# The Minerals, Metals & Materials Series

Camille Fleuriault · Joalet D. Steenkamp ·
Dean Gregurek · Jesse F. White ·
Quinn G. Reynolds · Phillip J. Mackey ·
Susanna A.C. Hockaday
Editors

# Advances in Pyrometallurgy

## Developing Low Carbon Pathways

TMS

Springer

*Editors*
Camille Fleuriault
Eramet Norway
Sauda, Norway

Dean Gregurek
RHI Magnesita Technology Center
Leoben, Austria

Quinn G. Reynolds
Mintek
Johannesburg, South Africa

Susanna A.C. Hockaday
Curtain University
Perth, WA, Australia

Joalet D. Steenkamp
Glencore XPS
Sudbury, ON, Canada

Jesse F. White
KTH Royal Institute of Technology
Stockholm, Sweden

Phillip J. Mackey
P. J. Mackey Technology Inc.
Kirkland, QC, Canada

ISSN 2367-1181 ISSN 2367-1696 (electronic)
The Minerals, Metals & Materials Series
ISBN 978-3-031-22633-5 ISBN 978-3-031-22634-2 (eBook)
https://doi.org/10.1007/978-3-031-22634-2



Cover Illustration: From Chapter "Ferronickel Production from Nickel Laterite via Sulfide Chemistry" Caspar Stinn et al., Figure 1: Distribution of iron (**a**), silicon (**b**), aluminum (**c**), manganese (**d**), magnesium (**e**), and nickel (**f**) in the laterite feedstock. Nickel is observed to be primarily distributed between iron-rich phases at a grade of around 1 wt% and magnesium-silicon-rich phases at a grade of around 3 wt%. Some manganese-rich phases are observed with elevated nickel contents on the order of 10 wt%. Silicon-rich magnesium-poor phases exhibit the lowest nickel trades at around 0.3 wt% or less (SEM/EDS map, scale bars: 500 μm). https://doi.org/10.1007/978-3-031-22634-2_25.

This Springer imprint is published by the registered company Springer Nature Switzerland AG
The registered company address is: Gewerbestrasse 11, 6330 Cham, Switzerland

# Preface

It is my pleasure to introduce the first installment of a new symposium series called "Advances in Pyrometallurgy", an exciting program sponsored by the Extraction and Processing Division and the Pyrometallurgy Committee of TMS. The theme of the first symposium is "Developing Low Carbon Pathways". Carbon intensive industries are at a crossroads: long-term manufacturing plans using pyrometallurgical processes all include decarbonization levers. We must solve the problem of fossil-based reduction and fossil-based power generation processes for metals production. As metallurgists, scientists, and engineers roll up their sleeves to face the greatest challenge of our generation, technologies enabling sustainable metals processing and its long-term vision develop at great speed.

The TMS community understands this challenge and embraces the opportunity. With this first symposium, we explore innovative and diverse strategies for the enablement of low carbon industries in the high-temperature metals and materials processing fields. In particular, the discussion highlights the potential of hydrogen as an alternative reducing agent for ironmaking, ferro, and manganese alloys smelting. It includes assessments of other alternatives to fossil carbon such as biocarbons, for the reduction of metal oxides but also manufacturing of electrodes and refractory. The symposium also covers novel energy efficiency and waste heat recovery concepts contributing to a lower footprint of production processes. In particular, renewable energies, such as solar power applied toward metallurgical practices, are systematically analyzed.

With this new symposium series, the Pyrometallurgy Committee wishes to illustrate how fundamental principles and advanced research translate to the production floor. Thus, the organizers favor a problem driven approach and the symposium will also include a keynote session focusing on the pathways taken to reduce carbon dependency within the industry, or directly for the industry.

I would like to thank the organizing committee for their contribution, invaluable input, and hard work. It has been a pleasure and a privilege to craft such an exciting program and to work with pyrometallurgy experts who are so passionate and knowledgeable about their field. I thank the authors for their excellent contributions, for the time spent writing the proceedings manuscripts, and for providing revisions throughout the peer-review process. Finally, I would like to thank the TMS staff for their support, especially Patricia Warren, Trudi Dunlap, Kelly Markel, Kelly Zappas, and Jeffrey Gnacinski. Additionally, many thanks to the TMS Programming Committee and the EPD Chair, Christina Meskers, for supporting the development of this new symposium.

Camille Fleuriault
Lead Organizer

**Advances in Pyrometallurgy: Developing Low Carbon Pathways Organizing Committee**

Camille Fleuriault
Joalet D. Steenkamp
Dean Gregurek
Jesse F. White
Quinn G. Reynolds
Phillip J. Mackey
Susanna A. C. Hockaday

# Contents

# About the Editors

**Camille Fleuriault** is senior project manager at Eramet Norway AS in Sauda, Norway. Her work focuses on identifying and enabling zero carbon strategies for the production of manganese alloys. She previously worked on developing innovative and environmentally friendly recycling processes for the secondary metals industry. She holds a B.S. in Geological Engineering and an M.Eng. in Mineral Engineering from the National School of Geological Engineering in Nancy, France, and a M.Sc. in Metallurgical Engineering from Colorado School of Mines, USA. She is chair of the TMS Pyrometallurgy Committee and a former *JOM* advisor for the same committee.

**Joalet D. Steenkamp** recently joined Glencore XPS in Sudbury, Ontario, Canada as Chief Metallurgist, Pyrometallurgy and Furnace Integrity. Her research focus areas are Furnace Containment, Furnace Tapping, and Reducing the Carbon-footprint of Ferrous Alloy Production. She holds a Ph.D. in Metallurgical Engineering from the University of Pretoria and an appointment as Visiting Adjunct Professor from the University of the Witwatersrand, both based in South Africa. Dr. Steenkamp has 13 years of experience in the South African industry/private sector (secondary steelmaking, ilmenite roasting and smelting, and manganese ferroalloy production) and 13 years in the research/public sector, of which the majority were spent at Mintek.

Dr. Steenkamp has been a TMS member since 2009. At TMS, she has been serving on the Pyrometallurgy

Committee since 2017 where she currently serves as Vice-Chair (2022–2024), served on the Industrial Advisory Committee (2018–2021) for which she was the inaugural Chair (2018–2020), serves on the Extraction and Processing Division Council as a representative for the Professional Development Committee (2021–2023), chaired the Organizing Committee for Furnace Tapping 2022, and now serves as a member of the organizing committee for the 2023 symposium Advances in Pyrometallurgy: Developing Low Carbon Pathways. In South Africa, she chaired the organizing committees for Furnace Tapping 2014 and 2018, the Schools on Manganese Ferroalloys Production in 2012, 2016, and 2020, and the School on the Production of Clean Steel in 2016. All of these events were hosted by the Southern African Institute of Mining and Metallurgy (SAIMM).

**Dean Gregurek** is a senior mineralogist in the RHI Magnesita Technology Center in Leoben, Austria since 2001. Dr. Gregurek received his M.Sc. degree at the University of Graz in 1995, his doctorate degree from the University of Leoben in 1999, and degree of assoc. prof. in 2019. Prior to RHI Magnesita, he worked for two years for Luzenac Europe in talc business. His current research interests and technical expertise are focused on chemical and mineralogical studies related to interactions between refractories, molten metals, and slags from pyrometallurgical furnaces. Dr. Gregurek has been a TMS member since 2012, *JOM* advisor (2014–2017), chair of the Pyrometallurgy Committee (2018–2020), and a co-organizer for the 7th–12th International Symposium on High-Temperature Metallurgical Processing (TMS Annual Meetings 2016–2021) and Furnace Tapping (TMS Annual Meeting 2022).

**Jesse F. White** holds a Ph.D. in Materials Science and Engineering from the KTH Royal Institute of Technology, an M.Sc. in Metallurgical and Materials Engineering from the Colorado School of Mines, and a B.S. in Metallurgical Engineering from the South Dakota School of Mines and Technology. He began his career in 1996 as a Process Engineer at the Kaiser Aluminum Mead Works. In 1997, he moved to Luleå, Sweden, and began as a Research Engineer at MEFOS working mainly in strip casting of steel. In 2002, he moved to Oslo, Norway and spent 5 years at Alstom as a Project Engineer designing, building, commissioning, and troubleshooting gas treatment systems for aluminum smelters around the world. Since 2007, he has been employed by Elkem in Kristiansand, Norway, starting out at Elkem Solar as a Research Engineer specializing in silicon refining, later moving to Elkem Technology, and since 2015 at Elkem Carbon. He is currently Technology Director at Elkem Carbon, supporting the production facilities in Brazil, China, Malaysia, Norway, and South Africa. In parallel, Dr. White is also currently an Affiliated Faculty Member of the Materials Science and Engineering Department at the KTH Royal Institute of Technology in Stockholm, where he teaches thermodynamics and conducts research in the areas of high-temperature experimental thermodynamics and metallurgical reactor design.

**Quinn G. Reynolds** holds an undergraduate degree in Chemical Engineering from the University of Kwazulu-Natal, a Masters in Engineering from the University of the Witwatersrand, and a Ph.D. in Applied Mathematics from the University of Cape Town. He has worked in the Pyrometallurgy Division at Mintek for the past 23 years. Mintek is a research institute conducting applied research and development to serve the extensive mineral processing and metallurgical industry in South Africa and worldwide. Dr. Reynolds' expertise includes mathematical and computational modelling of complex coupled phenomena in high temperature processes and in particular the application of high-performance computing and open-source modelling software to pyrometallurgy. His current areas of research include magnetohydrodynamic modelling of electric arcs, multiphysics fluid flow problems in furnace tapping and phase separation, combustion modelling for metallurgical processing, and discrete element modelling for particle flow problems. He has also performed extensive work in the characterization of the dynamic behavior of direct-current plasma arcs using high-speed photography and electrical measurement techniques.

**Phillip J. Mackey** obtained his B.Sc. (Honors) and Ph.D. degrees from the School of Metallurgy at the University of New South Wales, Australia. He then moved to Montreal, Canada to join Noranda to work on a new copper smelting process. As Pilot Plant Supervisor, Dr. Mackey helped develop the Noranda Process, which was first implemented at the Horne smelter in the early 1970s. Here he first learned the art and technique of tapping a high temperature melt—blister copper, copper matte, and slag. The Noranda Process, one of the most important copper smelting technologies of the twentieth century, also achieved early success in the United States, Australia, and China. Dr. Mackey was later instrumental in developing the Noranda Converting Process, of which he is a co-inventor, and which was installed at the Horne smelter in the late 1990s providing enhanced environmental performance. At Noranda and Falconbridge, he was involved with other initiatives, including new developments for processing nickel laterites and concluding technology agreements with other nations, notably Chile. He conducted due diligence studies on a range of projects around the world. He later formed his own consulting company, and this has led to a range of projects worldwide, including work on the development of a new nickel laterite project in Brazil.

**Susanna A. C. Hockaday** has 18 years of pyrometallurgical research experience in the non-ferrous industry. She joined Mintek in 2002 after obtaining her B. Chem. Eng. (Minerals Processing specialization) and M.Sc. in Extractive Metallurgy at the University of Stellenbosch in South Africa. During 2002 to 2010 she worked in the commercial projects group on various projects including the recovery of precious metals in liquid iron and the smelting of ores to produce design specifications of an industrial ferrochrome DC arc furnace. From 2011 to late 2015 she took a break from work and had two delightful children, now aged 11 and 8. From 2015 till

2021, she has been involved in research of new technologies for titanium metal production, chlorination of titanium dioxides in a fluidized bed, and the application of renewable energy in minerals processing. Since 2016 she has been enrolled at the University of Stellenbosch as a part-time Ph.D. student with the working title of "Solar thermal treatment of Manganese Ores". She acted as a work package leader responsible for €2 million of research toward advancement of solar thermal process heating technology in manganese ferroalloy production for the PRÉMA project. The PRÉMA project is funded by the European Union's Horizon 2020 Research and Innovation Programme under Grant Agreement No 820561 and is part of the SPIRE group of projects. Mrs. Hockaday was co-ordinator and the main author for Mintek's Roadmap for Solar Thermal Applications in Minerals Processing (STAMP). Mrs. Hockaday is a member of the South African Institute of Mining and Metallurgy (SAIMM) and part of the SAIMM technical programme committee. She was also head of the organizing committee for the SAIMM Colloquium on Renewable Energy Solutions for Energy Intensive Industry in 2020 and the Renewable Energy Solutions for Energy Intensive Industry Conference in 2021.

Mrs. Hockaday resigned from Mintek in June 2021 to move with her family to Perth, Australia. She has founded GamAesa as vehicle to continue her support of process innovation and renewable energy integrating in minerals processing.

# Part I
# Keynote

# Roadmap for Reduction of Fossil $CO_2$ Emissions in Eramet Mn Alloys

**B. Ravary and P. Gueudet**

**Abstract** Eramet produces materials useful for a low-emission society, in a resource-effective way, and for manganese (Mn) alloys, with a lower climate footprint than the industry average. Such high standards give increased competitivity because most stakeholders, in particular customers and investors, are interested in environmental-friendly production. This responsible strategy will eventually lead to improving profitability. Eramet has set goals for the reduction of emission of greenhouse gases from their production in the framework of the Science Based Target. In this paper, we present the strategy and a simplified roadmap to reach the target in Eramet Mn alloys activity. The roadmap is made a reality through actions and investments for industrial implementations. The reduction initiatives can be divided into four main areas, somewhat reflecting some sequences in time with overlap, from short term (2025) to long term (2040 and beyond): improvement of existing processes in resource and energy efficiency (2025), increase or introduction of biomass-based reductants to replace fossil carbonaceous materials (2030), carbon capture and usage (CCU) or storage (CCS) (2030), and development of innovative technologies (2040). All actions are rooted in scientific and techno-economic studies. Open innovation is necessary when developing technologies outside the core competence of the companies.

**Keywords** Manganese alloys · Climate · Roadmap · Bio-reductants · CCUS

## Introduction

The Paris agreement in 2015 targets to maintain global warming below 2 °C, compared to the pre-industrial age. Greenhouse Gas (GHG) emissions are to be

B. Ravary (✉)
Eramet Norway, Alfred Getz vei 2b, 7034 Trondheim, Norway
e-mail: benjamin.ravary@eramet.com

P. Gueudet
Eramet SA, 10 Boulevard de Grenelle, 75015 Paris, France
e-mail: pierre.gueudet@eramet.com



C. Fleuriault et al. (eds.), *Advances in Pyrometallurgy*, The Minerals, Metals & Materials Series, https://doi.org/10.1007/978-3-031-22634-2_1

reduced and countries signatories of the Paris agreement have set some targets for these reductions. The manganese (Mn) alloys business unit in Eramet has production facilities in the USA, Gabon, France, and Norway. The three Norwegian plants account for approximately 70% of the total production of Mn alloys in the business unit and were early in working on reducing their carbon footprint, following a strategy set by the broader Norwegian industry.

In its Intended Nationally Determined Contribution (INDC) to the Paris agreement, Norway committed to a 40% GHG emissions reduction in 2030 compared to the 1990 level, as well as achieving carbon neutrality by 2050. Carbon neutrality means that the same amount of GHG is stored and emitted so that the total net emissions are zero. As part of the European Green Deal, the European Union (EU) Commission proposed in September 2020 to raise the 2030 greenhouse gas emission reduction target, including emissions and removals, to at least 55% compared to 1990 [3]. On 14 July 2021, the European Commission adopted a series of legislative proposals setting out how it intends to achieve climate neutrality in the EU by 2050, including the intermediate target of at least 55% net reduction in greenhouse gas emissions by 2030. The package proposes to revise several pieces of EU climate legislation, including the EU Emissions Trading System (EU ETS), Effort Sharing Regulation, transport, and land use legislation, setting out in real terms the ways in which the Commission intends to reach EU climate targets under the European Green Deal [4].

The revised EU ETS Directive, which will apply for the period 2021–2030, will enable this through a mix of interlinked measures. EU ETS limits emissions from more than 11,000 heavy energy using installations (power stations and industrial plants, including metallurgical industry) and airlines. It covers around 40% of the EU's GHG emissions. In phase IV of the EU ETS, operators of installations subject to emissions trading may, upon request, receive a free allocation of emission allowances for the periods between 2021–2025 (first allocation period) and 2026–2030 (second allocation period). Allocations for 2021–2030 will have a major impact on the EU and Norwegian ferroalloy industry.

The roadmap for the Norwegian process industry [5] proposes a global vision and scenarios of technology development to achieve the Paris agreement goals in line with the EU expectations. Four technology breakthroughs are proposed: Carbon Capture and Storage (CCS) from both fossil and biogenic sources, increased use of hydrogen, increased use of biomass, use of zero-emission technologies and electricity, and circular economy. In addition, it illustrates the effects of potential new industries for producing sustainable fuels: E-fuel for aviation, ammonia for shipping, and advanced biofuels. The expected reductions linked to those technologies are presented in Fig. 1.

## Eramet's Climate Strategy

Decarbonization will soon become a "license to operate" and it is, therefore, a must for mines and alloys plants of Eramet Group. Having a low carbon footprint will be a condition to market our products, finance our projects and operations, and (retain

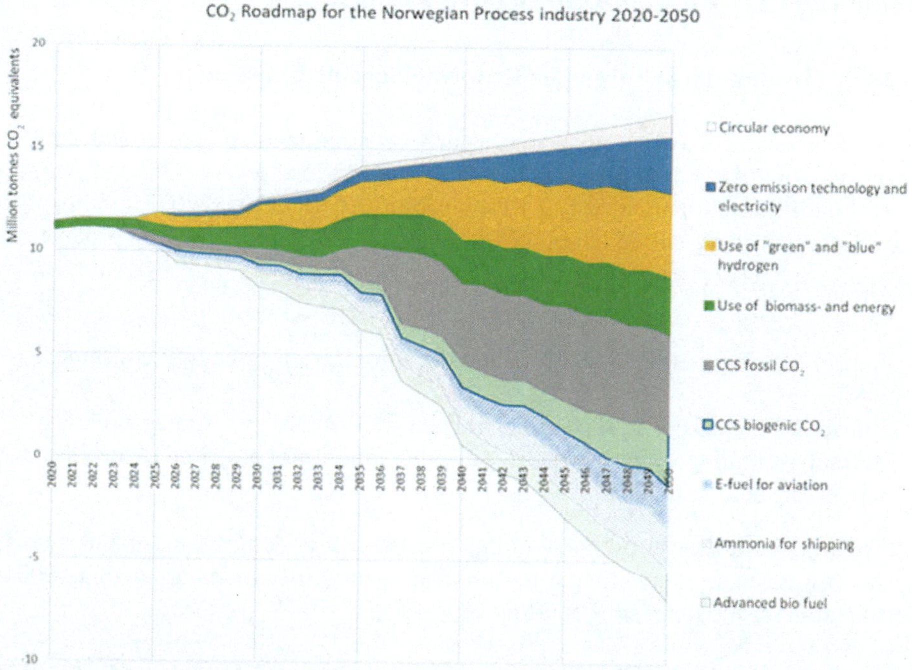


**Fig. 1** Expected reductions in GHG emissions from the implementation of different technologies in the Norwegian process industry [5]

and) recruit talents and partners. The reduction of $CO_2$ emissions requires action in different time frames from short-term adjustments of operation to developments requiring long lead times. The governance of our efforts considers the optimization of existing assets, the development of new technologies in partnership with peers, academics, and suppliers, as well as support for decarbonizing customers.

Eramet's answer to climate change is based on the following focus points:

- The reduction of $CO_2$ emissions on scopes 1 (internal emissions) and scope 2 (from the production of the electricity consumed).
- Helping customers and providers (scope 3 emissions) to reduce their GHG emissions, by offering products and solutions that mainly contribute to reducing the carbon footprint. This is reflected in one of the three pillars of the Group's strategy: "to expand the portfolio of activities towards energy transition metals".
- The promotion of circular economy.

## *Reducing $CO_2$ Emissions of Scopes 1 and 2*

### 2023: A Medium-Term Target for Reducing Specific Emissions

Eramet conducted a review to define a target for reducing scopes 1 and 2 $CO_2$ emissions, in 2018, based on technical and organizational levers. This led Eramet to include in its Corporate Social Responsibility (CSR), 2018–2023 roadmap, a significant reduction carbon target for the generated tons of $CO_2$ per ton product:

- Group Goal 2021 versus 2018: −26%, of which

– Impact of energy efficiency levers and decarbonization of energy consumed: −9.5%
– Impact of the business mix effect related to the Group's strategic choice to develop its mining activity, which is lower in emissions compared to the Group's processing activities: −16.5%

To structure all these progress initiatives, Eramet is deploying a management system for its energy and climate performance within its entities up to ISO 50001 certification for the main sites emitting $CO_2$.

### 2035: A Long-Term Objective Compatible with the Paris Agreements

Given the strong development of the mining activity, which is less carbon intensive than the pyrometallurgy activity, Eramet plans to meet its objective of reducing its specific $CO_2$ emissions much before 2023. The Group decided in 2020 to further accelerate the process through commitment to a Science Based Target, "well below 2 °C". Eramet is currently in "committed" approval status.

At constant perimeter, Eramet aims to reduce, in absolute value, its scope 1 and 2 $CO_2$ emissions by 40% in 2035 compared to 2019. This target requires activating all the levers identified, including those which are still at the R&D stage or at a pilot stage: Carbon Capture and Storage (or CCS), bio-reductants, electrification of mining activity, etc.

Eramet's carbon reduction trajectory thus depends on the Group's ability to develop multi-year, cross-functional projects on the following main axes:

- Decarbonization of purchased electricity (purchases, investments)
- Decarbonization of processes (bio-reductants and hydrogen)
- $CO_2$ capture and storage (CCS, in partnership)

The priority actions are:

- The development of CCS in partnership with other players: this is the most impactful action in terms of $CO_2$ savings and the costs are the main obstacle. We plan to develop a pilot and identify the least capital-intensive technologies.

- The use of bio-reductants in ore reduction: challenges for this level include finding biomass managed in a sustainable manner and compatible with the constraints of our processes (mechanical strength, polluting elements).
- The implementation of purchases and production of electricity from renewable sources coupled with the electrification of mines: the successful implementation of this lever is based, in parallel with the development of technical solutions, on a change of background culture (electric mining trucks for example) which requires long-term support.
- The improvement of the pre-reduction of ores and gradual introduction of hydrogen to this end.

This roadmap on scopes 1 and 2 is accompanied by a qualitative objective of reducing emissions in scope 3: Eramet is committed to encouraging its customers to reduce their own emissions.

### 2050: Carbon Neutrality on Scopes 1 and 2

Eramet targets carbon neutrality of its scope 1 and 2 emissions by 2050. This ambition relies on CCS and the use of bio-reductants, together with the implementation of disruptive technologies.

Most of the significant actions to reduce Eramet's carbon footprint take place over a medium to long- term horizon, the next few years being mainly devoted to confirming the potential gains through pilots.

The actions identified can only be implemented on the condition that the market reflects the investment costs in carbon and commodity prices. In this case, it would be a substantial increase in the price of carbon, and therefore, that of metals.

## Eramet Mn Alloys Roadmap

Eramet Mn Alloys will be the main contributor to reach Eramet Group's targets. Its plants have already among the lowest specific $CO_2$ emissions in the industry as illustrated for one line of its products in Fig. 2.

Eramet Norway, the Norwegian subsidiary of Eramet in Mn Alloys, was an early mover in its efforts to curb its carbon footprint and decided in 2017 to establish its "Climate and Environment Roadmap—Towards 2030 and 2050". The process started with a mapping of carbon emission reduction technologies (Fig. 3), based on their effectiveness and implementation. A Multi-Criteria Decision Analysis (MCDA) was applied to prioritize the levers, that are currently used in the roadmap for $CO_2$ reduction. A steering committee, including both representatives from Mn Alloys and Group, ensures a systematic follow-up and continuous update of our project portfolio.

The manganese alloys climate roadmap targets a reduction of emissions of 70% in 2035 through actions distributed in three main time frames (Fig. 4):

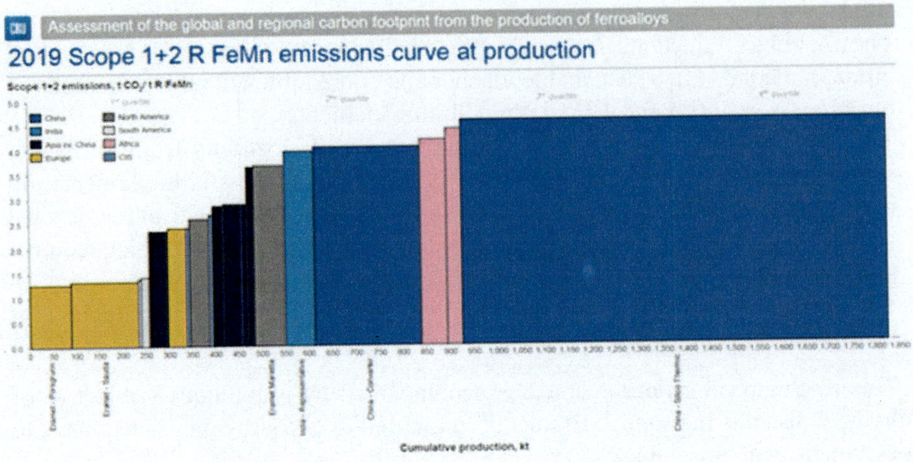


**Fig. 2** Specific emissions of $CO_2$ in refined ferromanganese as a function of production (kt) for different producers Specific emissions Eramet Porsgrunn and Sauda—approximately 1.3 $tCO_2$/t R FeMn [1]

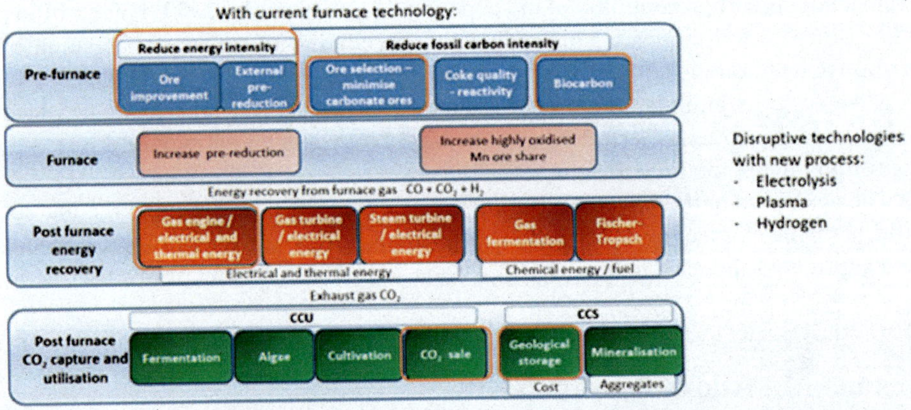


**Fig. 3** Overview of main technologies with current furnace technology (left) and disruptive technologies (right). Highlighted with orange frame—technologies in the Mn alloys roadmap

- Five-year perspective: reduce carbon consumption through improvement of operation of existing processes, promoting energy-saving chemical reactions, the pre-reduction of highly oxidized ores, and limiting the amount of carbonate in the charge [2].
- Ten-year perspective:
- Introduce a significant share of bio-reductant, also called biocarbon, to replace part of the current fossil reductants, coke, and coal.
- Capture $CO_2$ through CCS or Carbon Capture and Utilisation (CCU).

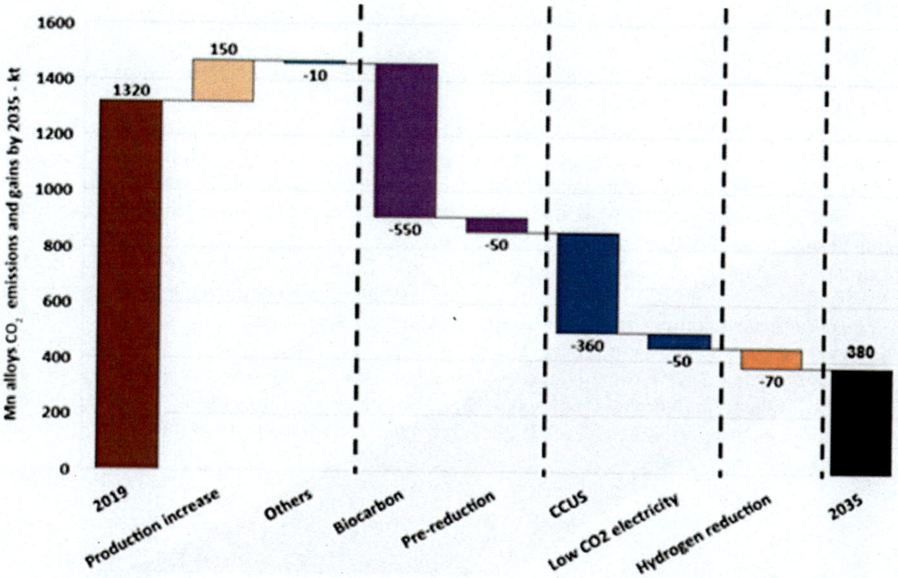


**Fig. 4** Main levers towards the 2035 target for Eramet Mn alloys. Low $CO_2$ electricity corresponds to moving from fossil-based electricity production to renewable (hydropower, wind, solar) or nuclear electricity as input for production (in scope 2)

- Longer term: Develop ground-breaking technologies, for instance using hydrogen, which may require drastic process changes and completely new production facilities.

Some details concerning the key R&D and industrial projects supporting the roadmap are presented in [6].

## Conclusion

Eramet has set objectives for the reduction of its $CO_2$ emissions using the Scientific Based Target framework. Eramet Mn Alloys will be the main contributor to reach these goals, following a climate roadmap, that is concretized through actions with different time frames. CCS and bio-reductants are its main levers in a 2035 perspective.

## References


1. CRU (2022) Assessment of the global and regional carbon footprint from the production of ferroalloys, Final Report, 1 June 2022
2. Davidsen J (2021) Reducing the $CO_2$ footprint from simn production by optimization of fluxes (September 12, 2021). In: Proceedings of the 16th international ferro-alloys congress (INFACON XVI) 2021. https://ssrn.com/abstract=3926019 or https://doi.org/10.2139/ssrn.3926019
3. European Commission (2020) 2030 climate and energy framework https://ec.europa.eu/clima/policies/strategies/2030_en. Accessed Dec 2020
4. European Commission (2021) Climate Action. EU Emissions Trading Systen (EU ETS). Revision for phase 4 (2021–2030) https://ec.europa.eu/clima/eu-action/eu-emissions-trading-system-eu-ets/revision-phase-4-2021-2030_en. Accessed Sept 2022
5. Prosess21 (2021) https://www.prosess21.no/
6. Ravary B, Bjelland KB, Valderhaug A, Messenlien AG (2021) Towards a climate-friendly ferroalloy industry in Norway (September 12, 2021). In: Proceedings of the 16th international ferro-alloys congress (INFACON XVI) 2021. https://ssrn.com/abstract=3926062 or https://doi.org/10.2139/ssrn.3926062

# Towards Net Zero Pyrometallurgical Processing with the ISASMELT™ and ISACYCLE™

**S. Nicol, T. Ryan, B. Hogg, and S. Nikolic**

**Abstract** With the growing global focus on reducing the environmental and social impacts of modern society, many smelters and recyclers are moving rapidly to decarbonise their processes. While existing and new solutions are required to optimise for reduced emissions across all reporting greenhouse gas (GHG) emission scopes, advancement of existing solutions can hold greater emissions-reduction potential. With a 37-year operating history and 25 global installations, across both primary and secondary (recycling) applications, ISASMELT™ and ISACYCLE™ technology is a mature, modern, and efficient smelting solution. The technology is well situated as it currently stands to achieve Net Zero status. The ability to decarbonise existing and new ISASMELT™ and ISACYCLE™ operations follows an emerging 'Emissions Optimisation' approach, which builds on significant energy-saving and emissions-reduction advancements derived from traditional profit-driven innovations. Scope 1 onsite direct emissions, and Scope 2 indirect emissions associated with electricity, heating, and cooling, can be minimised through the goal of attaining a low energy-use smelting operation. This can be achieved via aspects of the smelting process itself, including proprietary refractory design, advanced temperature control, waste heat capture, feed profile modifications, and oxygen enrichment. Further, scope 1 emissions can be eliminated through improvements such as alternative fuel substitution (i.e., hydrogen and sulphides) and off-gas processing. Finally, scope 3 emissions can be minimised through reduced maintenance, reduced consumables, and reduced equipment wear—by advancements including refractory design, lance design, and furnace control (for long campaign life and consistent production of high-quality products).

**Keywords** Net zero · Smelting · Greenhouse gas (GHG) emissions · Green metals · Critical minerals · Circular economy · Decarbonisation · Carbon · Smelting · Extractive metallurgy

---

S. Nicol (✉) · T. Ryan · B. Hogg · S. Nikolic
Glencore Technology, Level 10, 160 Ann Street, Brisbane, QLD 4000, Australia
e-mail: stuart.nicol1@glencore.com.au; isasmelt@glencore.com.au



C. Fleuriault et al. (eds.), *Advances in Pyrometallurgy*, The Minerals, Metals & Materials Series, https://doi.org/10.1007/978-3-031-22634-2_2

## Introduction

Extractive metallurgy plays a key role in mitigating environmental impacts associated with carbon dioxide and other greenhouse gas (GHG) emissions. By providing key materials to the world, metallurgists hold a great responsibility to decarbonise metallurgical systems and mineral processing operations. Specifically, the role of extractive metallurgists in providing 'critical minerals' is particularly crucial, as they are essential for the functioning of our modern lifestyles and global economy. Key minerals, such as the base metals copper, zinc, lead, and tin, are experiencing increasing global demand due to decarbonisation efforts, which involves the rapid rollout of renewable energy and electrification projects. To ensure these projects are constructed with green metals for a Net Zero future, extractive metallurgists must ensure decarbonisation efforts are made across all scopes of emission, and at all points of the critical minerals value chain.

Pyrometallurgical processes play a crucial role in unlocking metals for use in the supply chain. These processes chemically transform low-quality feeds to produce high-quality products for final refining and transport. Unlike other processing stages of the value chain, decarbonisation of pyrometallurgical processes requires more than just Scope 2 emissions reduction. Scope 2 emissions associated with purchased electricity, heating, and cooling, are comparatively simple to decarbonise. Scope 1 and 3 emissions reductions require metallurgical optimisation, due to the chemical processes involved in the smelting stage which traditionally involve the consumption and release of carbon as a fuel.

The ISASMELT™ technology is a mature, modern, and efficient smelting solution, well positioned for industry-leading emissions optimisation results, and Net Zero impact [8]. The technology was developed at Mount Isa, QLD to replace older, less efficient, and more emission-intensive technology [6]. Since its first installation, numerous developments in the technology have enabled even further reductions in gaseous emissions and fuel requirements. This progress has resulted in significantly lower smelting costs, with the Mount Isa Mines smelter becoming one of the lowest-cost operations in the world, despite its remote location in a developed country. The technology also significantly reduced the energy consumed by primary smelting, decreasing energy consumed at this stage of smelting by 93% [3]. Similar improvements have been realised at other smelters, such as the Ilo Smelter, with a 65% reduction in consumed fuel [13, 14].

Smelting operations have traditionally been optimised for profits, by maximising tonnage (revenue) and minimising costs (expenses). For 'Emissions Optimisation', existing smelting technologies must advance to also optimise their operation for reduced emissions. This is achieved by continuing to minimise energy usage, while also minimising and managing carbon consumption. The 'Traditional Optimisation' approach has driven the development of extremely efficient process outcomes. However, there remain additional areas for further advancement, development, and innovation to achieve a Net Zero smelter. These areas include feed modifications,

**Table 1** Optimisation parameters—traditional versus emissions optimisation

| | Manipulated variables | Measured variables | Outcomes |
|---|---|---|---|
| Traditional (cost-based) optimisation | • Furnace feed rate<br>• Feed profile modifications to manage impurities, size, moisture<br>• Oxygen enrichment<br>• Steady furnace control<br>• Advanced refractory design<br>• Advanced lance design<br>• Efficient ancillary equipment design (i.e., fans, blowers, conveying) | • Tonnage<br>• Campaign life and parts integrity<br>• Required smelting energy<br>• Furnace temperature<br>• Slag chemistry<br>• Product grades<br>• Electricity input<br>• Recycle rates | • Increased revenue through maximised tonnage and product grade/quality<br>• Decreased OPEX with minimised energy and consumables<br>• Decreased CAPEX with minimised maintenance and shutdowns |
| Emissions optimisation[a] | • Feed profile modifications to reduce carbon | • Direct, indirect, and embodied GHG emissions | • Reduced GHG emissions across scope 1, 2, and 3 |
| | • Alternative fuels substitution (i.e., hydrogen, pyrite concentrates) | | |
| | • Advanced waste heat capture system design | | |
| | • Off-gas processing | | |

[a] Additional to the traditional optimisation activities

materials substitution for fuels and reagents, further heat integration, and advanced gas processing (Table 1).

These 'Emissions Optimisation' activities will be explored with reference to the emission reporting scopes, to highlight the impact of emissions optimisation for an ISASMELT™ technology installation or modification. The following block flow diagram defines these emissions optimisation scope boundaries for the system (Fig. 1).

## ISASMELT™ and ISACYCLE™ Technology

ISASMELT™ furnaces are modern bath-smelting processes for the smelting of non-ferrous materials from both primary and secondary materials to produce various matte, slag, and metal products. The ISACYCLE™ furnace is an adaption of the ISASMELT™ furnace, to smelt secondary materials and various waste streams, such as electronic and municipal waste streams. The technology processes these streams

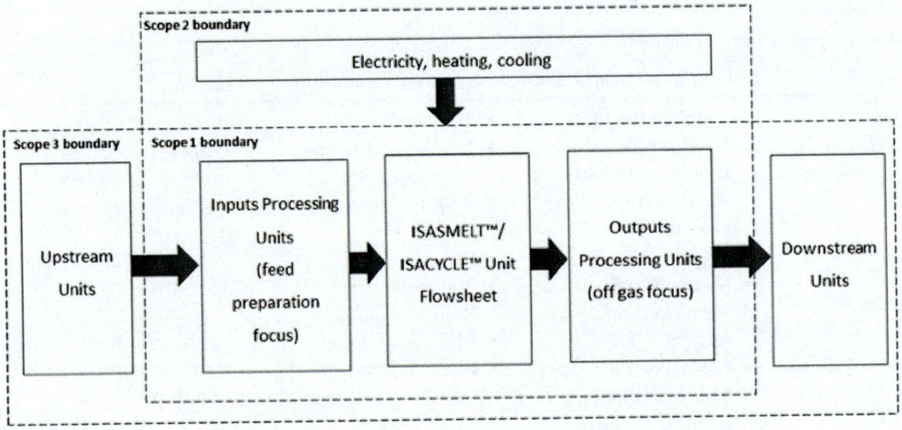


**Fig. 1** 'Emissions optimisation' scope 1, 2, and 3 emissions boundaries

to treat hazardous materials and circulate and recycle finite mineral stocks. Both furnaces can be constructed at any scale, to suit the required facility.

The ISASMELT™ top submerged lance (TSL) technology was developed at Mount Isa Mines (now part of Glencore) during the early 1980s [6] . It was identified that the smelter was running old technology, with new technology required to significantly reduce energy consumed in the smelter. The furnace technology was first tested in a 250 kg/h test rig in the 1980s [4]. The furnace technology was subsequently scaled-up to an operational demonstration plant, and finally a full-scale furnace. The technology resulted in 93% less energy being consumed in the primary furnace [3], and the smelter became one of the lowest-cost operations globally [3]. Due to the success of the ISASMELT™ at Mount Isa Mines, the technology has been installed at 25 sites around the world, with smelters using the technology to process nickel [5], lead [15], and copper [1, 2] concentrates and secondary materials.

The ISASMELT™ and ISACYCLE™ furnaces, depicted in Fig. 2, are cylindrical vessels with a flat roof. In an ISASMELT™ furnace, the vessel is refractory lined and regularly achieves campaign lives of 4 years of operation, without copper coolers [11]. A centrally located submerged lance injects air, oxygen, and fuel into a molten slag bath. This blast of air, oxygen, and fuel down the lance oxidises and violently agitates the liquid slag, to ensure a rapid reaction between this oxidised slag and feed materials. A frozen layer of slag forms on the outside of the lance and protects it from the aggressive environment in the furnace. The furnace products, slag, and metal or matte, can be tapped simultaneously or separately through water-cooled copper tapholes.

Table 2 presents key parameters and associated values.

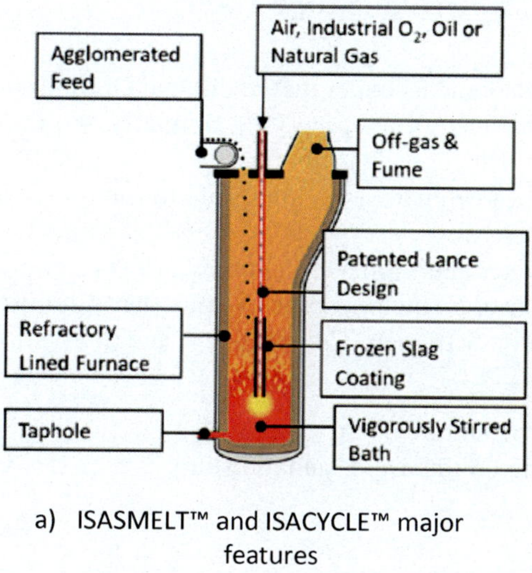


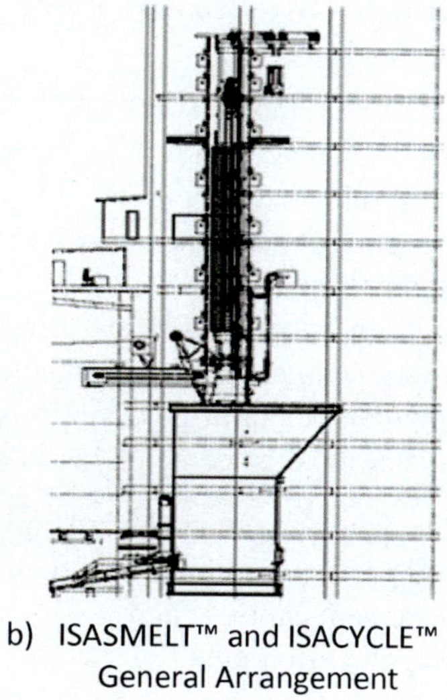


**Fig. 2** ISASMELT™ and ISACYCLE™ technology

**Table 2** ISASMELT™ and ISACYCLE™ operating parameters and conditions

| Parameter | Value |
|---|---|
| Facility sizes | 1 kg/h–200 t/h |
| Lance total flowrate | 10–71,000 $Nm^3/h$ |
| Furnace fuel/reductant | Coal, plastic, coke, secondary scrap (e.g., e-scrap) |
| Furnace trim fuel supply | Natural gas, diesel, pulverised coal, waste oil, hydrogen |
| Lance oxygen enrichment | 21 (air) to 92 vol% $O_2$ proven |
| Furnace availability | 92% (including re-brick and maintenance) |
| Furnace lining | Fully bricked, copper staves, and intensively cooled copper panels proven |
| Taphole types | Combined or separate metal/matte and slag tapholes proven |
| Furnace campaign | 4+ years proven by four separate ISASMELT™ licensees, 6+ years possible |
| Ramp-up to design capacity | 3 months proven |
| Furnace operation | Batch and Continuous proven, able to change back and forth during asset life |
| Feed size | <100 mm proven |
| Feed moisture | Up to 12 wt% proven |
| Feed delivery system to furnace | Vibrating or belt-style for coarse and/or wet feed<br>Pneumatic injection for fine, volatile, or low-density feed |

## Developments in Low Energy Smelting (Scope 1 and 2)

ISASMELT™ and ISACYCLE™ developments under the Traditional Optimisation approach, involve progress towards low-energy smelting [10]. Similarly, the Emissions Optimisation approach for the reduction and minimisation of scope 1 and scope 2 emissions, requires attaining a low-energy smelting state to minimise fuel requirements.

To attain a low-energy smelting state, and ultimately autonomous (or net positive energy) smelting, inefficiencies in the smelting process must be minimised. This involves metallurgical optimisation activities. These scope 1 and 2 emissions minimisation advancements include:

- Advanced refractory design and temperature control,
- Feed profile modifications, focusing on impurities and moisture,
- Waste heat capture; and
- Oxygen enrichment.

Recent advancements in the ISASMELT™ technology have led to a significant improvement in the energy efficiency of the furnace technology. A summary of the total energy savings available with the next generation of the technology is provided in Table 3, where the energy available from the feed is 11 MW. For other furnace technologies, 17 MW of energy from carbon-based fuels is required. Using modern ISASMELT™ Technology, there is a surplus of 7 MW available for the melting and smelting of other materials.

### *Advanced Refractory Design and Temperature Control*

Many advancements for low-energy smelting have been made in the most recent generation of ISASMELT™ furnaces. These developments were built on the success

**Table 3** Energy savings with modern ISASMELT™ designs (350 kt/a Cu from sulphide concentrates; the energy available from the feed is 11 MW)

| | Energy saving (MW) |
|---|---|
| Lower dusting rates (10% to 2%[a]) | 5 |
| Fully bricked furnace, with insulative lining[a] | 8 |
| Lower furnace temperature (1250 °C–1190 °C[a]) | 2 |
| Advanced slag control (including 20 °C temperature decrease) | 1 |
| Pneumatic injection of dried concentrate (9.5 wt% to 0 wt% moisture) | 8 |
| Oxygen enrichment (60 vol% to 95 vol% in Lance) | 2 |
| Total energy saving | 28 |

[a] Standard ISASMELT™ Design, comparison with alternative furnace technology shown

of its first generation, aiming to decrease capital cost, operating cost, and emissions through refractory design and temperature control. In this way, developments progressed with traditional optimisation at the forefront; but provided strong grounds for significant progress towards emissions optimisation.

Energy is lost through furnace walls in two ways. Firstly, through into the surrounding air, and secondly, to the cooling water used in water-cooled furnace components. Therefore, energy efficiency can be increased considerably by decreasing the heat lost to the surrounding air and cooling water system. For the ISASMELT™ furnace, the non-requirement for a complex and expensive water-cooling system also results in comparatively low heat losses from the furnace. Additional minor loss areas include the water-cooled splash block and water-cooled roof. The ISASMELT™ furnace is constructed with both an insulative and working lining. The working lining uses bricks resistant to slag attack, with the bricks slowly wearing over time. The insulative lining is constructed from bricks quickly corroded by the slag, but highly insulative. This layered lining approach effectively reduces heat lost from the furnace, improving the furnace's energy efficiency.

Smelting energy inefficiencies may also be a result of poor furnace temperature management. Successful furnace temperature management works to ensure that the minimum energy is used to produce a molten bath. This is achieved by minimising slag mass and temperature, through advanced process chemistry design and with chemical adjustments made online through advanced slag chemistry process modelling tools. Further, advanced furnace tapholes and advanced tapping designs are commonly installed on ISASMELT™ furnaces. This ensures molten material produced during low-energy smelting can be easily tapped from the furnace.

### *Advanced Waste Heat Capture*

Optimised ISASMELT™ systems under the traditional optimisation approach include heat recovery, to provide steam for heating on-site, or for export to other plant areas. However, there exist several opportunities for improved emissions optimisation outcomes. Typical ISASMELT™ practice includes integration of the furnace's water-cooled components, with the off-gas waste heat boiler (e.g., boiler tube roof), enabling the waste heat from these components to be recovered as well.

Additional waste heat capture systems are possible across the broader system boundary, including units upstream and downstream of the immediate ISASMELT™/ISACYCLE™ unit. These include waste heat within output processing units (e.g., acid plant units, scrubber). Further, excess generation of saturated steam can be converted to electricity, for use within the site. If any excess electricity was sold to a wider energy grid, it may be classified as a carbon offset activity.

## *Feed Profile Modifications*

In addition to the heat lost from the vessel, a large portion of smelting energy inefficiencies is due to the feed profile, both moisture and recycled materials. The recycles in the feed consume significant quantities of heat as they are heated to the furnace temperature. They then leave the furnace in the form of an enthalpy. Through minimisation of the quantity of these recycles, less energy is required to smelt the feed. This can be achieved by minimising the slag and dust recycles; which can be achieved with efficient smelting and low mechanical dusting, achieved with the existing ISASMELT™ technology.

Modern and advanced ISASMELT™ lance designs enable reduced moisture levels in feed, while minimising dusting rates. First-generation ISASMELT™ installations utilised concentrate agglomerated with water to produce a coarse feed with low mechanical dusting, resulting in <2 wt% of the feed reporting to the off gas as dust. This was a flexible, low cost, and easy to operate solution—resulting in several technical advances in process control and minor element processing (e.g., compact dust leaching plants). The dusting rate achieved with the agglomeration technology provided much lower dusting rates compared to other smelting technologies, which typically operates with more than 10–15 wt% dusting rates. However, the wet concentrate required additional energy to vaporise the water and heat the steam. Modern ISASMELT™ designs can inject dry concentrate down the lance to maintain low dusting rates (similar to, or less than, those achieved with water-based agglomeration techniques). The moisture in the concentrate can, therefore, be removed prior to injection, using a steam dryer, which may be operated on steam from the waste heat boiler.

## *Oxygen Enrichment*

Advanced and industrially proven lance designs have enabled up to 92 vol% oxygen to be used in the lance. A high oxygen enrichment decreases the nitrogen that is a 'spectator species' in the furnace, reporting to the off gas and resulting in lower energy efficiencies. This area also contributes significantly to low-energy smelting advancements. The production of industrial oxygen uses considerable quantities of electricity, but this can be obtained from a renewable source—potentially on-site (e.g., solar).

## Alternative Fuels and Off-Gas Management (Scope 1)

Scope 1 and 2 elimination activities extend on the traditional optimisation and emissions optimisation areas, and work to directly remove carbon inputs and manage carbon outputs. This includes:

- Alternative fuels substitution (i.e., hydrogen, pyrite concentrates); and
- Off-gas processing.

### *Alternative Fuel Substitution*

Depending on the feed material being smelted, reactions in the furnace either produce or consume energy. An energy-negative smelter requires fuel, traditionally derived from carbon-based sources. An energy-positive smelter requires coolant, which may consist of an energy-negative feed, recycled streams, or water. The balancing of energy-positive and energy-negative components presents an opportunity to minimise or eliminate carbon during smelting. Oxide feed materials, such as metallic scraps, are energy negative and require fuel and a reductant to produce a molten metal phase. For these materials, there are two avenues of carbon fuel substitution, which are sulphides or hydrogen.

Sulphides are energy positive, generating heat when oxidised to produce a molten matte and $SO_{2(g)}$. They are also easy to handle and safe to operate with, with many years of prior experience available in the industry. Useful sources include a variety of minerals processing waste streams, including Albion Process™ leach residues, base metal sulphide concentrates, pyrite tailings, and some types of scrap metals. The use of sulphide materials as an energy source for the processing of recycled copper-containing materials has been practiced at several ISASMELT™ operations. For composition and physical properties stability, the sulphide feed can be mixed and agglomerated with other feeds prior to smelting.

Lance injection of hydrogen into the ISASMELT™ furnace bath presents the second carbon-free solution. This promotes optimal bath agitation, which ensures combustion and reduction processes occur at greater than 98% efficiency. Additional advancements and further piloting to support this fuel type are required, including the latest valve trains for safe and controlled hydrogen flow.

Table 4 presents the range of fuel options available for secondary copper scrap, as an 'oxide feed material' example. This provides a comparison of fuel density between traditional carbon fuel types (coal), and the discussed sulphide and hydrogen options. It is noted that high-grade copper scrap, while having minimal oxygen, requires energy to melt and react in the order of 0.20 kWh/kg of scrap.

**Table 4** Fuels available for secondary copper smelting

| | Energy available[a] (kWh/kg) |
|---|---|
| Pyrite concentrate (Dry) | 1.10 |
| Albion process™ residue (Wet) | 1.10 |
| Dry copper concentrate (75% $CuFeS_2$, 25% $FeS_2$) | 0.60 |
| Shredded steel (dry) | 0.65 |
| Hydrogen | 27.00 |
| Coal | 8.00 |

[a] Includes the energy to heat and flux the reaction products (1180 °C)

### *Off-Gas Processing*

Finally, emissions capture and processing are required for any remaining carbon leaving the system in the furnace off gas. ISACYCLE™ and ISASMELT™ technology have been developed with gas control systems, which capture the energy from the off-gas stream while ensuring high concentration and containment of the main ($CO_2$ and $SO_2$) gas species. This efficient capture allows for potential use in combination with carbon capture, or other imminent processing techniques. Ultimately, this works to eliminate carbon dioxide emissions when processing primary, secondary (recycling), or other materials.

## Reduced Maintenance and Consumables (Scope 3)

Embodied emissions upstream and downstream of the smelting operation can be decreased by reducing the consumables required, which includes fuel and reductants—achieved through scope 1 emissions optimisation (low-energy smelting, alternative fuels). Additionally, scope 3 upstream and downstream emissions can be minimised through the following advancements:

- Maintenance and consumptions minimisation; and
- Low equipment wear design.

Further, reduction of required downstream processing is a significant scope 3 optimisation activity. Production of consistent-quality outputs helps recover critical minerals with minimal recycling and reprocessing. The slag output product provides a means of residual waste treatment and transformation, into a safe and usable product. These activities help enable the circular economy, reduce the need for more energy-intensive extraction, and eliminate emissions or pollution associated with irresponsible waste management.

## *Maintenance and Consumption Minimisation*

Every smelting operation requires maintenance and repair work due to either equipment wear or poor operation of equipment. This requires new materials to be purchased and installed, and old materials to be disposed of, resulting in additional upstream and downstream carbon emissions. To reduce these emissions, maintenance and repair works require optimisation through both component life extension and component simplification.

For ISASMELT™ and ISACYCLE™ operation, a maintenance and repair optimisation focus area is the replacement of working furnace bricks. Significant progress has been made to decrease the wear rate for both the lance and furnace brick components, with advanced ISASMELT™ operations requiring minimal maintenance and repairs compared to industry competitors. Across industry, several approaches have been taken to decrease refractory consumption per unit of metal produced, including increased furnace intensity, intensive water cooling, advanced process control, and designing for a stagnant bath at the vessel wall. Many smelting technologies focus on copper cooler advancements as an insurance solution to a process problem. Instead, ISASMELT™ developments have focused on process control. This has resulted in a variety of benefits for ISASMELT™ and ISACYCLE™ technology, including significantly longer campaign life while maintaining low furnace heat losses and operating costs. Advanced ISASMELT™ operations can achieve campaigns of over 6 years.

In addition, the ISASMELT™ technology maximises furnace intensity, thereby reducing construction and repair materials. The furnace technology uses less than 35% of the materials used by other furnace technologies for significant reductions in required upstream equipment, and associated scope 3 emissions [7]. Table 5 presents the advancements for decreased refractory consumption across industry with ISASMELT™'s comparative reduction for process control focus.

**Table 5** Technology and advancements used to decrease refractory consumption [7, 11, 12]

| Technology advancement | Refractory per unit of metal |
|---|---|
| Increased vessel intensity | 50% Reduction |
| Copper coolers | 50% Reduction (offset by increased fuel consumption) |
| Advanced process control | 80% Reduction with ISASMELT™[a] (20% Others) |
| Stagnant bath at vessel walls | 30% Reduction |

[a] Unique to ISASMELT™ furnaces

### *Low Equipment Wear Design*

Lances, burners, tuyeres, and electrodes are used in the furnace, to introduce reactants into the vessel and drive the chemical processes that are occurring. As such, these components are exposed to high temperatures and corrosive materials. A number of approaches are used to improve the maintenance and repair of these components, including:

- water cooling to extend component life [12],
- supersonic injection to move the aggressive reaction zone away from the walls [9],
- equipment simplification to reduce downtime and costs with replacements; and,
- advanced process control to manage the conditions in the vessel [11].

The ISASMELT™ lance is made from simple steel components, with a protective layer of frozen slag, which forms due to the cooler gases being injected down the centre of the lance. However, various conditions can result in the breakdown of this layer, leading to the bath corroding the lance. Early ISASMELT™ operations were able to achieve 1–2 days of operation per lance, but with advances in lance operation and design, a lance has now been demonstrated that can operate for more than 900 h of operation immersed in the bath. In addition, developments have been made to the lance design to simplify their construction, maintenance, and repair. These two developments have significantly reduced lance materials consumed by ISASMELT™ operations, and hence the associated scope 3 emissions.

## Conclusion

Smelting and recycling processes, required to supply key materials and critical minerals for a Net Zero future, currently emit carbon dioxide emissions across all emissions reporting scopes, as both a direct and indirect result of their operation. Developments and advancements for both ISASMELT™ and ISACYCLE™ technology are examples of how existing technologies can be adapted for primary and secondary (recycling) Net Zero smelting. This has been achieved by advanced refractory design, temperature control, waste heat capture, feed profile modifications, oxygen enrichment, alternative fuel substitution, off-gas processing, and reduced maintenance, consumables, and equipment wear. ISASMELT™ and ISACYCLE™ technologies are global, mature, and efficient smelting solutions for Net Zero impact. They are available as new solutions for construction, or through an upgrade to existing operations.

## References

1. Alvear G, Nikolic S (2013) ISASMELT for recycling of valuable elements contributing to a more sustainable society. In: Proceedings of REWAS 2013: enabling materials resource sustainability, annual meeting of the minerals, metals and materials society (TMS), San Antonio, Texas, United States of America, Mar 2013
2. Arthur PS, Hunt SP (2005) "ISASMELT™—25 years of continuous evolution. In: Nilmani M, Rankin WJ (eds) (2005) Floyd international symposium on sustainable development in metals processing. NCS Associates (Australia), pp 73–94
3. Arthur P, Butler B, Edwards J, Fountain C, Hunt S, Tuppuranien J (2003) The ISASMELT™ Process—an example of successful industrial R&D. In: Proceedings of metallurgical and materials processing: principles and technologies, Yazawa international symposium, annual meeting of the minerals, metals and materials society (TMS). San Diego, United States of America, 2–6 Mar 2003
4. Arthur PS, Alvear G (2011) Incremental scale up of ISASMELT™—the key to its success. In: Davis BR, Kapusta JPT (eds) (2014) Proceedings of the 50th annual conference of metallurgists of CIM, pp 203–214
5. Bakker ML, Nikolic S, Mackey PJ (2011) ISASMELT™ TSL—applications for nickel. Miner Eng 24(7):610–619
6. Coulter M, Fountain C (1989) The ISASMELT™ process does copper smelting. In: Non-ferrous smelting symposium. Port Pirie, South Australia, AusIMM, pp 237–240, Sept 1989
7. Gonzales T, Walter G, White M (2019) Comparison of smelting technologies. Copper 2019. Vancouver, Canada, 18–21 Aug 2019
8. Hogg B, Nikolic S, Voigt P, Telford P (2018) ISASMELT™ technology for sulfide smelting. In: Davis BR et al. (ed) extraction 2018, the minerals, metals & materials series (TMMMS) book series, Springer. https://doi.org/10.1007/978-3-319-95022-8_11
9. Kapusta J (2018) Sonic injection in sulphide bath smelting: an update. J SAIMM 118:1131–1139
10. Nicol S, Corrie D, Barter B, Nikolic S, Hogg B (2022) Adaptability of the ISASMELT™ technology for the sustainable treatment of wastes, TMS 2022. Anaheim, California, USA, 27 Feb–3 Mar 2022
11. Nikolic S, Hogg B, Voigt P (2019) Freeze lining refractories in non-ferrous TLS smelting systems, TMS 2019. San Antonia, Texas, USA, 10–14 Mar 2019
12. Taniguchi T, Matsutani T, Sato H (2006) Technological innovations in the mitsubish process to achieve four year campaigns, TMS 2006. San Antonio, Texas, USA, 12–16 Mar 2006
13. Walqui H, Noriega C (2007) Design, construction and start-up of the ilo smelter modernization project, SME, SME Annual meeting 2007. Denver, Colorado, USA, 25–28 Feb 2007
14. Walqui H, Noriega C, Partington P, Alvear G (2006) SPCC's 1,200,000 TPA Copper ISASMELT™, TMS 2006. San Antonio, Texas, USA, 12–16 Mar 2006
15. Errington WJ, Fewings JH, Keran VP, Denholm WT (1993) The Isasmelt lead smelting process. Trans Ins Min Metall Sec (C) 96(1987):1–6

# Part II
# Hydrogen

# Hydrogen, a Promising Carbon Substitute in Metallurgy?

**Juergen Antrekowitsch, Stefan Wibner, and Gustav Hanke**

**Abstract** Minimization of $CO_2$ emissions has become one of the main topics in our daily life. Especially industry, as one of the biggest emitters, has to contribute essentially to a global $CO_2$ reduction. When focusing on the metal production sector, one of the most promising solutions seems to be the utilization of hydrogen. Similar to carbon, hydrogen can act as both a reducing agent and an energy source. However, this looks easy only at the first glance. Various aspects such as efficient production, storage, transport, requirement for new processes concepts or a more difficult off-gas treatment underline that such a change will also be a future challenge. The paper gives an overview of hydrogen-production processes with some details about the ongoing development of methane pyrolysis as an effective way to generate hydrogen in huge volumes. Furthermore, it discusses where hydrogen can be used efficiently in metallurgical processes and in which areas alternatives have to be developed.

**Keyword** Environmental effects · Process technology · Pyrometallurgy · Hydrogen

## Hydrogen Production

In 2020, about one fifth of the hydrogen was produced as a by-product of petroleum refineries. About 60% are generated out of natural gas and another fifth out of coal (see pie-chart in Fig. 1) [1].

This production generates about 900 Mt of $CO_2$ emissions which indicates an urgent requirement for a change in hydrogen production technologies.

The most commonly used technology to produce hydrogen is steam reforming of hydrocarbons, methane in particular. The endothermic process is operated at an elevated pressure at temperatures of 700–800 °C in the presence of a catalyst. The overall reaction can be described as a combination of the reforming reaction and the water–gas-shift reaction (reactions 1 and 2) [2].

J. Antrekowitsch (✉) · S. Wibner · G. Hanke
Chair of Nonferrous Metallurgy, University of Leoben, Leoben, Austria
e-mail: Juergen.antrekowitsch@unileoben.ac.at


C. Fleuriault et al. (eds.), *Advances in Pyrometallurgy*, The Minerals, Metals & Materials Series, https://doi.org/10.1007/978-3-031-22634-2_3

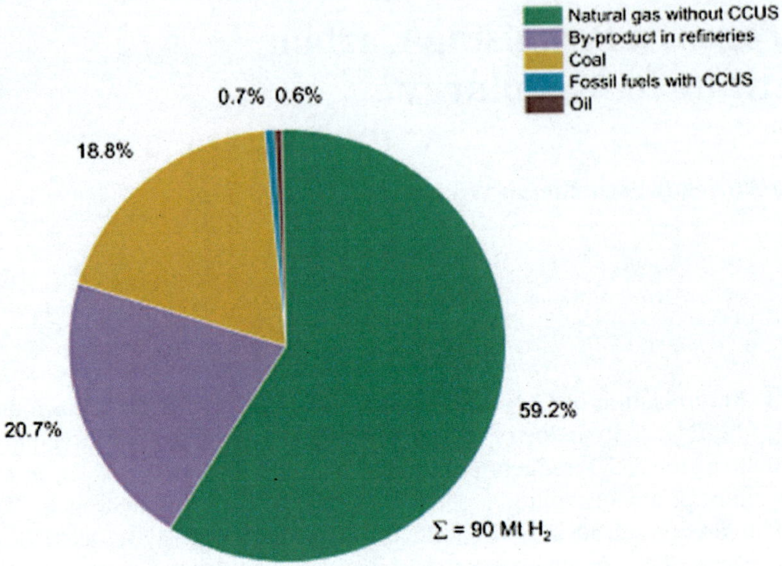


**Fig. 1** Sources for the production of hydrogen 2020 [1]

$$CH_4 + H_2O = CO + 3H_2 \tag{1}$$

$$CO + H_2O = CO_2 + H_2 \tag{2}$$

The overall reaction is:

$$CH_4 + 2H_2O = CO_2 + 4H_2 \tag{3}$$

Steam reforming of methane is considered a sophisticated technology with high production rates and good economic efficiency. One main disadvantage of this technology is the production of $CO_2$. For this, the production of hydrogen by electrolysis represents an interesting alternative. The decomposition of water is described in Eq. 4.

$$H_2O = H_2 + 0.5\,O_2 \tag{4}$$

However, much more energy is required per mol of hydrogen accounting for 286 kJ/ mol instead of 41 kJ/mol in case of steam reforming, taking only the chemical reaction into account. Due to the fact that most of the energy still comes from fossil fuels, this technology shows a rather high $CO_2$ footprint. Together with the high energy consumption, it shows a quite negative ecological impact as long as the energy is not provided by alternative $CO_2$ neutral energy sources [2].

Methane pyrolysis describes the one-step process of thermal cracking of methane in presence of a catalyst to form hydrogen and carbon according to Eq. 5.

$$CH_4 = C + 2H_2 \tag{5}$$

The technology uses catalysts such as different metals. This process is investigated in depth by the Chair of Nonferrous Metallurgy, University of Leoben, and is described in detail in the following chapter.

## Methane Pyrolysis

Methane pyrolysis describes the one-step process of thermal cracking of $CH_4$ in presence of a catalyst to form hydrogen and carbon according to Eq. 5.

A major advantage is the significantly lower amount of energy (38 kJ/mol $H_2$) required compared to water electrolysis and the lower $CO_2$ footprint compared to steam reforming. The mandatory heat for pyrolysis can be provided electrically or by combustion of fossil fuels. Transition metals such as iron, nickel or cobalt, activated carbon or metal oxides in solid form are used as catalysts in traditional $CH_4$ pyrolysis processes. The rapid deactivation of these catalysts due to the produced carbon has a highly detrimental effect on the overall efficiency. Several innovations including the utilization of molten catalytic alloys on which the hardly soluble pyrolysis carbon floats bear the potential to remedy this disadvantage.

For this reason, at the University of Leoben, a team is working on process and catalyst development for the decomposition of methane and natural gas in the molten metal reactor (MMR), where high $CH_4$ conversion rates have already been achieved with an innovative concept. In this context, the research and development of efficient liquid metal catalysts, which consist of different metal alloy systems, is essential. With regard to the experimental test setup, constant further development and scaling of the existing facilities are necessary. Therefore, the optimization of the gas introduction system and the solid separation devices, as well as the product gas treatment, are important sub-areas.

### *Experimental Setup of the Molten Metal Reactor (MMR)*

Methane pyrolysis on a laboratory scale at Montanuniversität Leoben, Chair of Nonferrous Metallurgy takes place in an inductively heated molten metal reactor, as shown schematically in Fig. 2 [2]. The catalytic metals or alloys are melted in a crucible with an inner diameter of 6.5 cm by induction heating. The process temperature is determined with two thermocouples (type K) located outside the bottom of the crucible and logged continuously during the experiment. A flange connects the crucible with an airtight stainless steel piping system. Methane (or natural gas) is introduced into the liquid catalyst through a lance made of $Al_2O_3$. The product gas flows along tubes and exits the experimental setup via a hot gas filter while the gas composition is measured simultaneously in a gas analyzer. The produced

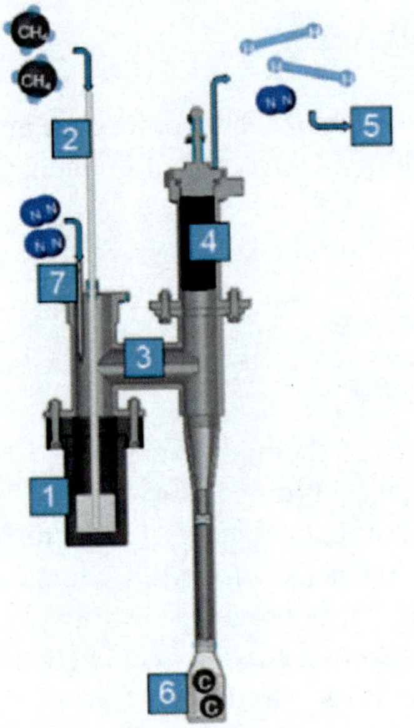


1. Crucible
2. Alumina lance
3. Piping system
4. Hot gas filter
5. Gas analysis
6. Glass bottle
7. Stainless-steel lance (carrier gas $N_2$)

**Fig. 2** Schematic representation of the molten metal reactor designed at the Montanuniversität Leoben [2]

solid carbon is separated gravimetrically from the product gas stream and collected in a glass bottle. Previous experiments demonstrated the formation of intermediates, mainly polyaromatic hydrocarbons, besides hydrogen in the process. To avoid condensation of these PAHs in the hot gas filter, the corresponding section of the piping system is heated to a temperature of 300 °C by heating wires. Additional flushing with nitrogen, which is introduced into the system by means of a stainless steel lance, results in a higher gas flow rate to avoid an accumulation of solid particles, especially in the horizontal pipe section [2].

## *Catalytic Metal Bath*

Some metals, for instance nickel, are active catalysts for the decomposition of methane, but have high melting points, resulting in high process temperatures. However, the active components can also be dissolved in other metals, some of which have low melting points, like tin, resulting in a lower melting point for the system. Copper is often used as a base metal due to its positive properties (catalytic effect, good alloyability, standard metal). The required process temperatures depend on the main metal used. They can be 600–900 °C for tin, 900–1200 °C for copper, and 1200–1600 °C for nickel. When selecting suitable alloys, it is important to ensure that they are thermally stable. The required energy for the pyrolysis process results mainly from the endothermic methane decomposition and from heat losses.

## Influence Parameters on the Formed Solid Carbon Fraction

With the molten metal reactor setup described above, carbon sample quantities in the 100 g range can be generated and subsequently analyzed.

The morphology of the carbon formed can be influenced by the following parameters:

- Temperature of the molten metal bath
- Composition of the chosen catalytic alloy

According to Keipi et al. [3], the morphology of the formed carbon in the thermo-catalytic decomposition of methane is depending on the reaction temperature and the selected catalyst as shown in Fig. 3.

It also has to be mentioned that a small amount of finely dispersed metal particles are carried out with the solid carbon fraction in the MMR. The proportion, as well as the composition of these impurities, depend on:

- Gas flow rate of the input-gas
- Liquid alloy composition
- Covering phase of the molten metal
- Gas injection technology
- Size of the gas space.

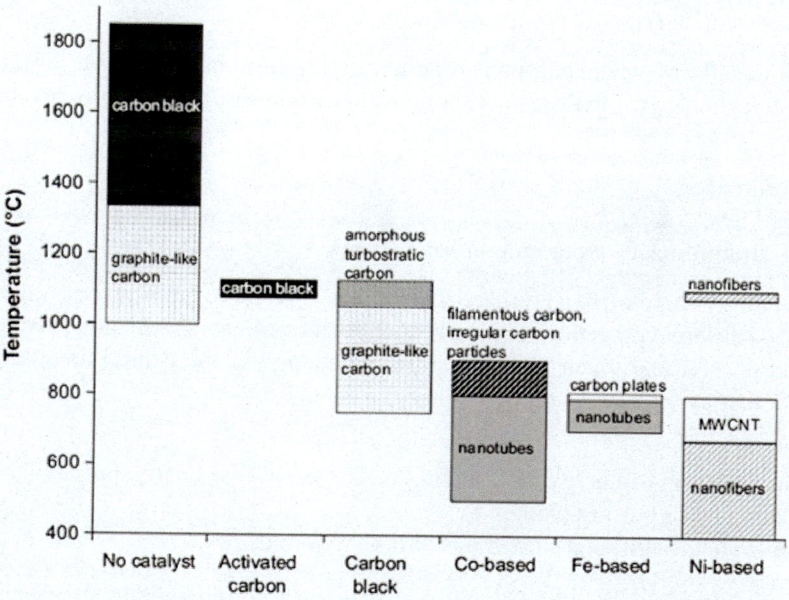


**Fig. 3** Main carbon products obtained as a function of temperature and catalyst employed in the thermo-catalytic decomposition of methane [3]

### Analysis of the Formed Solid Carbon Fraction

With regard to the exact analysis of the solid carbon fraction that is formed during the process in the MMR, the following analytical methods are used: Characterization of impurity concentration, microstructure, density, specific surface area, and porosity of produced solid carbon materials using scanning electron microscopy, X-ray diffraction, Raman spectroscopy, pycnometry, thermogravimetric analysis, and gas sorption measurements. Furthermore, the activation/functionalization of the produced solid carbon materials is carried out by thermo- and plasma-chemical methods.

## Hydrogen in Metallurgy

For metallurgy, hydrogen seems to be a good alternative to fossil carbon because it can replace its purpose as a reducing agent and also act as an energy source. However, burning hydrogen to deliver process energy is expensive and has an influence on the reduction ability of the gas.

### Metallurgical Processes Where Implementation of Hydrogen Is Difficult

Areas in metallurgy where carbon is used in its solid form have to face a substantial change of technology if hydrogen is applied as a carbon substitute. Typical examples are:

- shaft furnaces,
- rotary kilns,
- short drum furnaces (operating in solid stage).

Especially a standard blast furnace for ironmaking relies on a solid reducing agent supporting different procedures during reduction, movement of the charge, avoidance of liquid phases, and much more. Besides the change of the central metallurgical facility, other aspects such as:

- off-gas treatment
- additional form of energy (to compensate for the excess energy provided by the oxidation of carbon) as well as
- more sophisticated safety measures will be required.

Out of this, the metallurgical industry will face a tremendous change in the overall technology, more to say, a completely new process. Such an implementation will hardly be short term because companies cannot afford to scrap all their existing units and replace them with new technologies within a few years. This, in a

CAPEX-extensive industry sector such as metallurgy, is impossible to handle from an economic point of view. Examples are listed in the following:

- iron production via the blast furnace
- primary lead production in shaft furnaces
- secondary copper production in shaft furnaces
- steel dust recycling via waelz kilns
- cast iron production in cupola furnaces
- recycling of other by-products like lead slag or leach residues via rotary kilns

## *Processes Where a Replacement of Fossil Carbon by Hydrogen Is Easier to Realize*

For other process types, already implementing gaseous forms of reducing agents (e.g., natural gas), a substitution by hydrogen could be realized in a shorter period of time, keeping the general technology the same and just adapting the off-gas system and adding an additional source of energy could compensate the combustion of carbon. Examples are:

- vertical retorts that already fulfil the required type of material treatment or
- metal baths where a reduction by injected hydrogen gas bubbles could be realized.

This would be the case for:

- ironmaking in vertical retorts or fluidized bed reactors
- copper anode furnace
- primary lead production based on QSL, Kivcet or TSL
- tin smelting

Nevertheless, an important aspect that always has to be considered is how to realize a sufficient yield of the implemented hydrogen. Bubbles will quickly pass melts and time for reduction is limited. The same is true for retorts. This means the systems to recover hydrogen from the off-gas are highly important to allow the economic utilization of hydrogen.

## Charcoal as a Temporary Solution

In case a solid reducing agent should be replaced, causing a new process concept that would be only realized in long term, alternatives have to be taken into consideration to allow a $CO_2$ neutral production. One option would be charcoal, which per definition is $CO_2$ neutral, and is in general similar to the used fossil coke carrier.

Charcoal, especially if produced at higher pyrolysis temperatures or longer retention time, shows a high content of fixed carbon enabling substitution of fossil-based carbon in a wide range of applications. Some examples are [4]:

- Ferro-alloy-production in submerged arc furnaces
- Substitution of carbon carriers in rotary kilns
- Utilization as a foaming agent in electric arc furnace steel making
- Reducing agent in lead metallurgy (primary and secondary).

For iron production charcoal does not form an alternative due to its low mechanical strength allowing only small units as they are partly operated in Brazil. Also, required amounts would be hard to supply due to the limited availability of proper biomass.

Another difficulty is the high reactivity that limits the application of charcoal from biomass in different processes, causing losses and with this the requirement of higher specific amounts of reducing agent [4].

## Summary

Utilization of hydrogen is a possible solution to realize a $CO_2$ neutral metal production. However, depending on the type of process such a substitution is quite challenging and for some specific fields would require major changes in technology. Therefore, we face a step-wise approach that will allow a certain substitution short to mid-term while major changes cannot be assumed before the mid of the twenty-first century. Alternative options such as a substitution of fossil coke for biochar or the implementation of technologies based on electrical energy could help to close this gap but are also not always easy to realize regarding availability and economic feasibility.

## References


1. IEA (2021) Global hydrogen review 2021. Paris: IEA
2. Scheiblehner D et al. (2022) Hydrogen production by methane pyrolysis in molten binary copper alloys. Int J Hydrogen Energy. https://doi.org/10.1016/j.ijhydene.2022.08.115
3. Keipi T et al (2016) Thermo-catalytic decomposition of methane: the effect of reaction parameters on process design and the utilization possibilities of the produced carbonXE "carbon." Energy Convers Manage 126:923–934
4. Agirre Arisketa I, Griessacher T, Rösler G, Antrekowitsch J (2013) Production of charcoal as an alternative reducing agent from agricultural residues using a semi-continuous semi-pilot scale pyrolysis screw reactor. Fuel Processung Technol 106:114–121

# Use of $H_2$ in Mn-Ferroalloy Production

**Merete Tangstad, Trygve Schanche, and Faan de Preez**

**Abstract** The use of $H_2$ as a reductant in the iron and steel industry is an obvious choice towards carbon neutrality. For pyrometallurgical processes like Mn and Si, $H_2$ cannot be the only solution, as the stability of the oxides of these elements is higher compared to Fe. $H_2$ can, however, be used in the Mn-process together with other low $CO_2$ emission mitigations. In the Mn-ferroalloy process, $H_2$ can be used to reduce higher manganese oxides to MnO, and the last part of the reduction to metallic Mn can be done with biocarbon or with electrolysis. In studies from NTNU, the reduction of higher manganese oxides to MnO has been investigated with pure $H_2$ or with $CO/H_2$ mixtures. It is shown that the $H_2$ containing gases will give higher reduction rate compared to $CO/CO_2$ gases with the same reduction potential. The reduction rate will be increased in the order of 20–100%, and $H_2$ has a larger impact on the reduction rate for high oxygen pressures. In $H_2$ gas, it is seen that the final degree of reduction is higher than using CO gas, and this is believed to be due to the formation of metallic iron. The reduction rate in $H_2$ gases varies for various ores, and as for CO containing gases, the reduction rate of Comilog ore is faster than the Nchwaning ore, and activation energies of 23 kJ/mol versus 68 kJ/mol are, respectively, found for isothermal experiments in CO, $H_2$, and $CO_2$ gases.

**Keywords** Mn-ferroalloy process · Hydrogen · Prereduction

## Introduction

Mn-ferroalloys are mainly produced in Submerged Arc Furnaces (SAF) as many other ferroalloys. The most typical geometry is the circular furnace with 3 Søderberg electrodes. In the lower part of the furnace, the electrical energy is provided by the electrodes, and here the temperature is 1400–1600 °C. At these temperatures,

M. Tangstad (✉) · T. Schanche
Norwegian University of Science and Technology, Trondheim, Norway
e-mail: Merete.tangstad@ntnu.no

F. de Preez
North West University, Potchefstroom, South Africa



C. Fleuriault et al. (eds.), *Advances in Pyrometallurgy*, The Minerals, Metals & Materials Series, https://doi.org/10.1007/978-3-031-22634-2_4

the oxide materials are melted and reduced to metal, before it is drained from the furnace as metal and slag [1]. The slag/metal ratio is typically in the range of 0.5–1. The raw material mix, containing Mn-sources, fluxes, and carbon materials, is being fed to the charge by gravity at the top of the furnace, where the temperature is at its lowest, in the range of 200–600 °C. As slag and metal are being tapped from the furnace, the rest of the material in the furnace will descend while heated by the gas coming from the high-temperature zone. The high-temperature zone is called the cokebed (1400–1600 °C), due to a stagnant layer of carbon materials, while the low-temperature zone (200–1400 °C) is called the prereduction zone.

In the high-temperature zone, the following carbon-consuming reactions occur in the metal-producing reactions during High Carbon FeMn (HC FeMn) production [2]:

$$MnO + C = Mn + CO \tag{1}$$

$$FeO + C = Fe + CO \tag{2}$$

$$C = C(\text{carbon dissolves into the metal}) \tag{3}$$

Analysing the $CO_2$ emission, one can start to discuss the total carbon consumption per ton of metal produced. The typical composition of HC FeMn is 76–79% Mn, 7% C, and the rest is iron. As shown in Table 1, this requires 273 kg of carbon, however, only 203 kg will go into the gas phase as CO, and the rest will dissolve in the metal. There is, however, also some carbon consumption in the prereduction zone. The higher Mn- and Fe-oxides are reduced with CO gas to $CO_2$ (Rx. 4) and if this reaction happens above about 800 °C, the $CO_2$ will subsequently react with the carbon materials according to the Boudouard reaction (Rx. 5), and the sum of these two reactions (Rx. 6) will consume carbon [2, 3].

$$Mn_yO_X + CO = Mn_yO_{X-1} + CO_2 \tag{4}$$

$$\underline{CO_2 + C = 2CO} \tag{5}$$

$$Mn_yO_X + C = Mn_yO_{X-1} + CO \tag{6}$$

$CO_2$ from the decomposition of carbonates may also react according to the Boudouard reaction (Rx. 5), and hence the total reaction will be according to Rx. 7.

$$CaCO_3 + C = CaO + 2CO \tag{7}$$

**Table 1** Example of carbon consumption and subsequently $CO_2$ emission producing a metal with 79% Mn, a slag/metal ratio of 0.7, 40% MnO in the slag, 100 kg lime/ton metal, and 0% prereduction

| | | kg C | kg $CO_2$ |
|---|---|---|---|
| 790 kg Mn | $MnO + C = Mn + CO$ | 173 | 633 |
| 140 kg Fe | $FeO + C = Fe + CO$ | 30 | 110 |
| 70 kg C | C dissolved in metal | 70 | 0 |
| *C in cokebed zone* | | 273 | 743 |
| s/m = 0,7, 40%MnO | $Mn_3O_4 + C = 3MnO + CO$ | 59 | 217 |
| | $Fe_3O_4 + C = 3FeO + CO$ | 10 | 37 |
| 100 kg CaCO3/ton metal | $CaCO_3 + C = CaO + 2CO$ | 12 | *88* |
| *0% degree of prereduction* | | 81 | 341 |
| Total | | 354 | 1084 |

Based on industrial operation over 5 years and 4 different furnaces [3], it has been seen that the $CO_2$ from the reduction of $MnO_2$, $Mn_2O_3$, and $Fe_2O_3$ with CO gas, will not react with solid carbon, and hence the total oxygen content of the charge mix will not affect the total carbon consumption. The $CO_2$ from $Mn_3O_4$ and $Fe_3O_4$ to MnO and FeO, as well as the $CO_2$ from the decomposition of lime ($CaCO_3$), may or may not react with carbon to CO gas. If none of this $CO_2$ is reacting with C, it has been defined as 100% degree of prereduction, and if all of this $CO_2$ reacts with C, it has been defined as 0% degree of prereduction. Industrially, it was seen that the typical degree of prereduction was changing between 0 and 60%. The degree of prereduction is determined by the type and size of Mn sources as well as furnace operation.

So far we have discussed carbon consumption, as it is proportional to the final $CO_2$ emissions (with the exemption of the carbon dissolved in the metal). This is always the case for open furnaces where all CO gas will burn on the charge top. There are, however, some notes that must be included on this issue. For closed furnaces, the off gas will be a mixture of CO and $CO_2$ gas. If the gas is flared, again the carbon consumption will be proportional to the C consumption. As the off gas from Mn-alloy production may contain high amounts of CO (50–90%), this gas may be sold to other users of CO as fuel. When this gas is burned to $CO_2$ and emitted, the total amount of $CO_2$ emissions will again be proportional to the original carbon consumption. This $CO_2$ may, however, be reported as a part of the customer's $CO_2$ emissions and not the metal producer. From an environmental point of view, the total $CO_2$ emission will still be proportional to the carbon consumption in the metal-producing furnace, and the goal is to reduce this to a minimum.

In Table 1, the carbon consumption needed in the HC FeMn is shown. In the high-temperature zone, 273 kg of carbon per ton of metal is required, however, 70 kg of these will be dissolved in the metal, and hence 203 kg C or 743 kg $CO_2$ will be emitted from the reactions in this zone. If no prereduction is occurring, the additional carbon consumption in the prereduction zone is 81 kg C, that is, 341 kg $CO_2$ will be emitted. This means that with these operating conditions, 31% of the $CO_2$ emitted from the furnace will come from the prereduction zone.

One of the strategies to reduce the $CO_2$ emissions in today's processes is to increase the degree of prereduction in the furnace, that is, to reduce the extent of the Boudouard reaction, and hence reduce the total carbon and energy consumption. Figure 1 shows the carbon consumption and the $CO_2$ emission for the same charge as shown in Table 1 as a function of the degree of prereduction. As previously noted, the highest degree of prereduction seen industrially was in the area of 60%, and this would reduce the $CO_2$ emissions to 870 kg $CO_2$, that is, 19% reduction. To obtain such numbers, the operation is optimal under today's conditions. The main factor for this to happen is, first and foremost, good raw materials. If the raw materials could be developed further to obtain a 100% degree of prereduction, a total reduction of 31% $CO_2$ emission compared to 0% prereduction could be obtained.

One of the discussed paths to reduce $CO_2$ emissions is to use $H_2$ instead of solid carbon. MnO is a stable oxide and cannot be reduced with $H_2$, only with C as shown in reaction 1. FeO and the higher manganese oxides can, however, be reduced with $H_2$ according to reactions 8 and 9 as seen in Fig. 2.

$$Mn_yO_X + H_2 = Mn_yO_{X-1} + H_2O \tag{8}$$

$$FeO + H_2 = Fe + H_2O \tag{9}$$

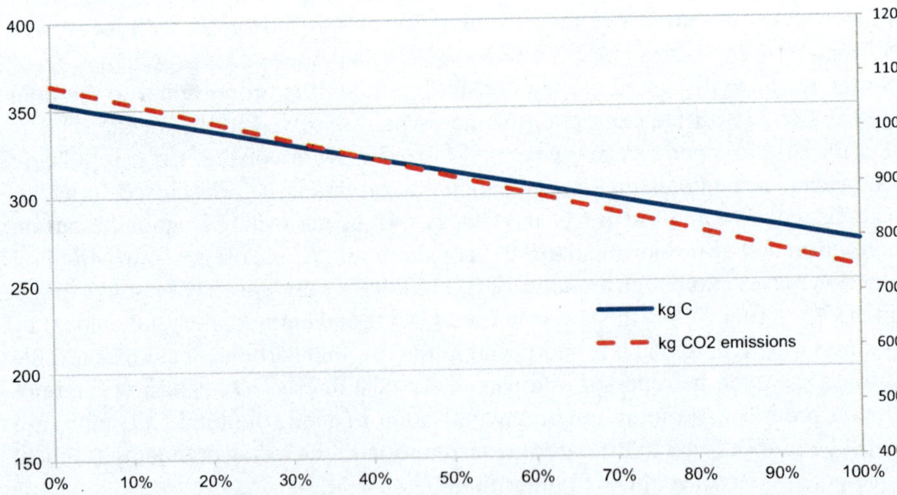


**Fig. 1** C consumption and subsequently $CO_2$ emissions for a given charge (slag/metal ratio = 0.7, %MnO in slag = 40%, 100 kg $CaCO_3$ per ton metal, all Mn sources are $MnO_2$, $Mn_2O_3$ or $Mn_3O_4$)

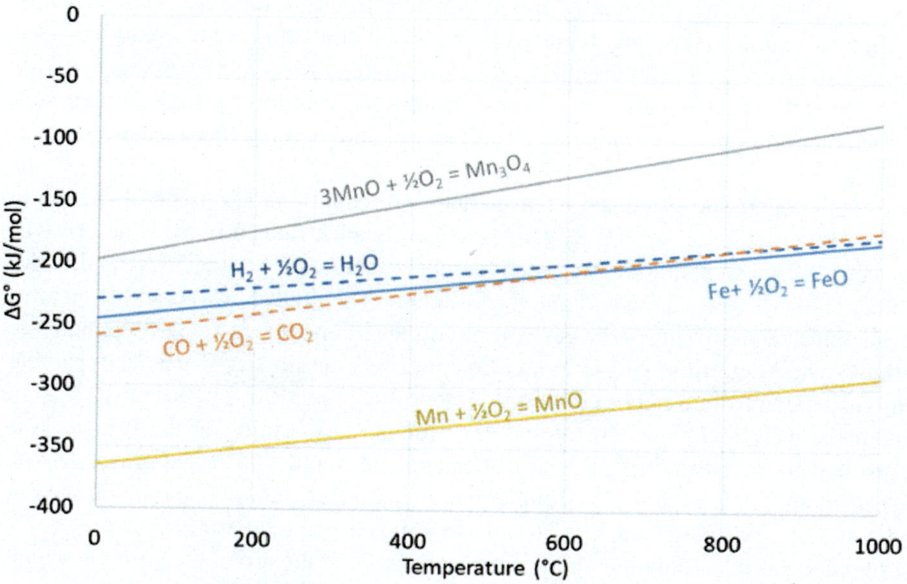


**Fig. 2** Ellingham diagram of Mn- and iron-oxides compared to $H_2/H_2O$ and $CO/CO_2$ gases (calculated from HSC Chemistry10)

In a process where the reactions of the prereduction zone could be separated from the high-temperature cokebed zone, e.g., in a prereduction unit, $H_2$ could be used to reduce the higher manganese oxides and the FeO without any Boudouard reaction. A more possible scenario is to operate a prereduction unit with no solid carbon, and hence both the CO from the furnace and/or additional $H_2$ may be used. In both cases, the total emitted $CO_2$ would now be down to 633 kg $CO_2$ per ton of metal produced (or 743 kg $CO_2$ if the iron is reduced to metallic iron in the prereduction unit). The conditions which are required to reduce the ore down to MnO and Fe are investigated in a number of projects [4–8]. This paper will sum up the work done at the Norwegian University of Science and Technology on the use of $H_2$ as a reductant for Mn-ores.

## Methods and Materials

Nchwaning and Comilog ore were the raw materials used in most of the studies, in addition to one study using UMK ore. UMK is a semi-oxidized ($Mn_2O_3$) carbonate ore. The main minerals are braunite I ($3(Mn,Fe)_2O_3.MnSiO_3$), carbonates (calcite ($CaCO_3$), dolomite ($CaMg(CO_3)_2$), and hausmannite ($Mn_3O_4$). Kutnahorite ($(Ca,Mn,Mg,Fe)(CO_3)_2$), bixbyite ($(Mn,Fe)_2O_3$), and hematite ($Fe_2O_3$) is also found but to a lesser degree [9]. Comilog is a high oxygen ore, where the oxygen level is close to that of $MnO_2$. The ore is low in iron and contains no carbonates. It contains approximately 5 wt% chemically bound moisture found in minerals such as lithiophorite and nsutite, as recently reported [5, 10, 11]. Nchwaning ore is a semi-oxidized ore, where the oxygen level is close to 1.5 correlating to $Mn_2O_3$. It contains insignificant amounts of chemically bound moisture but has a relatively high iron content and smaller amounts of carbonate. The chemical analyses for the ores are given in Table 2.

A TGA technique was used to investigate prereduction. The experiments were performed in an Entech VTF 80/15 vertical resistance tube furnace. The ore was placed in a high-temperature gas-tight stainless-steel (steel grade 253 MA) double-wall crucible (height 45 cm and diameter 4.8 cm). The double-wall crucible ensures preheating and premixing of the gas. The gas outlet is connected to an off-gas analyser (NDIR) which determines the CO and $CO_2$ concentrations in the off-gas. The crucible was suspended from a balance (Mettler Toledo PR2003DR, Switzerland,10 mg) to record the weight changes during the experiment. The heating of the furnace was controlled by a Eurotherm PID controller and measured with a calibrated S-type thermocouple. A K-type thermocouple placed in an alumina tube and positioned in the middle of the charge was used to measure the charge temperature.

Flexible gas inlet and outlet pipes were connected to the top of the crucible which allowed multiple gasses to mix before entering the crucible. Each gas had its own mass flow controller (Bronkhorst F-201C). The weight of the sample before and after the experiment was recorded to confirm the net weight loss obtained from the TGA data. After the experiment was complete the sample was allowed to cool in Ar.

The chemical composition of the raw materials and the prereduced ores was examined by X-ray fluorescence (Bruker AXS S4 Pioneer X-Ray fluorescence spectrometer) using the fused bead technique. Permanganometric titration (ASTM 465-11:2017) was employed to measure the excess oxygen above MnO which is expressed as $MnO_2$. The carbon concentration was analysed with a LECO (Combustion-IR) instrument and the $CO_2$ concentration was determined by assuming that all carbon in the ore is in the form of $CO_2$ (Fig. 3).

As seen in Table 3, the temperature profile and gas compositions vary in the different studies. For the isothermal experiments, two strategies were chosen. Lobo heated the Nchwaning pellets in Ar to the isothermal temperature. Lobo also roasted the pellets for 2 h at 800 °C to get good mechanical strength. It has, however, been seen that the heating conditions are affecting the subsequent prereduction rate [13], and Schanche and Davies hence chose to heat the ore in the gas species used during the isothermal reduction, to simulate an industrial process. The disadvantage of this method is that some of the reduction will occur during the heating. Also, in the non-isothermal experiments, where a fixed heating rate was used, two different strategies were chosen. Larssen chose to use the same gas atmosphere during the whole temperature range. It was, however, seen that CO would decompose to C and $CO_2$ at lower temperature in $H_2$ rich atmospheres, and hence Ngoy heated the ore to 500 °C in $CO_2$, before adding the reducing gas composition from 500 to 1000 °C. As seen by both Ngoy [14] and Biørnstad [9], Comilog ore would then decompose from MnO to $Mn_2O_3$ during the heating step, while Nchwaning would hardly change the oxidation degree.

**Table 2** Chemical composition of ores (average ± std. dev.)

| | Fe, tot (wt%) | Mn, tot (wt%) | $MnO_2$ (wt%) | MnO (wt%) | CaO (wt%) | $SiO_2$ (wt%) | $Al_2O_3$ (wt%) | $K_2O$ (wt%) | $CO_2$ (wt%) |
|---|---|---|---|---|---|---|---|---|---|
| Comilog [6] | 3.1 | 51.0 | 76.4 | 3.5 | 0.1 | 3.5 | 5.6 | 0.7 | 0.1 |
| Comilog [11] | 3.6 | 51.4 | 74.3 | | 0.1 | 3.3 | 5.1 | 0.8 | 0.15 |
| Nchwaning [6] | 10.0 | 46.4 | 34.6 | 31.6 | 5.9 | 6.7 | 0.5 | 0.0 | 3.0 |
| Ncwaning [8] | 8.7 ± 2.5 | 48.7 ± 3.3 | 38.8 ± 2.0 | | 5.9 ± 0.6 | 4.4 ± 0.1 | 0.4 ± 0.1 | 0.0 | 2.5 ± 0.5 |
| UMK [12] | 5.9 | 33.7 | 20.55 | | 17.0 | 7.3 | 0.4 | 0.06 | 17.0 |

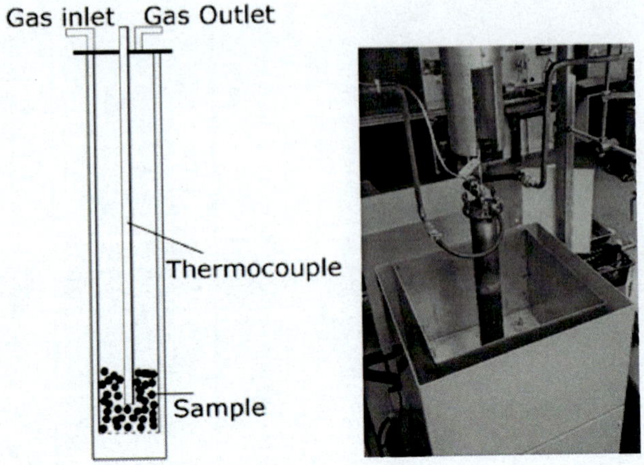


**Fig. 3** Sketch of the crucible used in the reduction experiments [6]

**Table 3** Experimental conditions of the studies discussed in this article

| | Lobo | Ngoy | Larssen | Davies | Schanche |
|---|---|---|---|---|---|
| Ores | Nchwaning | Nchwaning Comilog | Nchwaning Comilog | UMK | Nchwaning Comilog |
| | Pellets | Lumps | Lumps | Lumps | Lumps |
| Size (mm) | 10–14 | 9.5–16 | 11.2–15 | 9.5–16 | 9.2–15 |
| Gas | Various mixes of $H_2$, $H_2O$, CO, $CO_2$ | $CO/CO_2$, $CO/CO_2/H_2$ | $CO/CO_2$, $CO/CO_2/H_2$, $CO/CO_2/H_2O$ | $CO/CO_2$, $H_2/H_2O$, $H_2$ | $CO/CO_2$, $CO/CO_2/H_2$ |
| | Isothermal | Non-isothermal | Non-isothermal | Isotherma[a] | Isotherma[a] |
| Gas flow | 5 l/min | 2.85, 4.85 l/min | 4 l/min | 4 l/min | 4 l/min |
| Isothermal temperature | 950 °C | | | 700, 800, 900 °C | 605, 700, 790 °C (Nch) 365–605 °C (Com) |

[a] Parts of the reduction will occur before isothermal temperature is reached

## Results and Discussion

### *Reduction of Mn-Ores in $H_2$–$H_2O$ Gases Compared to CO–$CO_2$ Gases*

Figure 2 shows the reducing potential of $H_2/H_2O$ versus $CO/CO_2$ gases. For the same $H_2/H_2O$ ratio as $CO/CO_2$ ratio, the $CO/CO_2$ gas will have lower $O_2$ partial pressure below 800 °C and $H_2/H_2O$ gas will have the lowest $O_2$ partial pressure, and thus the highest reduction potential, above this temperature. It can also be seen that the difference in reduction potential is not very large above 800 °C; at 900 °C, the $O_2$

partial pressure for a 70/30 $H_2/H_2O$ gas is $5.5 \cdot 10^{-5}$ bar and for a 70/30 $CO/CO_2$ gas, it is $6.3 \cdot 10^{-5}$ bar. The overall picture is hence that CO and $H_2$ have a quite similar reduction potential in the temperature range of 500–1100 °C.

Figure 4 shows the difference between the reduction rate and the extent of the reduction in CO and the $H_2$ system for Nchwaning ore and Nchwaning pellets. Though the reduction potential is quite similar as discussed above, it is seen that the $H_2$ gas is reducing the ore to a higher extent than the CO gas. If, however, a mixture of $H_2$ and CO is used, the extent of the reduction will be in the same area as the CO gas. The higher extent of reduction is also seen by Davies [12] as shown in Fig. 5 when reducing the UMK carbonate ore. At 700 °C and 800 °C in CO–$CO_2$ gas, the carbonates are not yet decomposed as seen in Fig. 6, which explains the lower extent of mass loss at these temperatures. At 900 °C, where all the carbonates are decomposed, the extent of reduction is higher in $H_2$ compared to $H_2/H_2O$ and $CO/CO_2$ mixtures, which can be explained by the increased reduction potential in the $H_2$ gas as shown in Fig. 7. $H_2$ gas will, at equilibrium condition, reduce some iron to the metallic state, while the Mn will be in the (Mn,Fe)O phase. The 70/30 $H_2/H_2O$ and $CO/CO_2$ gas will be at the boundary line between the metallic Fe and FeO line. The XRD results, shown in Fig. 8, verifies that the metallic Fe is produced in $H_2$ gas. In the mixed gases, the metallic phase was not found. The difference in $\alpha$ when all FeO is reduced to Fe is 7%, and hence it seems like most of the FeO is reduced to the metallic state. Going back to the study of Lobo in Fig. 4 again, the difference between $H_2$ and CO cannot be explained by the difference in reduction potential, as the equilibrium state at this temperature and gas composition will now be metallic for both gas mixtures. Metallic iron was found by XRD for both gas mixtures, however, for the $H_2$ gas experiments, the monoxide was MnO, and for the CO gas, it was (Mn,Fe)O. This also indicates that it is the difference in the iron reduction that separates the extent of reduction.

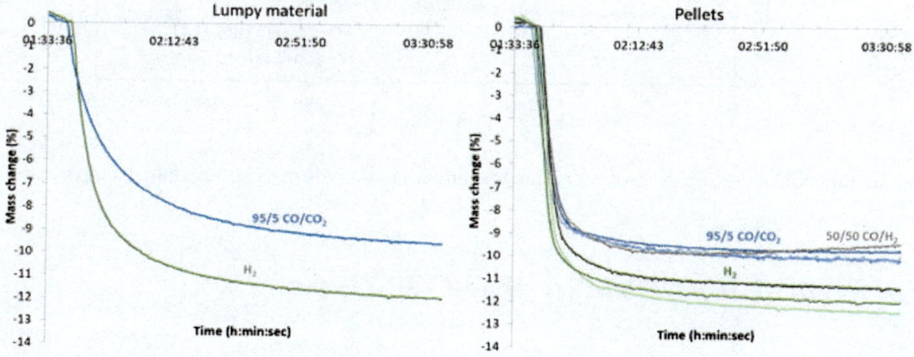


**Fig. 4** Comparison of reduction of lumpy material and pellets from Nchwaning ore with $H_2$ and 95%CO gas at 950 °C from Lobos study

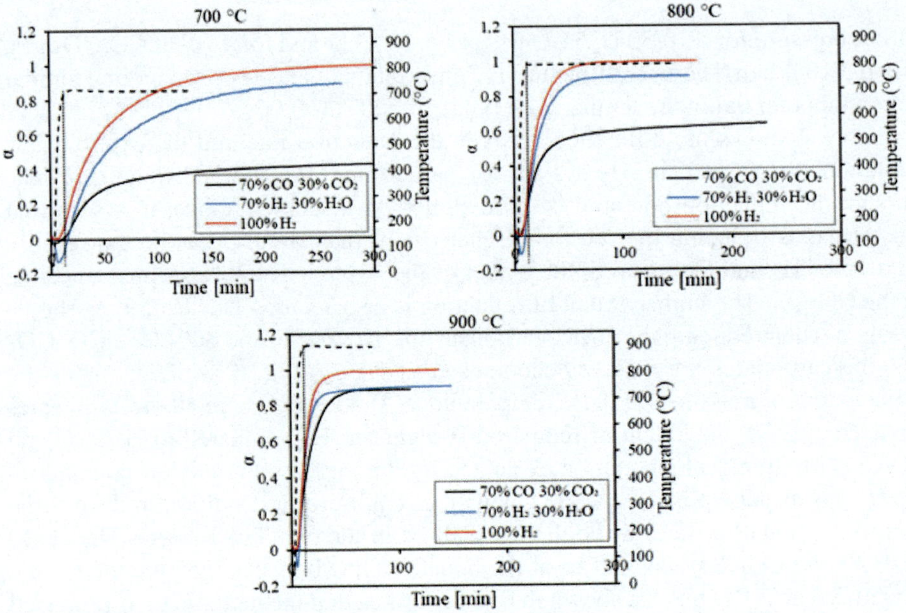


**Fig. 5** The extent of reduction and decomposition of carbonates in UMK ore at 700, 800, and 900 °C investigated by Davies [12]. Note a negative extent below 200 °C due to $H_2O$ condensation from the $H_2$–$H_2O$ gas

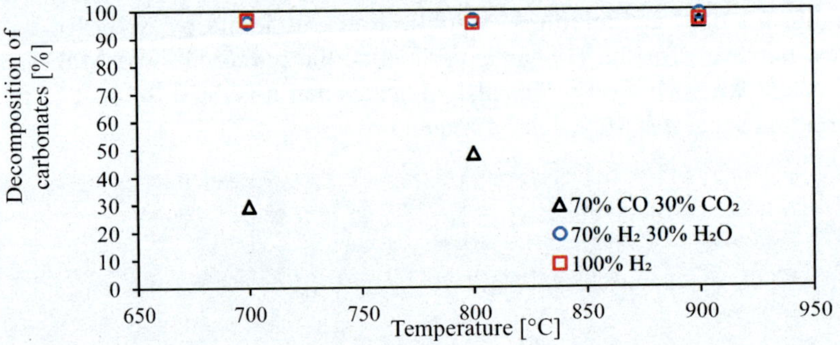


**Fig. 6** Fraction of the total wt% of carbonates decomposed in UMK ore as a function of temperature [12]

## *Reduction of Mn-Ores in $H_2$-$H_2O$-CO-$CO_2$ Gases*

Ngoy [14] investigated CO–$CO_2$–$H_2$ mixtures and compared them with CO–$CO_2$ mixtures with the same partial pressure of $O_2$ when in equilibrium, that is, the same reduction potential. The samples were heated to 500 °C in $CO_2$ before the reducing gas was added. Both Comilog and Nchwaning ore was investigated with two oxygen partial pressures. The mass loss is shown in Fig. 9. It is seen that the rate is higher in

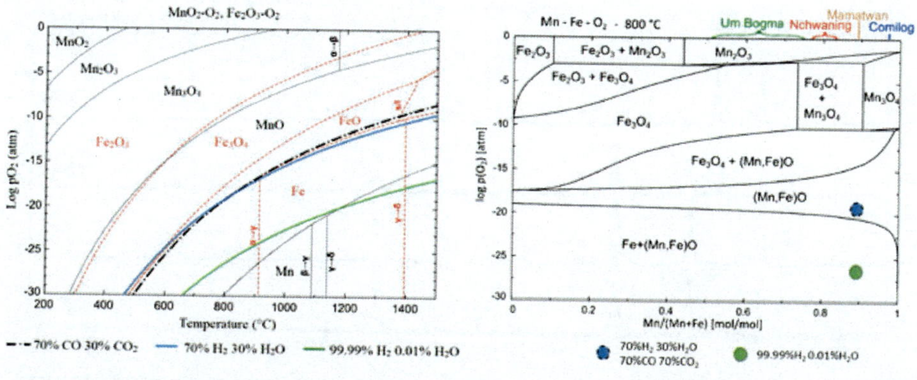


**Fig. 7** The Fe–O and Mn–O stability diagrams show the stable phases with varying oxygen partial pressure and temperature calculated by Factsage and HSC chemistry. The left show the superimposed Fe–O and Mn–O diagram, and the right shows solid solution Fe–Mn–O diagram [11, 12]

the $H_2$ containing gases and lower in $O_2$ containing gases as shown in Fig. 9a, c. At higher oxygen pressures, Fig. 9b, c, the $H_2$ containing gas reduces the Comilog ore faster. For the Nchwaning ore, the $H_2$ gas starts the reduction a bit later, however, the rate is faster when it has started. This experiment may be an outlier as it does not fit the results in the later work done. In the $H_2$ containing gas experiments, carbon deposition was seen both visually after the experiment, and on the weight curve, as shown by the increasing weight for Comilog ore. If carbon deposition occurred during the whole reduction path, it means that the $H_2$ containing gases will reduce the ore faster than the graph shows.

Larssen [6] investigated the prereduction of Comilog ore with increasing temperature for two oxygen partial pressures, with and without $H_2$, as seen in Fig. 10. Both for the high oxygen partial pressures experiments, the 50/50% $CO/CO_2$ and 44/12/44% $CO/H_2/CO_2$ experiments, and the low oxygen partial pressure experiments, the 80/20% $CO/CO_2$ and 70/12/18% $CO/H_2/CO_2$ experiments, the hydrogen has a strong effect on the reduction rate. It was also noted that the C content after the experiments was above1% for the $H_2$ containing gas and about 0.2–0.3 in the CO–$CO_2$ experiments, which again means that carbon was deposited during the experiments, and hence that the real weight loss from the reduction was even higher. In the low oxygen partial pressure experiment, one did not experience the typical temperature peak from high oxygen ores just below 600 °C, where the remaining $MnO_2$ would decompose to $Mn_2O_3$ [5]. This is explained by the fast reduction due to the low oxygen partial pressure and $H_2$ being present.

Similar experiments were performed using Nchwaning ore, however, they were influenced by carbon deposition [6]. For both oxygen partial pressures, the carbon deposition was initiated above 400 °C. This correlates well with previous reports where carbon deposition is initiated at 400 °C for this type of system, and the deposition extent is promoted by an increased CO concentration in the CO–$CO_2$ atmosphere [10]. The carbon deposition also increased with increased $H_2$ pressure as seen

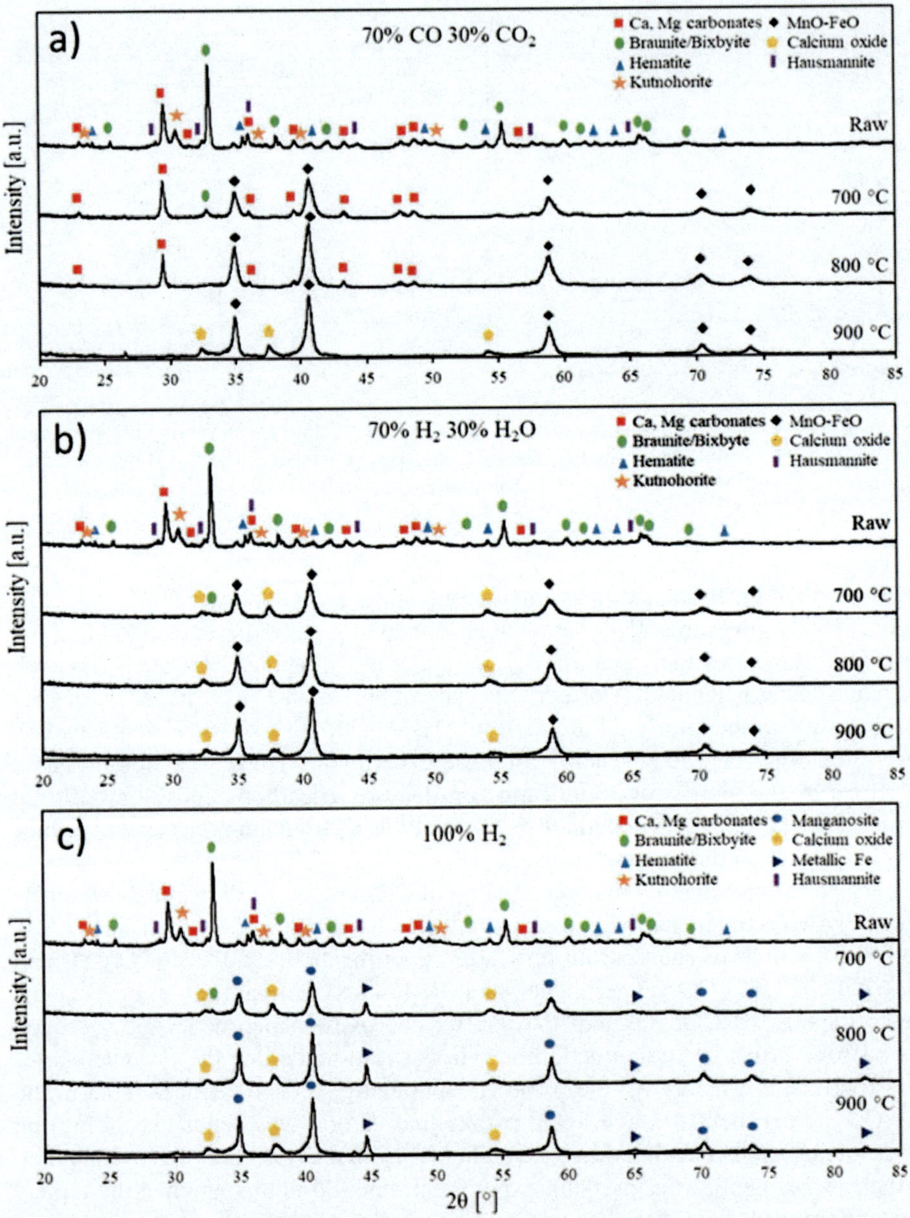


**Fig. 8** XRD results of the ore and prereduced in 70% CO 30% $CO_2$ (**a**), 70% $H_2$ 30% $H_2O$ (**b**), and 100% $H_2$ (**c**) at 700, 800, and 900 °C. Chemical formula of compounds: calcite ($CaCO_3$); magnesium carbonate ($MgCO_3$) braunite ($3(Mn,Fe)_2O_3{\cdot}MnSiO_3$); bixbyite ($(Mn,Fe)_2O_3$); hematite ($Fe_2O_3$); kutnahorite ($(Ca,Mn,Mg,Fe)(CO_3)_2$); manganosite (MnO); calcium oxide (CaO); hausmannite ($Mn_3O_4$) [12]

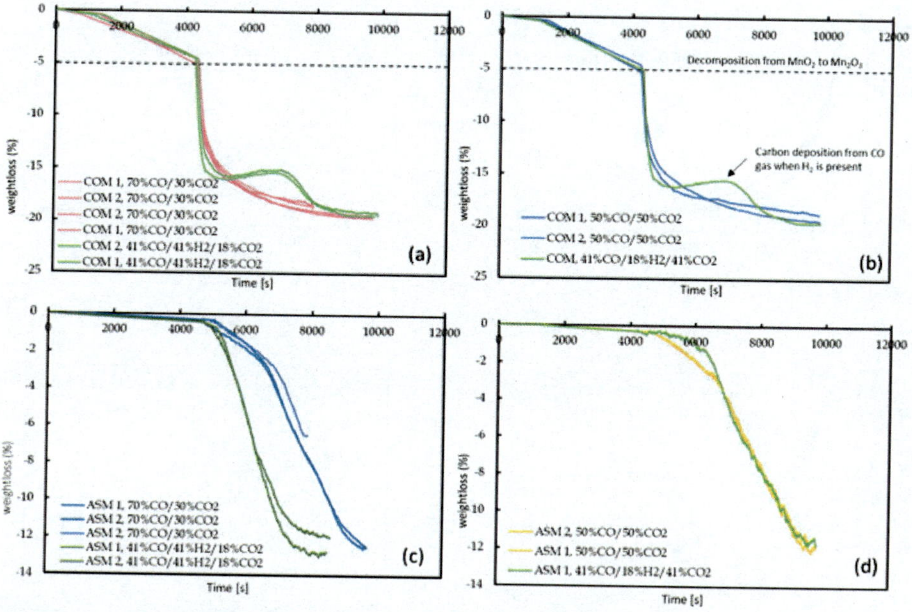


**Fig. 9** Comparison of mass loss for $H_2$ containing and $H_2$ free gas for Comilog (**a**, **b**) and Nchwaning ores (**c**, **d**). (**a**, **c**) have the same low oxygen pressure and (**b**, **d**) has a higher oxygen pressure in the gas. *Note* In (**c**) the two $H_2$ containing gas experiments have been shifted in time, to start the reduction at the same time as the $H_2$ free gases in (**c**) [14]

**Fig. 10** Weight loss for non-isothermal reduction of Comilog with $H_2$ containing and $H_2$ free gas composition, where the grey lines have high oxygen pressure, and the blue lines have low oxygen pressure [6]

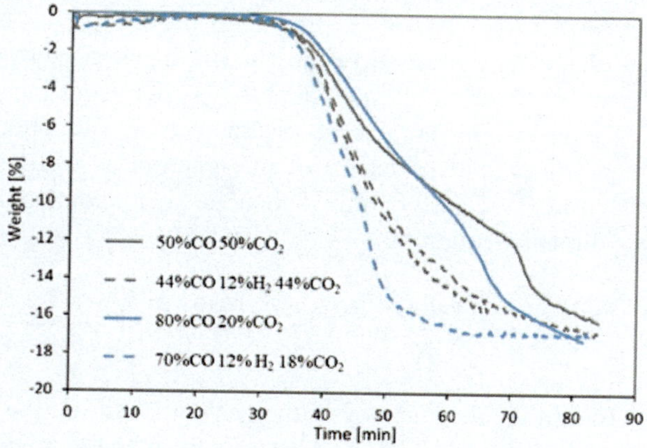


by Ngoy et al. [14]. Compared to Nchwaning ore, Comilog ore produces more $CO_2$ during the reduction with CO gas, and carbon deposition occurs to a lesser extent.

Schanche [8, 11] investigated the isothermal prereduction for Nchwaning and Comilog ore with and without $H_2$ present in $CO/CO_2$ gases as seen in Figs. 11 and 12. As for previous studies, the $H_2$ added to the $CO/CO_2$ gases increased the reduction rate, mainly in the area of 30–50%, with the exemption of high oxygen gas mixture

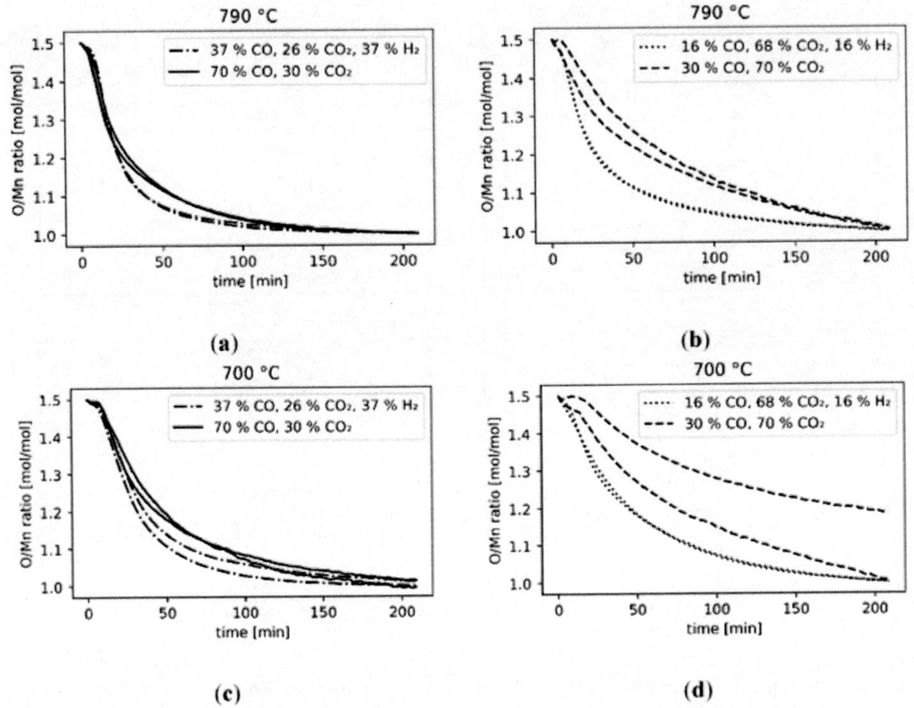


**Fig. 11** The effect of hydrogen at constant oxygen partial pressure at 790 °C for **a** low $pO_2$, and **b** high $pO_2$(Nchwaning ore) [11]

and Nchwaning ore, where the $H_2$ in the gas mixture more than doubled the rate. Ngoy also found in his study that the $H_2$ in the gas increased the rate by 20–40%.

The reduction rate can be assumed to be a function of the reduction potential, temperature, and the extent of the reaction as well as the presence of $H_2$ in the reduction gas. Based on this, Schanche [11] modelled the reaction rate according to the following equation:

$$\frac{d\alpha}{dt} = k(T)f(\alpha) = k_0 \exp\left(-\frac{E_a}{RT(t)}\right)(1-\alpha)^x \quad (10)$$

To find the dependency of the gas composition, the logarithm of $k_0$ is plotted as a function of ln(1-$CO_2$). The parameters *m* and *k* are determined based on the slope and *y*-intercept of the linear regression between the points as shown in Fig. 13. Based on the correlation, the total equation describing the reaction rate is given in Eq. (10).

$$\frac{d\alpha}{dt} = k \exp\left(-\frac{E_a}{RT(t)}\right)(1-\alpha)^x p^m_{(1-CO_2)} \quad (11)$$

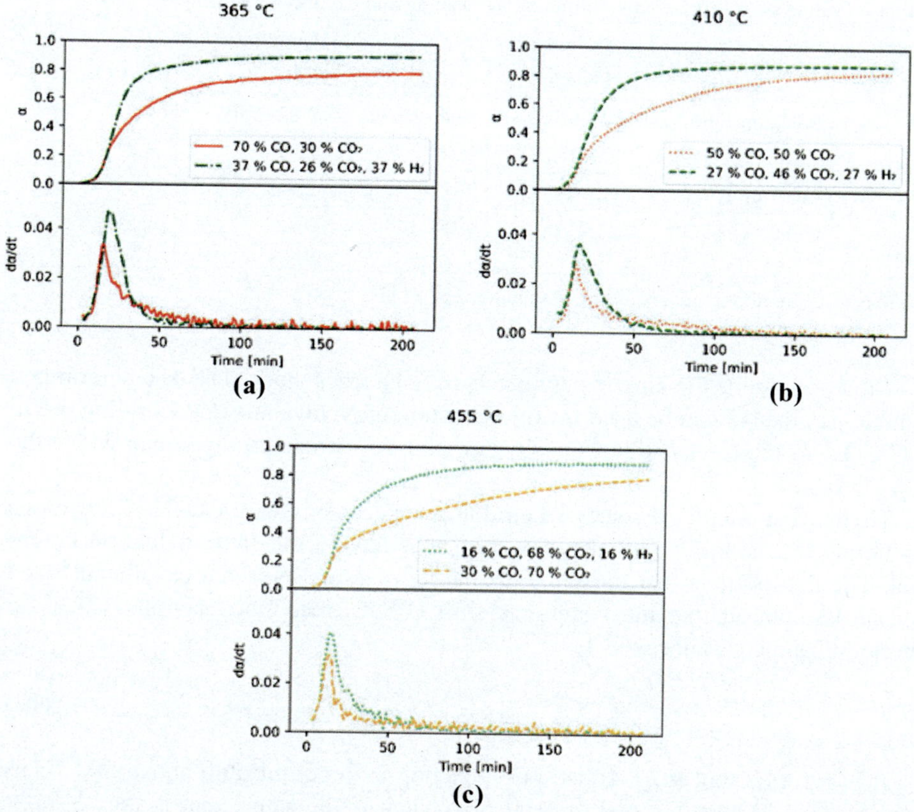


**Fig. 12** The fractional conversion curves for Comilog ore; **a** hydrogen free gas mixtures at 410 °C, **b** hydrogen containing gas mixtures at 410 °C, **c** low $pO_2$ at 365 °C, **d** medium $pO_2$ at 410 °C, and **e** high $pO_2$ at 455 °C [11]

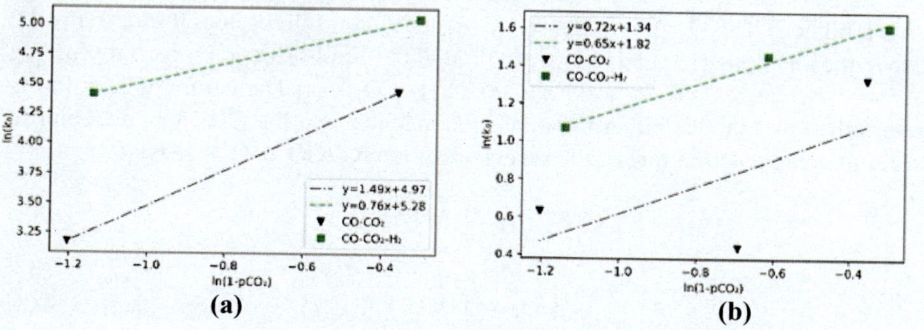


**Fig. 13** The dependency of reaction rate on gas composition for **a** Nchwaning ore and **b** Comilog ore [11]

**Table 4** Values obtained from modelling of Nchwaning and Comilog ore using Eq. 11

| Parameter | | Nchwaning | | Comilog | |
|---|---|---|---|---|---|
| | | $CO/CO_2$ | $CO/CO_2/H_2$ | $CO/CO_2$ | $CO/CO_2/H_2$ |
| $k$ | Rate constant | 75.9 | 69.4 | 3.82 (5.05) | 6.18 |
| Ea | Activation energy | 68.1 | 68.1 | 23.2 | 23.2 |
| $x$ | Reaction order | 2 | 2 | 2 | 2 |
| $m$ | Order of 1-$CO_2$ | 1.21 | 0.38 | 0.72 (0.82) | 0.65 |

Numbers in parenthesis are excluding the medium $pO_2$ [11]

Table 4 contains the kinetic parameters in Schanches work [11]. The determined kinetic parameters can be used for the reduction of Nchwaning and Comilog ore in $CO/CO_2$ gas mixtures and in $CO/CO_2$ gas mixtures containing hydrogen where the $H_2/CO$ ratio is 1.

There are a couple of issues when doing experiments with CO–$H_2$–$CO_2$ gases assuming the same $O_2$ partial pressure, that is having the same reduction potential. The first issue is that the same $O_2$ partial pressure has been calculated based on an assumption that the water–gas shift (WGS) reaction, reaction (12), is in thermodynamic equilibrium.

$$CO_2 + H_2 = CO + H_2O \tag{12}$$

It is reported that [6, 11] the WGS reaction is in equilibrium above 500 °C as seen in Fig. 14. As the total amount of oxygen is the same, that is the $\frac{CO+H_2}{CO_2+H_2O}$ ratio is constant in the gas, the reduction potential of the gas is quite constant even if the WGS reaction is not in equilibrium. The second, and more serious issue, is that C is being deposited according to the reverse reaction (5) ($2CO = C + CO_2$), especially when $H_2$ is present. This could be due to the WGS reaction which will give a higher $CO/CO_2$ ratio when $H_2$ is added. On a microscale, it could also be due to the reactions (13) and (14), even though the equilibrium $CH_4$ partial pressure is $5.10^{-5} bar$ at 800 °C for a 60/20/20 $CO/H_2/CO_2$ gas. The carbon deposition is determined by kinetics rather than thermodynamics, and the $CH_4$ may decompose faster than the reaction rate of the reverse reaction 5 ($2CO = C + CO_2$).

$$2CO + 2H_2 = CH_4 + CO_2 \tag{13}$$

$$CH_4 = 2H_2 + C \tag{14}$$

To summarize the impact of $H_2$ in $CO/CO_2$ mixtures, it is generally seen that $H_2$ will increase the reduction rate for the same reduction potential, that is, the same or similar $O_2$ partial pressure. This is believed to be due to the mass transfer in the gas phase. As seen in Fig. 15, the diffusion coefficient of $H_2$ in $CO_2$ is many times higher

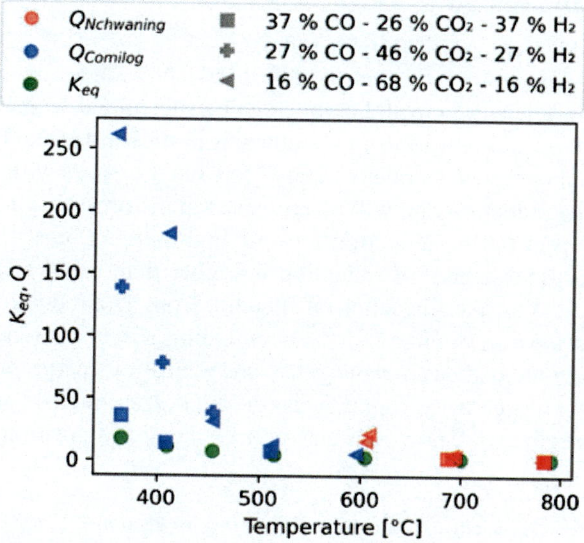


**Fig. 14** Shows the reaction quotients (Q) from the final gas composition of the hydrogen containing experiments and the equilibrium constant, Keq, from the water–gas shift reaction, as a function of temperature. The equilibrium constants were calculated using HSC Chemistry [11]

than the diffusion coefficient of CO in $CO_2$ and may hence explain the increased reaction rate. It has been shown in all the discussed studies that the prereduction of Mn-ores is not in equilibrium and hence the increased mass transfer with $H_2$ will increase the reaction rate.

As a final short-point, it can be mentioned that the decrepitation during the reduction in $H_2$ containing gases has not changed compared to $H_2$ free gases [11].

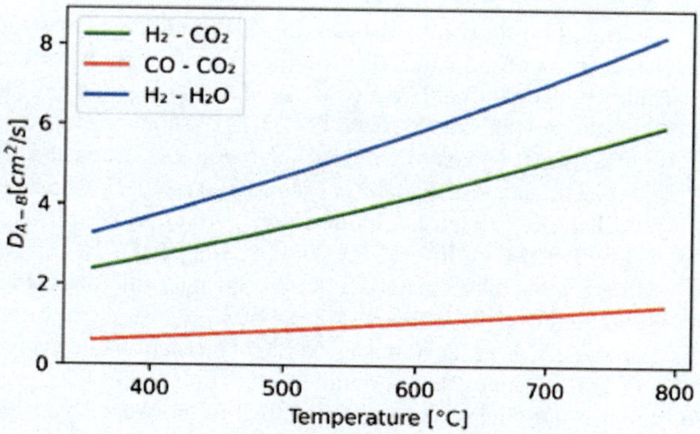


**Fig. 15** The calculated binary diffusion coefficients for $H_2$–$H_2O$, $H_2$–$CO_2$, and CO–$CO_2$ [11]

## Conclusion

Today, solid carbon is used to partly prereduce the Mn-ores in the production of Mn-ferroalloys. Solid carbon can be substituted with $H_2$ and hence reduce the environmental footprint. In 4 studies, it is shown that the $H_2$ containing gases will give higher reduction rate compared to $CO/CO_2$ gases with the same reduction potential. The reduction rate will be increased in the order of 20–100%, where $H_2$ has a larger impact on the reduction rate for high oxygen pressures. In $H_2$ gas, it is seen that the final degree of reduction is higher than using CO gas, and this is believed to be due to the formation of metallic iron. The reduction rate in $H_2$ gases varies for various ores, and as for CO containing gases, the reduction rate of Comilog ore is faster than the Nchwaning ore, and activation energies of 23 kJ/mol versus 68 kJ/mol are, respectively, found for isothermal experiments. For the carbonate UMK ore, the presence of $CO_2$ in the gas reduces the decomposition rate of the carbonate.

**Acknowledgements** The Norwegian Research Council and the Norwegian ferroalloy industry are acknowledged through the SFI Metal production (237738), the Thanos project (309475), the Reduced $CO_2$-project (280968), FME High Eff (257632) and Gasferrosil (224950). Stephen Lobo is acknowledged for agreeing to include unpublished work regarding the prereduction of Mn-ores.

## References

1. Tangstad M, Bublik S, Haghdani S, Einarsrud KE, Tang K (2021) Slag properties in the primary production process of Mn-Ferroalloys. Metall Mater Trans B 52(6):3688–3707. https://doi.org/10.1007/s11663-021-02347-8
2. Olsen S, Tangstad M, Lindstad T (2007) Production of manganese ferroalloys. Tapir, Trondheim
3. Ishak R, Tangstad M (2007) Degree of prereduction without coke consumption in industrial furnaces. The Indian Ferro Alloy Producers' Association, New Delhi, India, pp 268–280
4. Larssen TA, Tangstad M, Kero IT (2018) Gaseous reduction of Mn ores in CO-$CO_2$ atmosphere. In: Davis BR, Moats MS, Wang S, Gregurek D, Kapusta J, Battle TP, Schlesinger ME, Alvear Flores GR, Jak E, Goodall G, Free ML, Asselin E, Chagnes A, Dreisinger D, Jeffrey M, Lee J, Miller G, Petersen J, Ciminelli VST, Xu Q, Molnar R, Adams J, Liu W, Verbaan N, Goode J, London IM, Azimi G, Forstner A, Kappes R, Bhambhani T (eds) Extraction 2018. The Minerals, Metals & Materials Series. Springer International Publishing, Cham, pp 1093–1101. https://doi.org/10.1007/978-3-319-95022-8_88
5. Larssen TA, Senk D, Tangstad M (2021) Reaction rate analysis of manganese ore prereduction in CO-$CO_2$ atmosphere. Metall Mater Trans B 1–14
6. Larssen TA, Tangstad M (2022) Effect of moisture, hydrogen, and water-gas shift reaction on the prereduction behavior of Comilog and Nchwaning manganese ores. Metall Mater Trans B. https://doi.org/10.1007/s11663-022-02511-8
7. Ringdalen E, Gjøvik JE, Larssen TA, Tangstad M (2021) Pretreatment of manganese ores in different gas-atmospheres—a method to reduce energy consumption and $CO_2$ emissions in Mn-alloy production. In: Proceedings of the 16th international ferro-alloys congress (INFACON XVI) 2021. https://doi.org/10.2139/ssrn.3930059
8. Schanche TL, Tangstad M (2021) Prereduction of Nchwaning ore in $CO/CO_2/H_2$ gas mixtures. Minerals 11(10):1097. https://doi.org/10.3390/min11101097

9. Biørnstad O (2020) Decripitation of Cimilog, Assmang and UMK manganese ores during prereduction. MSc Thesis, Norwegian University of Science and Technology
10. Larssen TA, Senk D, Tangstad M (2020) Reduction of manganese ores in $CO$-$CO_2$ atmosphere. Metall Trans B. https://doi.org/10.1007/s11663-020-02018-0
11. Schanche T (2022) Pretreatment of manganese ores in $CO/CO_2/H_2$ atmospheres. PhD thesis, Norwegian University of Science and Technology
12. Davies J, Tangstad M, Schanche T, de Preez F (2022) Pre-reduction of united manganese of Kalahari ore in $CO/CO_2$, $H_2/H_2O$, and $H_2$ atmospheres. Metall Mater Trans B. Is submitted
13. Berg KL (2000) Kinetics of manganese ore reduction by carbon monoxide. Metall Mater Trans B 31B:477–490
14. Ngoy D, Sukhomlinov D, Tangstad M (2020) Pre-reduction behaviour of manganese ores in $H_2$ and CO containing gases. ISIJ Int 60(11):7

# Development of Fossil-Free Technologies for the Metallurgical Industry—Swerim Pilot and Industrial Experiences

**Ida Heintz, Elsayed Mousa, and Guozhu Ye**

**Abstract** The metal industry alone contributes to about 8% of the $CO_2$ emissions globally. For a sustainable and carbon-neutral metal industry, the use of $H_2$ and biocarbon for the substitution of fossil carbon is essential. Swerim has in recent years made great efforts to support the metal industry for a smooth C-neutral transition. This paper will highlight the projects using biocarbon and $H_2$ in different metallurgical processes including the installation of an $H_2$-electrolyser which could be connected to various pilot test facilities in Swerim including an EAF, a DC furnace, and a Universal Converter of 6–10 tons in scale. The Swerim $H_2$-electrolyser has the capacity to produce 100 $Nm^3$ of $H_2$/h. Our unique gas processing equipment for the separation of $CO_2$ and production of $H_2$ in combination with our demo facilities enable us to develop high TRL demonstration of different CCUS opportunities for the steel and metal industry. The use of $H_2$ and biocarbon in metallurgical applications will also be discussed.

**Keywords** Metal industry · $CO_2$-neutral · Pilot and demonstration · $H_2$-metallurgy · Sustainability · Fossil free

## Introduction

The metal industry alone contributes to about 8% of the $CO_2$ emission globally. For a sustainable and carbon-neutral metal industry, the use of $H_2$ and biocarbon for the substitution of fossil carbon is essential. Swerim, as a leading institute for metallurgical research and technology, and with its unique pilot testing facilities,

I. Heintz · E. Mousa · G. Ye (✉)
Swerim AB, Box 812, 97125 Lulea, Swerim, Sweden
e-mail: guozhu.ye@swerim.se

I. Heintz
e-mail: ida.heintz@swerim.se

E. Mousa
e-mail: elsayed.mousa@swerim.se


C. Fleuriault et al. (eds.), *Advances in Pyrometallurgy*, The Minerals, Metals & Materials Series, https://doi.org/10.1007/978-3-031-22634-2_5

has in recent years made great efforts to support the metal industry for a smooth C-neutral transition.

Swerim plant test facilities include a 5 MVA EAF furnace of a capacity of 6–10 tons that has been used for smelting of DRI/HBI including those produced by $H_2$, conventional EAF using biocarbon and smelting reduction for the recovery of residues from metal and steel industries. Swerim also has a DC furnace of 3 MW which has been used for the development of new metallurgical processes for ilmenite smelting, ferroalloy production, and processing of slag and dust from metallurgical industries. Spent batteries and catalysts have also been processed using the DC furnace. The Swerim universal converter, with a capacity of 5–6 tons of metal, can simulate all kinds of converter processes, ferrous and nonferrous. New smelting processes aiming for low $CO_2$ nonferrous metal production processes have also been pilot-tested.

In the 90s, the Swedish and European iron and steel industry had also installed specific pilot plants such as EBF of LKAB and a Single Belt Strip Caster for the development of ultralow $CO_2$ steelmaking processes and breakthrough energy saving technologies.

Some of these as well as ongoing $H_2$- and biocarbon-based projects and efforts will be highlighted in this paper.

## Development of $CO_2$ and $H_2$ Technology for the Metal Industry

Hydrogen has been identified as one of the key enablers in the industrial transformation in Sweden and globally. Hydrogen can be a parameter in more or less all research areas to develop a fossil free and circular metallurgical industry. Swerim has worked with hydrogen-related questions for a long time in metallurgy, material development, and combustion. We have a unique position and know-how in the field of hydrogen in metallurgical applications, covering the entire value chain of metals from raw materials to recycling, as well as the different steps of hydrogen applications, such as production, use, recovery, and storage. All those steps need to be considered when evaluating hydrogen applications, and energy efficiency is key in each and every one of them, in order to secure a cost and environmentally-efficient process.

### *Pilot Test Facilities*

At Swerim in Luleå, there are unique pilot and demo facilities and gas infrastructure (Table 1) available for research and demonstration for decarbonization research. It includes an alkaline electrolyser with a capacity of ca 100 $m^3/h$ at 15 bar (Fig. 1)

**Table 1** Swerim's demo and pilot gas infrastructure

| Gas supply | Flow to separation ($Nm^3/h$) | Flow to gas processing ($Nm^3/h$) | Status |
|---|---|---|---|
| BFG | 800 for high P<br>2500 for low P | 50 $CO_2$ (240 in 2023)<br>40 $H_2$ (100 in 2023) | |
| BOF | 400 | $CO_2 + H_2$ (2023) | |
| Hydrogen | | 100 $H_2$ (Electrol)<br>Higher volumes under discussion | Connection to pilot and demo |
| LPG | As required | As required | |

including a hydrogen infrastructure that can supply all other pilot equipment with hydrogen for tests in different applications (Fig. 2). Furthermore, gasses from SSAB such as Blast Furnace Gas (BFG) at different pressures and later during 2023, a pipeline for 400m$^3$/h of Basic Oxygen Furnace gas (BOF) can be supplied.

**Fig. 1** Swerim's alkaline electrolyser, capacity 100 m$^3$/h

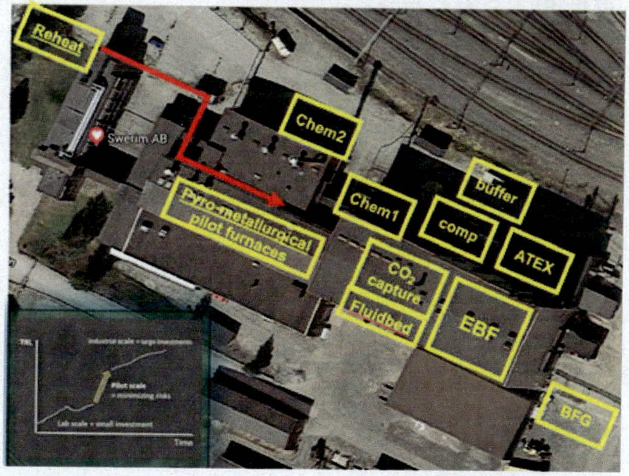


**Fig. 2** Swerim's metallurgical pilot facilities are all connected to hydrogen and metallurgical gas infrastructure

The hydrogen infrastructure includes pipelines to our pilot furnace and demo halls. Hydrogen use in metallurgical applications can be developed in pilot trials infrastructure in our fix pilot equipment such as:

- EBF (Experimental Blast Furnace)
- ECF (Experimental Combustion Furnace)
- Walking beam furnace
- Chamber furnace
- Universal converter
- Pilot AC furnace
- Pilot DC furnace
- Fluidized bed reactor.

The ATEX, demo halls offer the opportunity to demonstrate other applications, from metallurgical as well as different value chains. In the Chem stations, tanks with potential hydrogen carriers or correlating by-products can be stored during the demonstration of concepts.

## *Example of the Project of Interest for Fossil-Free Metal Industry*

The gas infrastructure has been developed in a series of projects funded by the EU framework Horizon 2020 and Horizons Europe. The first project, Stepwise [1] demonstrated $CO_2$ separation in combination with hydrogen production from BFG for power production using SEWGS (sorption enhanced water gas shift) technology. The role of Swerim has been to construct and conduct the demonstration trials using a single column unit. Following Stepwise was FReSMe [2], where methanol production using the $CO_2$ and $H_2$ from the shifted BFG and hydrogen from the electrolyser was demonstrated at TRL6.[1] Further development of the technology and concept is carried out in pilot trials in the $C^4U$ [3] project where BFG is shifted and used for reheating application in a chamber furnace. The use of BFG in a walking beam furnace, combined with post-combustion $CO_2$ separation, is also demonstrated within the project. To fully develop the concept into TRL7[2], the INITIATE project [4], where the concept to produce ammonia for Urea production from BOG will be demonstrated, at an equivalent of 5 t (Urea)/d.

The hydrogen infrastructure and the operational experience from procurement, site acceptance tests, and control system of the electrolyser give us a unique opportunity for experience exchange for other actors' future electrolyser investment projects. To better understand the electrolyser system at Swerim, a dynamic model using

[1] TRL6 (technology readiness level 6) = System/process prototype demonstration in an operational environment.

[2] TRL7 = Integrated Pilot System Demonstrated: System/process prototype demonstration in an operational environment (integrated pilot system level).

Modelica has been developed and used. The model can be adapted to other electrolyser systems with different pressures, control systems, and other equipment, as well as different electrolyser types, to help gain valuable information for future implementations by an independent party.

The ATEX demo halls give the possibility to demonstrate new and developing technologies and evaluate how they could be integrated into metallurgical contexts, for example, in combination with off gases from EAF. Other ongoing projects at Swerim include hydrogen use in zinc fuming together with Boliden and Linde Gas, co-funded by Swedish Energy Agency where pilot trials will be conducted. The theoretical $CO_2$ savings potential is 80,000 $CO_2$ tons per annum.

### *Future Opportunities*

Hydrogen in industrial applications is nothing new. It is already used in chemical industries to produce, for example, methanol and ammonia, and in a metallurgical context, it is found as a component in gases such as natural gas for DRI production, and in metallurgical off-gases such as blast furnace and coke oven gas. All mentioned hydrogen-rich gases also contain carbon, which will produce carbon oxides in combustion or reaction. The focus for future metallurgical processes is, therefore, to use hydrogen as a separate gas, not in a mixture, but also to find industrial symbiosis with chemical industries for CCU or CCUS solutions. The different uses of hydrogen that we foresee are: reducing agent, reheating applications, system perspective, and other fuel applications. Future potential ideas include recovery of oxidic residues and slags, battery recycling, and reduction of nonferrous ores as well as concepts of CCU and CCS with future metallurgical off-gases.

## Use of Biocarbon for the Metal Industry

Biomass is another feasible approach to mitigate the fossil $CO_2$ emissions from the metallurgical industry which are based on fossil coke and coal. The biomass products could be introduced into the shaft furnaces such as blast furnace and/or cupola furnace by top charging in the form of briquettes to replace fossil coke and/or injection through tuyeres to replace the fossil pulverized coal (PC), and hence reduce the fossil $CO_2$ emission. The feasibility of replacing the traditional top charging coke and PC with biocarbon will be discussed in the following section.

## *Briquettes Containing Biocarbon*

Cold-bonded briquettes can contribute not only to recycle the steel mill residues, but also to the charging of pre-treated biomass into the blast furnace. The blast furnaces in Nordic countries (Sweden and Finland) are working with almost 100% pellets and the residues are recycled via cement-bonded briquettes. The charging rate of the briquettes is approximately 100–120 kg/tHM (Hot metal). The mechanical strength of the briquettes is the most critical factor that is significantly affected by adding pre-treated biomass to the briquetting blend and it is required to add more cement to maintain the required strength.

In the Bioagglomerate project [5], full-scale trials have been conducted by charging 39–64 kg/tHM of bio-briquettes containing 1.8 wt% of torrefied pelletized sawdust. The cement ratio in the briquetting mixture was increased from 8% in the reference mixture of standard briquettes to 10–12% in bio-briquettes. The trials were followed by the evaluation of the blast furnace's efficiency and process stability. It was reported that the gas utilization was higher during bio-briquette charging periods without a change in pressure drop up to 64 kg/tHM charging of bio-briquettes which indicates the sustained shaft permeability. The rate of blast furnace dust generation and dust composition was not changed significantly. Moreover, the top gas composition did not indicate a release of hydrocarbon from the torrefied pelletized sawdust in connection to the charging of bio-briquettes. As shown in Fig. 3, the total $CO_2$ emission was reduced by about 33–40 kg/tHM when using 64 kg/tHM bio-briquettes.

In the steel meets forest project, the partial replacement of cement in cold-bonded briquettes with biomass lignin was studied [6]. Lignin is a byproduct from the Kraft pulp mills and it is abundant and renewable, therefore, it has attracted more and more attention for potential applications such as replacing cement in cold-bonded blast furnace briquettes. Cement was replaced with lignin at a rate of 0, 10, 25, 50, and 100%. The mechanical strength measured by the standard tumbler index (TI) for the briquettes produced with a replacement ratio of up to 50%. The replacement of

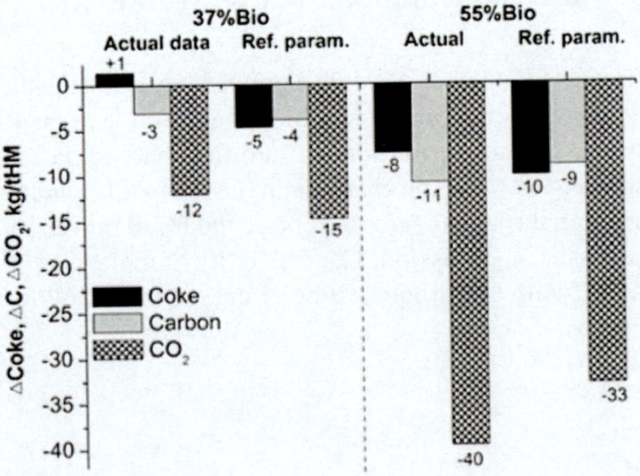


**Fig. 3** Differences (kg/tHM) in calculated coke, carbon, consumption, and $CO_2$ emission between the reference period and test periods [5]

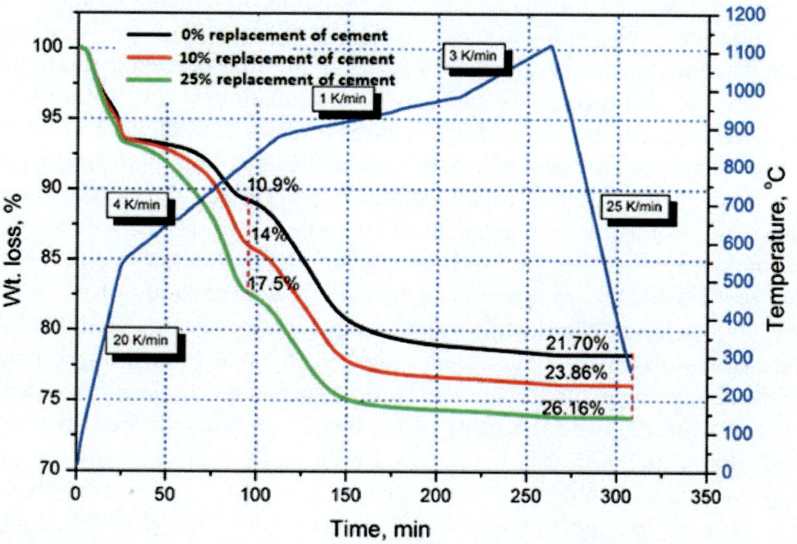


**Fig. 4** Reduction curves of briquettes prepared at different substitution levels of cement with lignin [6]

up to 25% of cement with lignin demonstrated adequate mechanical strength with TI of 74–82.5%. By increasing the replacement ratio of cement with lignin to 50%, a sharp decrease in the mechanical strength occurred and the TI dropped to 18% and most briquettes disintegrated. This indicated the possible replacement of up to 25% of cement with an equal percentage of lignin without impairing the strength of the briquettes. The reduction rate was studied using thermogravimetric analysis under an inert atmosphere and temperature profile simulated blast furnace conditions, as shown in Fig. 4. The reduction rate of briquettes increased by increasing the substitution ratio of cement with lignin, which indicated the positive contribution of lignin not only in reducing the cement ratio in the briquettes and consequently slag generation, but it can also contribute to the reduction of iron oxides and consequently reduce the coke consumption.

## *Biocarbon Briquettes for Cupola*

In the foundry industry, the cupola furnace is considered the most efficient melting unit for the production of cast iron. The cupola is a robust technology and a flexible melting unit with a relatively small footprint for larger melting capacities. Foundry coke is the source of carbon and energy in the cupola process. The furnace allows to efficiently carburize the metal and adjust the silicon content to the required level in cast iron. Although the cupola furnace has undergone modifications and developments to improve energy utilization and reduce fossil greenhouse gas (GHG) emissions, the demands for further decrease of GHG emissions challenge the foundries using this key melting process. The environmental regulations in Sweden aim to reach zero net fossil emission by 2045 which requires short and middle-term solutions to secure this technology for cast iron production and the automotive industry. In collaboration with the Volvo Group Truck Operations, Swerim has developed biocarbon briquettes with promising cold mechanical strength and energy density for cupola furnace implementation [7]. The study focused on the evaluation of different additives and binders for the compaction of biocarbon and the study of the behavior of renewable materials at high temperature on a laboratory scale to build the foundation for upscaling of briquettes. The effect of charcoal ratio in the briquettes, particle size, binder type, binder ratio, moisture content, and compaction pressure on the mechanical strength and reactivity of the biocarbon briquette was investigated. A combination of molasses and hydrated lime showed promising results regarding the energy density and cold mechanical strength; however, cracks were initiated by heating the biocarbon briquettes up to 850 °C and kept for 1 h under load (25 kg/briq.) in a gas media simulating the cupola furnace shaft gas (34 vol% CO, 7 vol% $CO_2$, and 59 vol% $N_2$). Further development is ongoing in the project Bio4Cupola via collaboration between Swerim, Volvo Group Truck, Envigas, and Carbomax aiming to develop and produce biocarbon with sufficient hot strength for Cupola furnace implementation.

## *Biocarbon Injection*

Injection of biomass to the shaft furnaces (blast furnace and/or cupola furnace) through tuyeres provides a flexible option for its utilization to replace the injected pulverized coal (PC) without affecting the top charging burden. In this case, mechanical strength is not a prerequisite meanwhile the devolatilization, gasification, combustion characteristics, and heating value are the major parameters affecting the Raceway Adiabatic Flame Temperature (RAFT) and consequently the furnace efficiency and stability. The replacement rate of PC by biomass materials like charcoal, torrefied and raw biomass was theoretically investigated by means of static heat and mass balance model. It was reported that 155 kg/tHM of pulverized coal could be replaced by 166.7 kg/tHM of charcoal. The replacement ratio of highly volatile biomass, like torrefied and raw wood pellets, is much less. Torrefied biomass can replace 22.8%, while raw wood pellets can replace only 20% of the injected PC. Further increase in the injection rate of highly volatile bio-based materials requires oxygen enrichment otherwise, the RAFT temperature decreases, and the top gas temperature increases [8, 9]. In our recent study [10] in collaboration with Envigas (biocarbon producer in Sweden), three grades of biochar, produced from the pyrolysis of sawdust, were evaluated for replacing PC. The total carbon content was 79.2,

93.4, and 89.2% in biochar 1, 2, and 3, respectively, while it was 81.6% in the reference PC. For each type of biochar, 6 cases were designed at different injection levels from 30 kg/tHM to 143 kg/tHM, which represent 100% replacement of PC in the applied case, while the top charged coke is fixed in all cases as reference. The oxygen enrichment, RAFT, and TGT are fixed for certain cases, and have been calculated by heat and mass balance-based model (MASMOD) in other cases to identify the optimum level of biochar injection. The MASMOD calculation showed that as the injection rate of biochar 1 and biochar 2 increased, the RAFT increased by ~190 °C, while TGT decreased by ~ 45 °C at 100% replacement of PC with biochar. By optimizing the moisture content of biochar and the oxygen enrichment in the blast, it is possible to reach 100% replacement of PC without much affecting the RAFT and TGT. Biochar 3 was able to replace 100% of PC without deteriorating the RAFT or TGT. The full replacement of PC with biochar enables the reduction of fossil $CO_2$ emission by 34%.

## Circularity, Sustainability, and Future Metallurgy

The circularity of the process and product value chain is an essential part of a sustainable system aiming for zero or low $CO_2$ emission. For instance, recycling steel scrap for steelmaking, using an EAF, consumes only ¼ of the energy needed for ore-based steelmaking. If the energy used in the EAF is green electricity and the added carbon is biocarbon, it becomes a fossil-free steelmaking approach. In most cases, recycling and use of secondary resources will not just save primary natural resources, but it will also reduce the overall energy needed as the production value chain will be much shortened and simplified. The mining part will be excluded and for scrap-based smelters, the mineral processing, the coke plant, the agglomeration plant, etc., could also be saved. In addition to these, the related offgas and water cleaning systems will be drastically minimized.

The development of sustainable recycling technologies is a major part of the Swerim R&D strategy. In the past 30 years, numbers of recycling processes have been developed using our pilot facilities. Some of these will be described to illustrate how we develop the new processes and the role of the pilot plant. The metallurgical principle for recycling secondary raw materials including wastes and residues from metal industries, applied at Swerim, is shown in Fig. 5.

Secondary resources including scrap, e-scrap from production and consumers, ashes, dust, sludge, and slags from metal and incineration industries generally consist of four major fractions: (a) stable oxides, i.e., CaO and $Al_2O_3$, (b) less stable oxides/metals which could be controlled to metal or to slag (c) volatile elements, i.e., Zn, Pb, Na, K, Li, etc., and (d) organic fraction such as plastics in e-scrap or textile, etc.

Based on the characterization of the materials and thermodynamic consideration, the most important process parameters such as process temperature, oxygen potential, slag chemistry, and in some instances also pressure, will be identified and the targeted

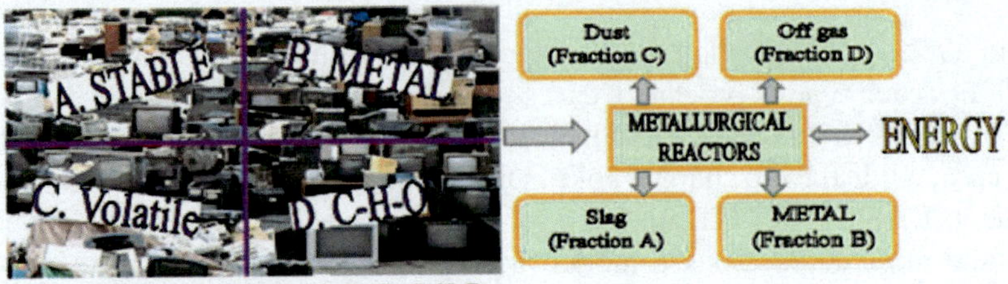


A: CaO, SiO2, **Al2O3**, MgO, **Li2O**
B: **Ni, FeO, MnO**, V2O3, Cr2O3, P2O5, **Cu, Co,** Mo
C: Zn, Pb, Na, K, Cl, **F**, Cd, Hg, **Li**
D: C-H-O, plastics/textile/fluff, **C**

**Fig. 5** Metallurgical principle for recycling [11]

products determined. A process concept will be developed and eventually tested in our laboratory or pilot plant. During metallurgical processing, under the preferred conditions, Fraction A will end up in the slag phase. Co, Ni, and Cu in Fraction B will easily be managed to the metal phase while Fe, Mn, Cr, and V can be controlled to the slag or to the metal by controlling the reduction potential. Fraction C will, in most cases, end up in the dust fraction, while Fraction D will be in the offgas as CO, $H_2$, $CO_2$, and $H_2O$.

Using the principle illustrated in Fig. 5, number of recycling processes have been developed using the 3 MW pilot DC furnace at Swerim. Figure 6 lists some of the materials tested and concepts developed.

In one of the most recent projects, Li-ion battery wastes were processed in the 10 tons pilot EAF to recover Li without compromising a high yield of recovery of Co and Ni as illustrated in Fig. 7. The recovery yield of Co and Ni is as high as 98–99%, whereas most of the Li could be recovered to the dust fraction as $Li_2CO_3$.

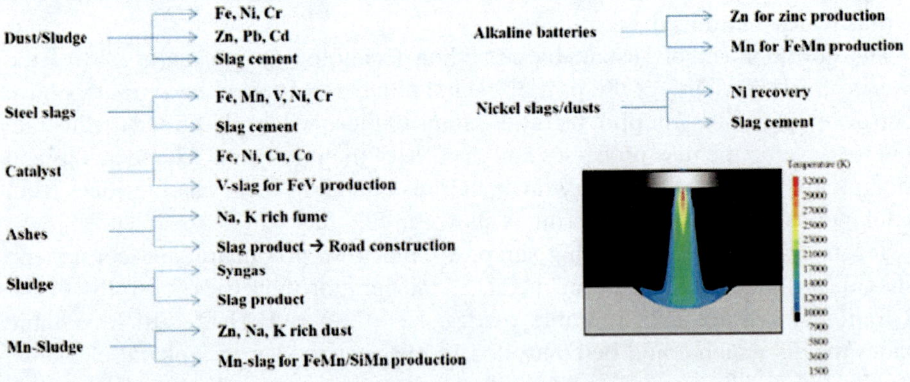


**Fig. 6** Concepts developed using the 3 MW DC furnace at Swerim [11]

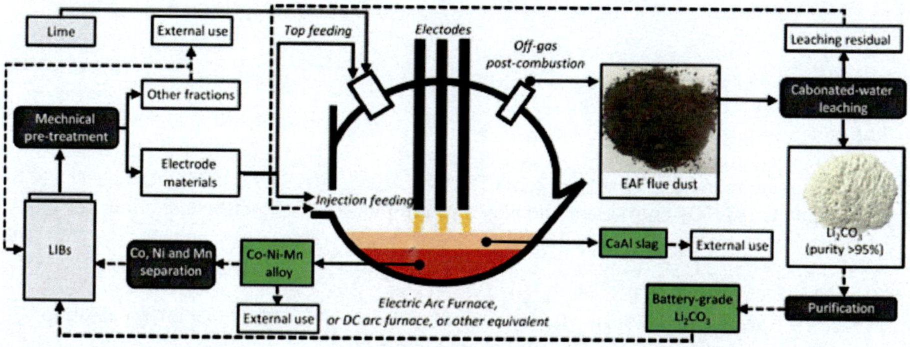


**Fig. 7** Recovery of Li, Co, Ni, and Mn from spent LiBs [12]

It has been shown in this section that circularity is one of the most important pillars of a fossil-free and sustainable society. Pyrometallurgy is of essential importance and pilot testing is inevitable for developing recycling processes.

## Concluding Summary

The development of future metallurgical processes will strongly be driven and guided by the Paris Agreement. The transition for fossil-free processes has already started. In Sweden, electrification of the energy-intensive process industries, including the metal industry, was already started some 10 years. A good example is the well-known Hybrit process. The Swerim experiences shown in this paper have demonstrated the strong need for pilot and demonstration test infrastructure and the importance of pilot testing for the development of new fossil-free metallurgical processes.

Transition from a C-based to an H-based steel industry has already been initiated. The same will be happening for the ferroalloy and nonferrous metal industry, as well as for the recycling industry. At Swerim, we already see a strong demand from the metal sectors for pilot testing, most with the aim to minimize or zero $CO_2$ processes, and valorization of metallurgical byproducts and residues.

It can be foreseen that demand for $H_2-$ and biocarbon-based processes will continue and the need for pilot testing will increase in the coming years. Pilot and demonstration plants with well-functional infrastructure are thereby of great importance.

**Acknowledgements** The authors thankfully acknowledge the Swedish Energy Agency (Energimyndigheten) for the financial support of the research projects Slag reduction with hydrogen (project number 49830-2), Bioagglomerat (project number 39150-1), and Bio4Cupola (project number P50878-1).

We also thankfully acknowledge the European Union's Horizon 2020 research innovation programme for the financial support of Stepwise (grant agreement no. 640769), FReSMe (grant agreement no. 727504), $C^4U$ (grant agreement no. 884418), and INITIATE (grant agreement no. 958318).

## References

1. https://www.stepwise.eu/
2. https://cordis.europa.eu/project/id/727504
3. https://c4u-project.eu/
4. https://www.initiate-project.eu/
5. Mousa E, Lundgren M, Ökvist LS, From L, Robles A, Hällsten S et al (2019) Reduced carbon consumption and $CO_2$ emission at the blast furnace by use of briquettes containing Torrefied sawdust. J Sustain Metall 5(3):391–401
6. Mousa EA, Ahmed HM, Wang C (2017) Novel approach towards biomass lignin utilization in ironmaking blast furnace. ISIJ Int 57(10):1788–1796
7. Mousa E, Kazemi M, Larsson M, Karlsson G, Persson E (2019) Potential for developing biocarbon briquette for foundry industry. Appl Sci 9:5288
8. Wang C, Mellin P, Lövgren J, Nilsson L, Yang W, Salman H et al (2015) Biomass as blast furnace injectant–considering availability, pretreatment and deployment in the Swedish steel industry. Energy Convers Manage 102:217–226
9. Wang C, Larsson M, Lövgren J, Nilsson L, Mellin P, Yang W et al (2014) Injection of solid biomass products into the blast furnace and its potential effects on an integrated steel plant. Energy Procedia 61:2184–2187
10. Mousa E, Sjöblom K (2022) Modeling and optimization of biochar injection into blast furnace to mitigate the fossil $CO_2$ emission. Sustainability 14:2393
11. Ye G (2020) Technology advances in pyrometallurgy with focus on recent development of sustainable processes—experiences of Swerim pilot plant activities. TMS 2020, February 23–27, San Diego, California, USA
12. Hu X, Mousa E, Ye G (2021) Recovery of Co, Ni, Mn, and Li from Li-ion batteries by smelting reduction—Part II: a pilot scale demonstration. J Power Sources 483:229089

# Investigation of High-$H_2$ Reducing Gas Delivery Through Shaft-Level Tuyeres With Computational Fluid Dynamics

**Tyamo Okosun, Samuel Nielson, Orlando Ugarte, and Chenn Q. Zhou**

**Abstract** As both a significant contributor to the steelmaking process and the largest single source of $CO_2$ emissions in integrated steelmaking, the blast furnace is a critical area of focus for decarbonization efforts. Generally, the replacement of coke with injected fuels such as natural gas or syngas has been a key pathway for reducing emissions. These injectants provide higher concentrations of $H_2$ reducing gas, decreasing reliance on CO reactions, but they are limited by their endothermic impacts on flame and reducing gas temperature. One potential method of circumventing these limitations is the use of shaft-level tuyeres for high rates of hot reducing gas delivery, in a similar fashion to the direct-reduced ironmaking process. In this paper, the impacts of such a system are investigated with Computational Fluid Dynamics to predict the reaction rates, coke replacement ratios, and carbon emissions of the blast furnace process under the proposed operating scheme.

**Keywords** Blast furnace · Hydrogen · Computational fluid dynamics · Coke rate

## Introduction

Steelmaking, one of the most critical industries to the functioning of modern society, contributes roughly 7% of global carbon dioxide emissions [1]. As the world attempts to address the clear and increasing impacts of greenhouse gas emissions, the steelmaking industry must identify methods by which their carbon emissions can be controlled and reduced. The industry has already made great strides in process efficiency and electrification, but there remain key areas in which significant gains can be made. One such area is the blast furnace (BF), a heavy fossil fuel-dependent reactor that remains the dominant method of ore reduction across the world. In North America, BF ironmaking provides the majority of pig iron production and remains the largest single fossil fuel consumer in integrated steel mills [2].

---

T. Okosun (✉) · S. Nielson · O. Ugarte · C. Q. Zhou
Center for Innovation Through Visualization and Simulation, Purdue University Northwest, 2200 169th Street, Hammond, IN 46323, USA
e-mail: tokosun@pnw.edu



C. Fleuriault et al. (eds.), *Advances in Pyrometallurgy*, The Minerals, Metals & Materials Series, https://doi.org/10.1007/978-3-031-22634-2_6

Metallurgical coke represents the primary fuel source for the blast furnace, and simultaneously the largest source of $CO_2$ generated from the process. Coke combustion generates high-temperature carbon monoxide, which is utilized in reduction reactions to remove oxygen from iron ore and melt the iron into liquid form. BFs are a mature manufacturing process, and technologies including hot blast, oxygen enrichment, and fuel injection have significantly decreased fuel rates and $CO_2$ emissions per ton of production since their inception. In particular, injected fuels have replaced a substantial fraction of coke in the process, and the use of natural gas in North America (due to relatively low fuel costs) has also been beneficial to $CO_2$ emissions. Natural gas use reduces emissions by replacing coke with hydrogen-bearing fuels. This shift enables the furnace to rely more heavily on $H_2$-based indirect reduction reactions, provided there exists the necessary thermal energy. The introduction of $H_2$-based chemical reactions can also result in increased productivity [3].

However, there are operational limitations to the rate at which hydrogen-bearing fuels can be supplied to the blast furnace, particularly the quenching effects of natural gas (and hydrogen rich reductants generally) in the furnace. Without mitigating measures, raceway temperatures decline as more natural gas is delivered to the furnace [4]. The corresponding mitigation techniques generally decrease top gas temperature simultaneously with increased raceway flame temperature, placing a cap on their effectiveness. Hydrogen injection presents similar limitations, both promoting the water gas reaction ($C + H_2O \rightarrow CO + H_2$) as in natural gas injection and increasing $H_2$ reduction reactions, which are endothermic in nature ($Fe_2O_3 + 3\ H_2 \rightarrow 2\ Fe + 3\ H_2O$, for instance, requires ~ 100 kJ), resulting in an even greater thermal deficit [5]. Replacing coke with gaseous fuels can also result in high pressure drops within the furnace. This occurs since a reduced coke rate leads to thinner coke layers in the burden, and therefore smaller coke slits in the cohesive zone. Additionally, gas volumes can increase with higher H2 use in the bosh gas.

Despite some desire to engage in these carbon-free $H_2$ reduction reactions, the aforementioned limitations (in combination with economic factors) have thus far kept carbon-based fuels firmly in place as the selection of choice for BF operation. However, as the industry looks to decarbonize, BFs must be prepared to adapt, and the use of $H_2$-based reducing gas at higher rates requires further consideration. In particular, methods by which the thermal quenching effects could be circumvented or managed are necessary. This research focuses on one such method, the delivery of hot reducing gas into the furnace above the cohesive zone for gas-solid reduction reactions.

Researchers have explored the concept of delivering hot reducing gas directly into the furnace shaft above the cohesive zone, with some focus on the concepts of recycling BF top gas, delivering pre-combusted natural gas, and delivering coal or natural gas-derived syngas [6–9]. Most earlier studies have focused primarily on modeling the overall impact of the concept with simplified physics, with more recent explorations endeavoring to understand the impact of gas flow and fluid dynamics on the chemical reactions occurring in a shaft tuyere operating scheme [10, 11]. Reducing gases could be generated from a variety of sources, including waste gasification, top gas recycling, natural gas partial combustion, and direct hydrogen injection, with varying levels of $CO_2$ emissions mitigations achievable across this range. Regardless of the source, the delivery of reducing gas above the cohesive zone would reduce the endothermic impacts of high-$H_2$ fuels, with the potential to further address the issue by preheating the gas before delivery [12, 13]. An enhanced understanding of the impacts of shaft-level tuyere gas delivery into a typical North American blast furnace (including potential process operation impacts such as gas penetration, gas species profiles, potential existence of a second cohesive zone, and others) will present useful information for operators and engineers considering the option as a method of $CO_2$ emissions reduction.

Experimental testing of a proposed modification to blast furnace operation can often present a myriad of technical concerns, including the potential to interrupt production. Laboratory-scale testing is effective at establishing key physics, but it can be difficult to assess process concerns that may arise under industrial-scale operation. Computational fluid dynamics (CFD) modeling is a useful, relatively rapid, and low-cost tool to determine the feasibility and impacts of proposed changes to multi-phase reacting flow systems. CFD modeling of the BF process has been widely applied by researchers seeking to understand, evaluate, and optimize the selection of operating conditions. In particular, the researchers at the Steel Manufacturing Simulation and Visualization Consortium (SMSVC) at Purdue University Northwest's (PNW) Center for Innovation through Visualization and Simulation (CIVS) have performed a range of simulation studies on BF performance for industrial operations using a combination of commercial CFD codes and in-house solvers. These solvers are applied in this research to simulate shaft-level reducing gas delivery and analyze its impacts on furnace operation.

## Methodology and Computational Models

The complex conditions within the blast furnace involve multi-phase reacting flow at varying timescales for gas and solid flow. To this end, the CFD modeling techniques applied in this research to the simulation of the blast furnace process utilize three coupled component models to capture the physics of combustion, chemical reduction reactions, heat transfer, and multi-phase flow occurring within the furnace. These component models handle simulation of (1) the blowpipe and tuyere downstream of the elbow and bustle pipe, (2) the raceway region outside the tuyere within the furnace, and (3) the shaft region of the furnace, including burden layers, cohesive zone, and all corresponding reactions. Simulations are performed with a combination of commercial (ANSYS Fluent 2021R2) and in-house solver codes, depending on the need for flexible geometry and mesh creation. The component models pass information to one another primarily in the direction of material flow, with the tuyere model determining the gas inputs for the raceway model, and the raceway model determining the conditions observed in the furnace bosh for the shaft model. Each of the component models (simulating hot blast flow, coke and auxiliary fuel combustion, reducing gas generation, reduction reactions, and cohesive zone location) employ the Semi-Implicit Method for Pressure-Linked Equations (SIMPLE) to discretize and solve the Navier–Stokes equations for mass, momentum, energy, and species, represented in a general form by Eq. (1).

$$\nabla \cdot (\rho \phi u) = \nabla \cdot (\Gamma_\phi \nabla \phi) - \nabla \cdot (\rho u' \phi') + S_\phi \tag{1}$$

where $\phi$ represents the property to be transported, either mass ($\phi = 1$), momentum ($\phi =$ velocity), or energy ($\phi =$ enthalpy). Additionally, $\rho$ is density, $u$ is velocity, $u'$ and $\phi'$ represent the fluctuating components of velocity and the transported property due to turbulence, and $S_\phi$ is a source term. Standard RANS turbulence models are used, and finite rate/eddy dissipation modeling is utilized to model the impact of turbulence on gas reactions.

The blowpipe and tuyere model is handled with the commercial solver ANSYS Fluent 2021R2, owing to the ease of geometry modifications for various tuyere designs, gas injection locations, lance shapes, and more. Key phenomena represented in this region with the simulation of the aforementioned Navier–Stokes equations are the turbulent mixing of hot blast and injected fuel as well as combustion reactions. Turbulence is handled with the $k$–$\varepsilon$ realizable turbulence model, and species transport modeling is used to predict chemical reactions, with the Eddy Dissipation/Finite Rate model providing for the influence of turbulence. Boundary conditions for this model utilize mass flow inlets for gas delivery (hot blast, natural gas, and other injected fuels), a pressure outlet at the tuyere outlet with blackbody radiation specifications to match the expected raceway temperature, and assumed heat fluxes on the blowpipe (HTC of ~10 $W/m^2K$) and tuyere walls (HTC of ~3000 $W/m^2K$) corresponding to natural convection of air around the insulated blowpipe or forced convection of cooling water through the tuyere channels, with both fluids beginning at ambient temperature.

Modeling of the raceway region is handled with an iterative methodology employing Fluent and an in-house CFD solver to simulate a slice of the furnace centered on a single tuyere and raceway, with assumed radial symmetry. First, a cavity in the coke bed outside the tuyere outlet is formed with a two-phase interpenetrating Eulerian cold flow model in ANSYS Fluent 2021R2, wherein blast air is driven into a granular phase bed representing the coke. Once a raceway envelope has formed, the phase volume fractions are frozen and exported as a coke bed porosity distribution for use in the in-house solver to simulate gas, coke, and other injected fuel combustion, as well as secondary chemical reactions. The tuyere region model outputs provide inlet conditions for both the cold flow and combustion simulations in the raceway region. Other boundary conditions include a pressure outlet at the furnace bosh, assumed heat losses to cooling on the furnace walls, and an assumed coke bed temperature based on current operating conditions. Gas temperatures, reaction rates, and flow patterns are calculated in this solver, and both gas density changes due to temperature and the rates of solid-to-gas mass transfer over the raceway region are then exported back to the formation solver so the raceway shape can be updated. This is repeated until a converged raceway envelope is achieved. Chemical reactions are handled with the combined Eddy Dissipation–Finite Rate Arrhenius model, with

**Table 1** Chemical reactions modeled in the raceway region

| Reaction | No. | | Chemical Equation | $A_S$ [1/s] | $B_i$ ($m^3$/(kg*s)) | Act. energy $E_i$ (J/mol) |
|---|---|---|---|---|---|---|
| Natural gas combustion | R1 | Eddy-Diss./Finite Rate | $CH_4 + 2O_2 \rightarrow CO_2 + 2H_2O$ | $1.6 \times 10^{10}$ | N/A | $1.081 \times 10^5$ |
| CO combustion | R2 | Eddy-Diss./Finite Rate | $2CO + O_2 \rightarrow 2CO_2$ | $7.0 \times 10^4$ | N/A | $6.651 \times 10^4$ |
| $H_2$ combustion | R3 | Eddy-Diss./Finite Rate | $2H_2 + O_2 \rightarrow 2H_2O$ | $5.4 \times 10^2$ | N/A | $1.255 \times 10^5$ |
| Coke oxidation | R4 | 1st order finite rate | Coke + $O_2 \rightarrow CO_2$ | N/A | $1.225 \times 10^3$ | $9.977 \times 10^4$ |
| Boudouard reaction | R5 | 1st order finite rate | Coke + $CO_2 \rightarrow 2CO$ | N/A | $7.351 \times 10^3$ | $1.380 \times 10^5$ |
| Water gas reaction | R6 | 1st order finite rate | Coke + $H_2O \rightarrow CO + H_2$ | N/A | $1.650 \times 10^5$ | $1.420 \times 10^5$ |

Arrhenius rate (for gas): $w_{\mathrm{Arr}} = A_s \rho^2 Y_{\mathrm{Fuel}} Y_{\mathrm{O2}} \exp\left(-\frac{E_s}{\mathrm{RT}}\right)$, where $A_s$ is a rate constant and $E_i$ is the activation energy

1st Order finite rate: $\dot{m}_{c,i} = -\pi d_p^2 \rho Y_{\mathrm{gasspecies}} B_i \exp\left(-\frac{E_i}{RT_p}\right)$, where $B_i$ is a rate constant and $E_i$ is the activation energy

the limiting rate being the lower of the two calculated. Table 1 summarizes the reactions included in the raceway model along with the reaction rates utilized for the ED–Finite Rate and 1st Order Finite Rate reactions. Additional in-depth detail on both the tuyere and raceway region models can be found in previous publications [14, 15].

Once a bosh gas profile is generated by the raceway model, it can be used as the inlet conditions for gas flow in a simulation of the blast furnace shaft to predict iron ore reduction, coke consumption, gas composition and temperature, burden composition and temperature, solution loss, and cohesive zone position in the furnace, among others. The computational domain is again a slice of the furnace, assuming radial symmetry, from below the bosh or belly region up to the furnace top. Generally, this symmetry around the furnace is deliberately maintained in operation, with the minor variations between tuyeres quickly becoming uniform as gas flows through the packed bed. Under some conditions, operators may utilize plugged tuyeres to increase or maintain tuyere velocity, such as when production rates are low, however, these scenarios are atypical and usually avoided. A quasi-steady assumption is applied for the gas phase flow, as it ascends through the burden in seconds, in comparison to the hours-long descent of the burden phase. The general transport equation for the gas and burden phases in the shaft region is listed below, with the burden phase utilizing a zeroed diffusion term since transport of properties is achieved solely by the movement of the burden itself. Detailed descriptions of the conservation equations for each phase can be found in previous publications [16, 17].

$$\nabla \cdot (\rho \phi u) = \nabla \cdot \left(\Gamma_\phi \nabla \phi\right) + S_\phi \tag{2}$$

Top gas pressure is set fixed at the shaft model outlet boundary. The burden descent velocity is assumed at simulation initialization, with a distribution along the furnace radius. The enthalpy of charged burden material is defined by the assumed material charge temperature. As previously mentioned, raceway region bosh gas results are applied as inlet conditions for gas flow at the bottom of the shaft region. Heat losses from the furnace shell and cooling apparatus are specified by convection boundary conditions calculated based on gas temperature and cooling water temperature, with an adjustable effective heat transfer coefficient accounting for convection between gas and the refractory, conduction within the staves, and convection between staves and cooling water which can be calibrated for each furnace simulated.

Chemical reactions incorporated in this model are listed in Table 2, with kinetics in the model based on research conducted by Tsay et al. [18]. CO, $CO_2$, $H_2$, $H_2O$, and $N_2$ are modeled in the gas phase, and $Fe_xO_x$, Fe, CaO, and MgO are modeled in the solid phase. The prediction of the furnace cohesive zone is performed on an iterative basis by updating the upper and lower boundaries of the zone based on isotherms of burden temperature. The upper boundary is defined by the softening temperature of the ore pellets, while the lower boundary is marked by the liquidus temperature, a variable based on the specific chemistry and selection of ore pellets and/or sinter. Within this region, the porosity of the ore layers in the furnace drops precipitously, resulting in significant resistance to penetrating gas flow. Details on specific assumptions for reaction kinetics and reaction models for gas–solid heat and mass transfer, and considerations for burden layer structure can be found in additional detail in previous publications [17, 19, 20].

**Table 2** Reactions included in the base CFD shaft model

| Reaction | No. | Chemical equation |
|---|---|---|
| Indirect reduction of iron oxide by CO | R1 | $3Fe_2O_3(s) + CO(g) \rightarrow 2Fe_3O_4 + CO_2(g)$ |
| | R2 | $Fe_3O_4 + CO(g) \rightarrow 3FeO(s) + CO_2(g)$ |
| | R3 | $FeO(s) + CO(g) \rightarrow Fe(s) + CO_2(g)$ |
| Indirect reduction of iron oxide by $H_2$ | R4 | $3Fe_2O_3(s) + H_2(g) \rightarrow 2Fe_3O_4 + H_2O\ (g)$ |
| | R5 | $Fe_3O_4 + H_2(g) \rightarrow 3FeO(s) + H_2O(g)$ |
| | R6 | $FeO(s) + H_2(g) \rightarrow Fe(s) + H_2O(g)$ |
| Boudouard reaction | R7 | $C(s) + CO_2(g) \rightarrow 2CO(g)$ |
| Water gas reaction | R8 | $C(s) + H_2O(g) \rightarrow CO(g) + H_2(g)$ |
| Flux decomposition | R9 | $MeCO_3(s) \rightarrow MeO(s) + CO_2(g)$ (Me = Ca, Mg) |
| Water gas shift reaction | R10 | $H_2(g) + CO_2(g) \rightarrow H_2O(g) + CO(g)$ |
| Direct reduction of liquid FeO | R11 | $C(s) + FeO(l) \rightarrow Fe(l) + CO(g)$ |

## Baseline Blast Furnace Simulation and Model Validation

Direct validation of the models employed to simulate blast furnace operation under industrial operating conditions is a difficult proposition, as many of the predicted parameters are effectively impossible to measure. The model has been applied to simulate multiple scenarios using inputs matching those in real-world operation with the aim of comparing simulated blast furnace performance to key operational data from industry furnaces. Previous publications have documented good agreement between modeling results and industrial data for top gas temperature profiles, raceway flame temperature, top gas chemistries, gas utilization, furnace pressure drops, and coke consumption rates. The difference between simulated and measured data for these values was generally observed to be lower than 5%, with good agreement in trends [19, 20].

The furnace utilized in this research was a mid-sized North American blast furnace operating with 24 tuyeres and tuyere level NG injection. The baseline operation conditions for this study were defined based on typical production scenarios for this furnace, with a wind rate of ~270,000 $Nm^3$/hr, hot blast oxygen enrichment of ~29% by volume, hot blast temperature of ~1450 K, and a NG injection rate of ~100 kg/thm (metric tons of hot metal). The simulation models were applied to this furnace under matching conditions to validate predictions against industrial data, the results of which have been previously published [21–23]. A comparison of key operational data for these conditions with CFD predictions is provided in Table 3. CFD results align to within 1% of the measured industrial values for coke rate and raceway flame temperature, and to within roughly 5% for top gas temperature and gas utilization. Figure 1 details the gas temperature and species distributions, burden layers, and cohesive zone within the furnace for these operating conditions, providing a point of comparison for analyses of the impacts of employing shaft tuyeres.

**Table 3** Baseline operating scenario for typical NA BF and Comparison of CFD predictions with industrial measurements for established operating conditions [21–23]

| | CFD predictions | Industrial values |
|---|---|---|
| Coke ratekg/thm | 392 | 390 |
| CO utilization (%) | 50.8 | ~50 |
| $H_2$ utilization (%) | 51.5 | ~50 |
| Avg. top gas temperature | 389 K (116 °C) | 383 K (~110 °C) |
| Raceway flame temperature | 2187 K (1914 °C) | 2173 K (~1900 °C) |

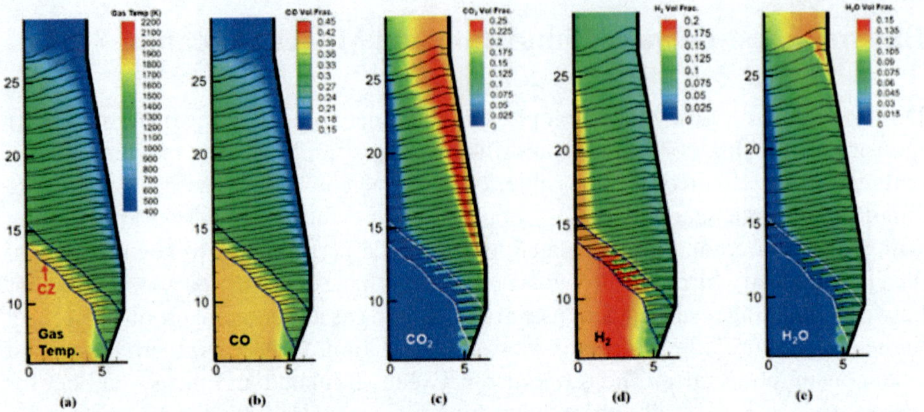


**Fig. 1** Cross-section views of contour distributions in the shaft region for **a** gas temperature; **b** CO vol. fraction; **c** $CO_2$ vol. fraction; **d** $H_2$ vol. fraction; and **e** $H_2O$ vol. fraction

## Reducing Gas Delivery Through Mid-Shaft Tuyeres

To capture the introduction of reducing gas through mid-shaft tuyeres in the existing CFD model, source terms accounting for the modification of mass, energy, and species were introduced to the corresponding governing equations. Gas flow in the shaft region of the furnace is represented with a simplification of the Navier–Stokes equations resulting in a potential flow analogous model, with pressure representing velocity potential [16, 17]. This simplification reduces computation time by a factor of up to four and produces identical results to the standard SIMPLE discretization of the Navier–Stokes equations under normal operation. However, in the case of the newly implemented shaft tuyeres, it requires an alternative approach to distribute gas flow from the perpendicular tuyere jet in the mid-shaft region. To address this limitation, a plume penetration sub-model was developed to approximate the plume penetration depth into the packed bed and the distribution of mass, energy, and species source terms required.

A simplified model of the furnace environment around the shaft tuyere was created in ANSYS Fluent, capturing a full radius slice of the BF. Gas was supplied through the bottom of the domain at 2 m/s, matching the average gas flow velocity in the furnace near the wall in the baseline case, with a species composition reflective of the furnace bosh gas composition. An additional inlet was specified at a diameter of six inches to reflect the largest potential dimensions of a tuyere which would be implemented in an industrial furnace. Hydrogen gas was supplied through this inlet at 1200 K and a velocity of 50 m/s, matching the flow rate conditions for a 15 kg/thm injection rate. The interior of the blast furnace domain was defined as a porous media with a porosity of 0.4, matching the assumed porosity of the ore layers above the furnace cohesive zone. The velocity, density, and species concentrations were extracted from this simplified model to develop a profile of the gas flux due to the mid-shaft tuyere flow. A curve fit against the data extracted from this sub-model

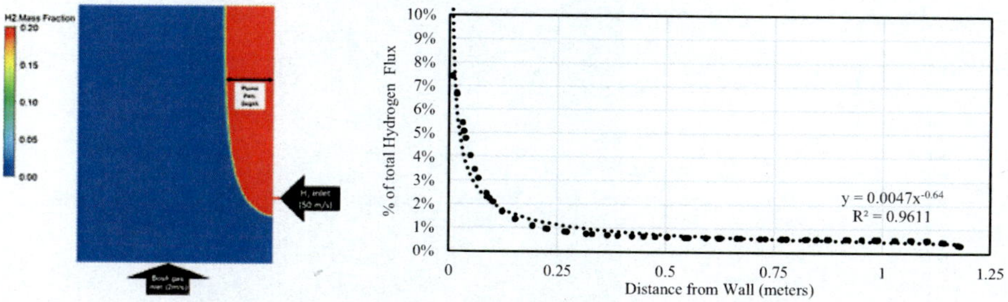


**Fig. 2** Contour of $H_2$ mass fraction in the simplified domain (left) and $H_2$ flux distribution (right)

was then utilized to define the distribution of mass, momentum, and species source terms at the mid-shaft tuyere level to be used in the full furnace model. Figure 2 shows the distribution of hydrogen gas in the simplified model and the corresponding data extraction and curve fitting for the reducing gas flux applied to the full furnace model.

The plume penetration sub-model applies a flux distribution for reducing gas across a specified plume penetration depth, 1.2 m in this case. The distribution is then normalized to ensure summation to unity over the depth of jet penetration into the bed, and the normalized flux distribution is then used to weigh the source terms for all cells in the injection region between the wall and the plume depth. Physically, this results in a greater portion of the influx from the mid-shaft tuyere being located near the wall, with a gradual decay along the furnace radius towards the center (and end of the penetration depth). Figure 3 shows an example of how the mid-shaft tuyere specification is applied to the CFD shaft model. This sub-model requires calibration for each new furnace and a specific set of operating conditions. Changes to mid-shaft tuyere injection velocity will impact plume penetration depth, as will changes to the bosh gas flow rate. This sub-model was implemented in the existing solver with flexible inputs to allow for ease of modification so that future work can investigate the impact of plume penetration depth on the performance of mid-shaft gas injection.

A range of scenarios was explored with this model to investigate the impacts of hydrogen gas delivery at an injection rate of 15 kg/thm through mid-shaft tuyeres at conditions otherwise matching the baseline for this blast furnace. Figure 4a–f details the impacts on gas temperature distributions in the furnace with varying mid-shaft tuyere injection positions. Figure 4g–l also details the gas species distributions observed in the shaft under these same conditions. Gas temperatures rapidly decline in the region of the gas plume formed by the mid-shaft tuyeres, resulting in a significant region of cooler temperatures next to the staves and furnace wall. Lower temperatures injection temperatures, as observed in Fig. 4a–c, result in slightly larger cool regions near the furnace walls compared to higher injection temperatures (Fig. 4d–f). Raising the mid-shaft tuyere injection height also slightly increases the radial size of the cooled region, while lower tuyere locations result in a shifting of the cohesive zone towards the center of the furnace. These impacts correlate directly with the plume of hydrogen gas observed in Fig. 4g–l. Reducing gas reactions between hydrogen and iron ore are endothermic, driving this rapid decline in gas temperatures in conjunction

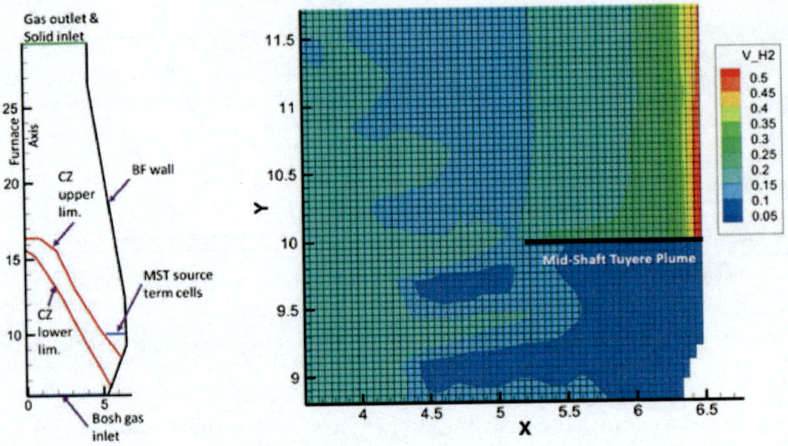


**Fig. 3** Overview of the computational domain (left) and zoom focused on inlet source for reducing gas through mid-shaft tuyere (right)

with the heat losses due to cooling in the furnace staves. It should be noted that the flat bottom of the reducing gas plume is an artifact of the methodology chosen to implement mid-shaft tuyere gas delivery into the existing potential flow model via source terms.

The key performance metrics for the furnace for each of these scenarios are compared with a baseline without mid-shaft tuyere gas delivery in Table 4. Positioning mid-shaft tuyeres deeper in the furnace correlates with higher top gas temperatures, and lower coke rates, with higher injection temperatures increasing the impact observed on both variables. Gas utilization for CO is largely unaffected by the addition of mid-shaft injection, experiencing a slight decline of an average of 0.6 percentage points. Hydrogen utilization shows a much more significant impact of mid-shaft injection, with an average decline of 13 percentage points relative to the baseline. The maximum coke rate reduction observed is −1.5% for injection at 10 m with an injection temperature of 1500 K.

This total benefit is somewhat limited, and would likely be insufficient to justify implementation. However, based on the examination of the $H_2$ utilization, gas plume location, and gas temperature distribution, it appears that a significant amount of the supplied reducing gas is being wasted, instead passing directly through the burden and exiting the furnace with the top gas. This hypothesis can be further explored by examining in greater detail the reduction rates of iron ore with hydrogen gas. Figure 5 explores the reduction rates for a scenario with the mid-shaft tuyere at 10 m and an injection temperature of 1250 K. The yellow line marks the boundary of the mid-shaft tuyere plume, as defined by a $H_2$ volume fraction of 0.2. It is clear that mid-shaft injection results in a shift in the position of the secondary reduction reaction region, sending it further down the furnace and reducing the amount of reduction observed near the furnace wall at heights above 13 m. This correlates with the lower gas temperatures near the wall with mid-shaft injection, as there is less

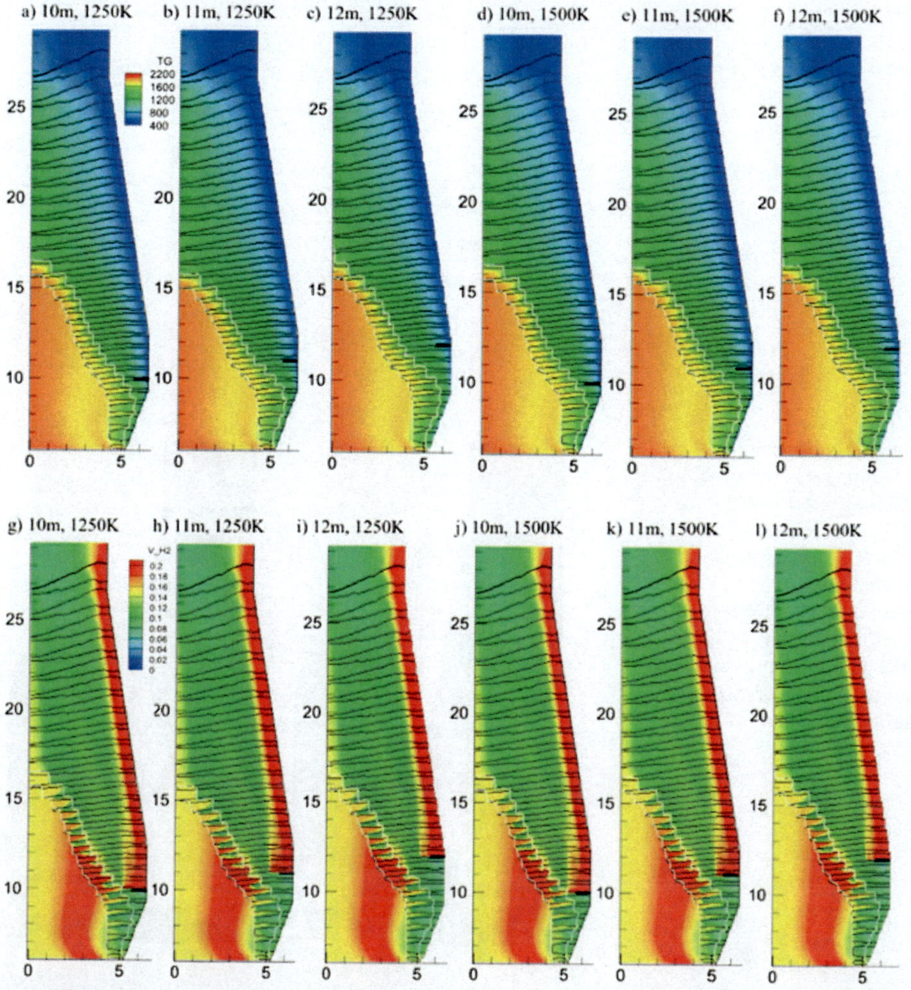


**Fig. 4** Comparison of gas temperature (**a–f**) and hydrogen volume fraction (**g–l**) contours at various injection temperatures and mid-shaft tuyere heights. Position of source cells marked with black bar

thermal energy available to drive the reduction reactions. Additionally, a spike in reduction is observed near the injection location at 10 m where available thermal energy is high enough to initiate reduction reactions.

This decline in available energy and reduction reactions in the mid-shaft tuyere plume region can also be observed by plotting the distributions of gas temperature and species distributions across the ore layers at different heights within the furnace. These results are analogous to the analysis of in-burden probe data, although they can present detail at greater depths within the burden when compared to real-world operational data. Figure 6 details the radial distributions at 10, 15, 20, and 25 m of gas temperature (on lines as shown in Fig. 6b), $H_2$ reduction reaction rate, $H_2$

**Table 4** Summary of top-level results for $H_2$ injection at 15 kg/thm through mid-shaft tuyeres

| MST height (m) | Inj temperature (K) | CO Util. (%) | $H_2$ Util (%) | TGT (°C) | Coke rate (kg/thm) | % change coke rate |
|---|---|---|---|---|---|---|
| Baseline (no MST) | | 51.39 | 52.17 | 129.6 | 391.9 | – |
| 10 | 1250 | 50.87 | 39.55 | 136.6 | 386.4 | −1.4% |
| 10 | 1500 | 50.89 | 39.99 | 137.2 | 385.9 | −1.5% |
| 11 | 1250 | 50.86 | 38.85 | 133.9 | 387.4 | −1.1% |
| 11 | 1500 | 50.68 | 39.63 | 134.8 | 386.4 | −1.4% |
| 12 | 1250 | 50.68 | 38.10 | 134.1 | 389.0 | −0.7% |
| 12 | 1500 | 50.55 | 38.45 | 134.1 | 388.7 | −0.8% |

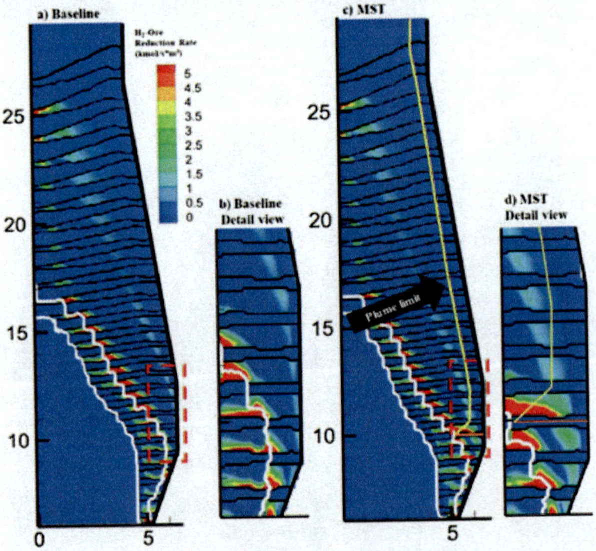


**Fig. 5** Hydrogen ore reduction rates for **a** the baseline and **c** 5 kg/thm of $H_2$ injected at 10 m and 1250 K. Detailed views of the mid-shaft tuyere injection region for **b** the baseline and **d** the MST case. MST plume limit is shown in yellow, with an orange line marking source term cells

volume fraction, and $H_2O$ volume fraction in the furnace for the best case scenario of 15 kg/thm $H_2$ delivery at 1250 K at a mid-shaft tuyere height of 10 m. At the shaft tuyere height of 10 m, gas temperatures in the final 40% of the furnace radius are roughly up to 250 K lower than the baseline case. This difference lessens as gas rises up the furnace shaft, however, it also spreads further across the furnace radius, with at least some impacts observed at 70% of the radius at 15 m and 90% of the radius at 20 m. The sharp spike in $H_2$ reduction reaction rate observed at 10 m near the furnace wall is a clear indication of reduction thanks to the gas delivered by the shaft tuyeres, however, the reduction rate declines rapidly with falling gas temperatures near the wall, and has almost entirely vanished by the time the gas reaches 20 m. This is due to the lack of requisite thermal energy to drive the reaction in the near-wall region higher in the furnace, exacerbated by the endothermic nature of the $H_2$ reduction

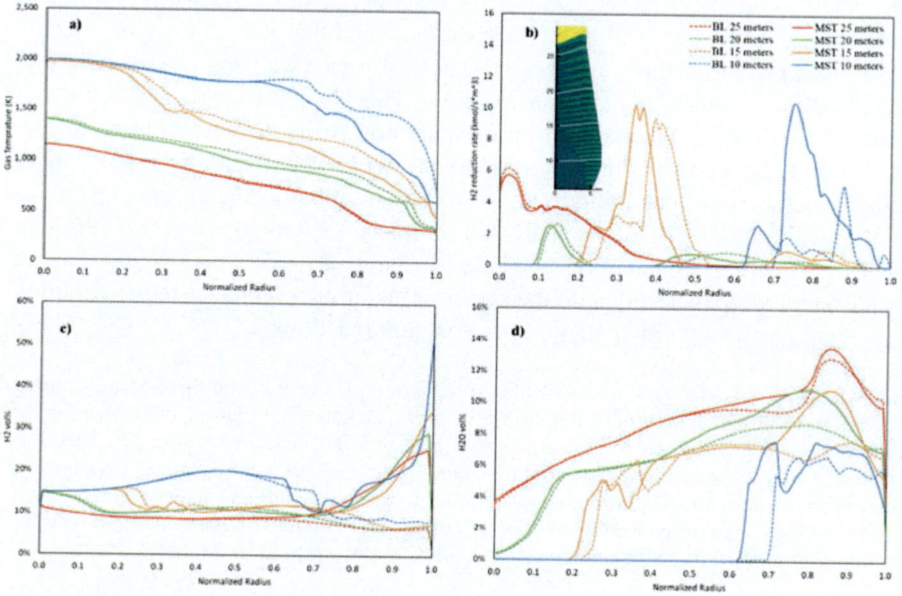


**Fig. 6** Radial distributions of **a** gas temperature, **b** $H_2$ reduction reaction rate, **c** $H_2$ volume percentage, and **d** $H_2O$ volume percentage, comparing the baseline scenario to 15 kg/thm $H_2$ delivery at 1250 K with mid-shaft tuyeres located at 10 m

reactions. As observed in the gas species contours, a significant amount of $H_2$ gas is wasted in this scenario, passing through the burden without reacting and leaving through the top gas. $H_2O$ volume fractions indicate a minor increase in water vapor (a product of the reduction reactions) near the furnace wall, heralding the minor benefits observed in coke consumption rate due to the additional reducing gas.

Further modification of operating conditions such as maximizing injection temperature, increasing plume penetration with higher injection velocity, and locating the tuyeres as low as possible in the furnace may improve the observed benefits to performance. Perhaps most significantly, as the $H_2$-iron ore reduction reaction is endothermic and provides a net cooling effect to the gas and burden near the furnace wall, a reduction in cooling water flow through the furnace staves (and by extension a reduction in cooling heat flux) may enable higher levels of reduction due to the additional available energy without overheating the staves.

## Conclusions

A modification to existing CFD models for simulation of blast furnace operation was developed to model the impacts of reducing gas delivery through mid-shaft tuyeres. A range of scenarios for $H_2$ delivery through these tuyeres at an injection rate of 15 kg/thm were investigated, with findings indicating a relatively minor improvement in the coke rate of 1.5%. Investigation of the reactions, gas species distributions, and temperature distributions within the furnace indicated that the majority of injected $H_2$ remained in a low-temperature region near the furnace wall and lacks the thermal energy to support additional reduction reactions. This phenomenon may be countered by decreasing the cooling load by lowering water flow to the staves and increasing the reducing gas temperature. Future research work will focus on these variables in an attempt to define stable operating conditions which establish a greater benefit from the use of mid-shaft tuyeres. Additionally, higher injection velocities achieved through higher reducing gas temperatures and smaller diameter tuyeres may be explored to provide better gas penetration into the center of the furnace, enhance the distribution of reducing gas, and promote additional reduction reactions.

**Acknowledgements** The authors would like to thank the members of the Steel Manufacturing Simulation and Visualization Consortium (SMSVC) for their support of this applied research. The authors acknowledge the SMSVC ironmaking project technical committee for providing operating conditions, geometry, validation data, and collaborative project guidance during model development and scenario research. Finally, the authors would like to thank the staff and students at the Center for Innovation through Visualization and Simulation (CIVS) at Purdue University Northwest for their support and input to the simulation research work.

## References

1. US Energy Information Administration IEO2021 issues in focus: energy implications of potential iron- and steel-sector decarbonization pathways, p 26. https://www.eia.gov/outlooks/ieo/IIF_industrial/pdf/IIF_industrial.pdf
2. World Steel (2020) Steel statistical yearbook 2020 concise version. World Steel Association, p 27. https://worldsteel.org/wp-content/uploads/Steel-Statistical-Yearbook-2020-concise-version.pdf
3. Geerdes M, Chigneau R, Kurunov I, Lingiardi O, Ricketts J (2015) Modern blast furnace ironmaking: an introduction. IOS Press, Amsterdam, The Netherlands, pp 59–75
4. Pistorius PC, Gibson J, Jampani M (2017) Natural gas utilization in blast furnace ironmaking: Tuyère injection, shaft injection and prereduction. In: Wang S, Free M, Alam S, Zhang M, Taylor P (eds) Applications of process engineering principles in materials processing, energy and environmental technologies. The Minerals, Metals & Materials Series. Springer, Cham. https://doi.org/10.1007/978-3-319-51091-0_26
5. Nogami H, Kashiwaya Y, Yamada D (2012) Simulation of blast furnace operation with intensive hydrogen injection. ISIJ Int 52(8):1523–1527. https://doi.org/10.2355/isijinternational.52.1523
6. Jampani M, Gibson J, Pistorius PC (2019) Increased use of natural gas in blast furnace ironmaking: mass and energy balance calculations. Metall Mater Trans B 50:1290–1299. https://doi.org/10.1007/s11663-019-01538-8
7. Hirsch A et al (2013) New blast furnace process (ULCOS). Report EUR 25085 EN, European Commission, Brussels
8. Kozub VN, Andronov VN, Popov NN, Bachinin AA, Tikhomirov EN (1971) Blast-furnace operation with injection of hot reducing gases at the Azovstal works. Steel USSR 1:8–10
9. Babich AI, Gudenau HW, Mavrommatis KT, Froehling C, Formoso A, Cores A, García L (2002) Choice of technological regimes of a blast furnace operation with injection of hot reducing gases. Rev Metalurgia 38(4):288–305

10. Yu X, Hu Z, Shen Y (2021) Modeling of hydrogen shaft injection in ironmaking blast furnaces. Fuel 302:12. https://doi.org/10.1016/j.fuel.2021.121092
11. Dong Z, Wang J, Zuo H, She X, Xue Q (2017) Analysis of gas-solid flow and shaft-injected gas distribution in an oxygen blast furnace using a discrete element method and computational fluid dynamics coupled model. Particuology 32:63–72. https://doi.org/10.1016/j.partic.2016.07.008
12. de Castro JA, de Medeiros GA, de Oliveira EM, de Campos MF, Nogami H (2020) The mini blast furnace process: an efficient reactor for green pig iron production using charcoal and hydrogen-rich gas: a study of cases. Metals 10:22. https://doi.org/10.3390/met10111501
13. van der Stel J, Louwerse G, Sert D, Hirsch A, Eklund N, Pettersson M (2013) Top gas recycling blast furnace developments for 'green' and sustainable ironmaking. Ironmaking Steelmaking 40(7):483–489. https://doi.org/10.1179/0301923313Z.000000000221
14. Okosun T (2018) Numerical simulation of combustion in the ironmaking blast furnace raceway. Ph.D. Dissertation, Purdue University, West Lafayette, IN, USA
15. Okosun T, Silaen A, Zhou C (2019) Review on computational modeling and visualization of the ironmaking blast furnace at Purdue University Northwest. Steel Res Int 90(4). https://doi.org/10.1002/srin.201900046
16. Fu D, Chen Y, Zhao Y, D'Alessio J, Ferron K, Zhou C (2014) CFD Modeling of multiphase reacting flow in blast furnace shaft with layered burden. Appl Therm Eng 66(1–2):298–308. https://doi.org/10.1016/j.applthermaleng.2014.01.065
17. Fu D (2014) Numerical simulation of ironmaking blast furnace shaft. Ph.D. Dissertation, Purdue University, West Lafayette, IN, USA
18. Tsay QT, Ray WH, Szekely J (1976) The modeling of hematite reduction with hydrogen plus carbon monoxide mixtures: Part I. The behavior of single pellets. AIChE J 22(6):1064–1072. https://doi.org/10.1002/aic.690220617
19. Zhou CQ (2012) Minimization of blast furnace fuel rate by optimizing burden and gas distribution. Final Technical Report to U.S. Department of Energy (DOE). https://doi.org/10.2172/1053052
20. Fu D, Chen Y, Rahman MT, Zhou C (2012) Validation of the numerical model for blast furnace shaft process. In: Proceedings of AISTech 2012, Atlanta, GA, USA, May 7–10, p 11
21. Okosun T, Nielson S, D'Alessio J, Ray S, Street S, Zhou C (2020) On the impacts of pre-heated natural gas injection in blast furnaces. Processes 8(7):771. https://doi.org/10.3390/pr8070771
22. Nielson S, Okosun T, Damstedt B, Jampani M, Zhou C (2021) Tuyere-level syngas injection in the blast furnace: a computational fluid dynamics investigation. Processes 9(8):1447. https://doi.org/10.3390/pr9081447
23. Okosun T, Nielson S, Zhou C (2022) Blast furnace hydrogen injection: investigating impacts and feasibility with computational fluid dynamics. JOM 74:1521–1532. https://doi.org/10.1007/s11837-022-05177-4

# Hydrogen Plasma Reduction of Iron Oxides

Check for updates

**Dierk Raabe, Matic Jovičević Klug, Yan Ma, Ömer Büyükuslu, Hauke Springer, and Isnaldi Souza Filho**

**Abstract** We present a study about the reduction of hematite ores by hydrogen plasma. The transformation kinetics and chemical composition are studied over several intermediate states. We find that the reduction kinetics depends on the balance between the input mass and arc powder. For an optimized input mass-arc power ratio, the reduction is obtained within 15 min exposure to the hydrogen plasma. Micro- and nanoscale chemical and microstructure analysis show that the gangue elements partition to the slag oxides, revealed by energy dispersive spectroscopy and atom probe tomography. Si-enrichment was observed in the interdendritic fayalite domains, at the wustite/iron hetero-interfaces and in the oxide particles inside iron. With proceeding reduction, however, such elements are gradually removed from the samples so that the final iron product is nearly free of gangue-related impurities.

**Keywords** Iron and steel · Pyrometallurgy · Sustainability

## Extended Abstract

The global steel industry produces more than 1.85 billion tons of steel every year. This qualifies steel as the most important metallic material, both in terms of volume produced and environmental impact. On the one hand steels are profound enablers of a sustainable future, for instance through their use in wind power plants, weight-reduced vehicles, safe infrastructures, and magnetic materials. On the other hand the primary synthesis of steels is based on the reduction of iron oxides by fossil carbon sources, releasing in average about 2 tons of $CO_2$ per ton of steel produced into the atmosphere. This makes iron and steel production the largest single cause of global warming [1].

This presentation is an introduction to the most important pending basic research questions associated with producing steel more sustainably, particularly with lower $CO_2$ emissions, placing focus on hydrogen-containing plasma based reduction

D. Raabe (✉) · M. J. Klug · Y. Ma · Ö. Büyükuslu · H. Springer · I. Souza Filho
Max-Planck-Institut für Eisenforschung, Max-Planck-Str. 1, 40237 Dusseldorf, Germany
e-mail: d.raabe@mpie.de


C. Fleuriault et al. (eds.), *Advances in Pyrometallurgy*, The Minerals, Metals & Materials Series, https://doi.org/10.1007/978-3-031-22634-2_7

methods. More specific, we present recent results about the melting and reduction of (a) commercial hematite ore pellets and of (b) partially direct reduced hematite ore pellets by hydrogen-containing plasma atmospheres [2, 3]. Using a state-of-the-art array of advanced characterization methods, we document the transformation kinetics and chemical composition over several intermediate states, from macroscopic dimensions down to the atomic scale [2]. We find that the overall oxide reduction kinetics is determined by the arc power, input mass, arc dynamics, hydrogen content, and chemical composition of the mineral feedstock. Under optimized conditions full reduction of hematite into iron is obtained within 15 min exposure to the hydrogen-containing plasma.

Energy dispersive spectroscopy, atom probe tomography, and micro- and nanoscale chemistry and microstructure analyses demonstrate that the gangue components partition to the slag oxides. Si-enrichment was found in the iron oxide particles, in the wüstite/iron hetero-interfaces, and in the interdendritic fayalite domains. However, as the reduction process continues, these components are gradually eliminated from the samples, resulting in a practically impure final iron product.

In a second set of experiments we charge hematite ores that were partially reduced via hydrogen-based solid-state direct reduction [4] into the plasma furnace for subsequent melting and reduction using a hydrogen-containing plasma [3]. The findings reveal that an optimal transfer point between these two reduction methods (direct reduction, plasma reduction) occurs where their chemical efficiency in hydrogen consumption is about equal. The reduction of hematite through magnetite into wüstite via direct reduction is efficient and fast, but it becomes very sluggish and inefficient when iron forms at the outermost layers of the iron ore pellets [3].

## References

1. Raabe D, Tasan CC, Olivetti EA (2019) Strategies for improving the sustainability of structural metals. Nature 575:64–74
2. Souza Filho IR et al (2021) Sustainable steel through hydrogen plasma reduction of iron ore: process, kinetics, microstructure, chemistry. Acta Mater 213:116971
3. Souza Filho IR et al (2022) Green steel at its crossroads: Hybrid hydrogen-based reduction of iron ores. J Clean Prod 340:130805
4. Kim SH et al (2021) Influence of microstructure and atomic-scale chemistry on the direct reduction of iron ore with hydrogen at 700 °C. Acta Mater 212:116933

# Hydrogen Plasma-Based Reduction of Metal Oxides

**Halvor Dalaker and Even Wilberg Hovig**

**Abstract** Hydrogen is a candidate to replace carbon in metal production, as it can reduce some metal ores (e.g., iron ore). However, for other oxides, such as those of manganese and chromium, the situation is much more challenging. As the exotic species found in hydrogen plasma are much more reactive than molecular hydrogen, the use of hydrogen plasma can improve hydrogen reduction for all these oxides. Here, using a plasma arc melter, samples of $Fe_2O_3$, $Cr_2O_3$ and MnO have been exposed to hydrogen plasma. Reactions between oxides and hydrogen have been observed in all cases, producing metallic iron, chromium, and manganese, hinting that plasma technology can play a part in sustainable metal production, allowing for carbon free production of chromium and manganese. The paper also discusses possible reaction mechanisms.

**Keywords** Hydrogen · Plasma · Manganese · Chromium · Steel · Sustainability

## Introduction

The metallurgical sector contributes to climate change through substantial $CO_2$ emissions. Some of these emissions are related to electricity production and can be mitigated by a change to a more renewable energy mix. Others result from carbon acting as a reducing agent in converting oxides to metal, with $CO_2$ as an unavoidable by-product. Sustainability therefore requires some combination of [1]:

- $CO_2$-capture and storage
- The use of $CO_2$-neutral biogenic carbon
- An entirely different and carbon free process, for example electrowinning.

H. Dalaker (✉)
SINTEF, Trondheim, Norway
e-mail: halvor.dalaker@sintef.no

E. W. Hovig
SINTEF, Oslo, Norway



C. Fleuriault et al. (eds.), *Advances in Pyrometallurgy*, The Minerals, Metals & Materials Series, https://doi.org/10.1007/978-3-031-22634-2_8

Hydrogen is thought by many (e.g. [2]) to play a big role in decarbonising the metallurgical sector, as it can replace carbon in the reduction of some metal oxides, for example iron:

$$FeO_x + xH_{2(g)} \rightarrow Fe + XH_2O_{(g)}.$$

Some metal oxides, however, are too stable to be reduced by gaseous molecular hydrogen under normal conditions. Exciting the hydrogen to the plasma state will convert some of the molecular hydrogen into exotic species—such as monoatomic hydrogen, H, rotationally excited hydrogen, $H_2^*$, and ions such as $H_2^+$ and $H^+$—that are more reactive towards metal oxides [3]. These species have the potential to react with metal oxides under conditions where molecular hydrogen would not [3], so if the reactive power of these species can be utilized, hydrogen-based extractive metallurgy of many more metals would become possible.

This is illustrated in Fig. 1, which shows an Ellingham diagram calculated using Factsage [4]. Each line shows the $\Delta G$ for the reaction between different elements and oxygen as a function of temperature. The relative position of these lines corresponds to the relative stability of the different oxides. The uppermost dashed line represents the reaction between hydrogen and oxygen to form water, while the very top line represents the reaction between MnO and oxygen to form $Mn_2O_3$. Since the curve for the reaction $O_2 + 4MnO = 2Mn_2O_3$ lies above that of the reaction $O_2 + 2H_2 = H_2O$, hydrogen can be used in prereduction of higher order manganese oxides to MnO. Indeed, prereduction of manganese using hydrogen has received some attention [5, 6]. The reaction between metallic manganese and oxygen however, $O_2 + 2Mn = 2MnO$, lies below the hydrogen line, and reduction of MnO to metallic manganese by molecular hydrogen is thermodynamically unfavourable. The authors are not aware of any reports that metallic manganese has been produced from MnO with hydrogen.

The situation is similar for the chrome/chrome-oxide system, in that the $Cr_2O_3$ is more stable than $H_2O$ across most relevant temperatures. However, at temperatures below approximately 2500 °C, $Cr_2O_3$ is less stable than MnO, and metallization

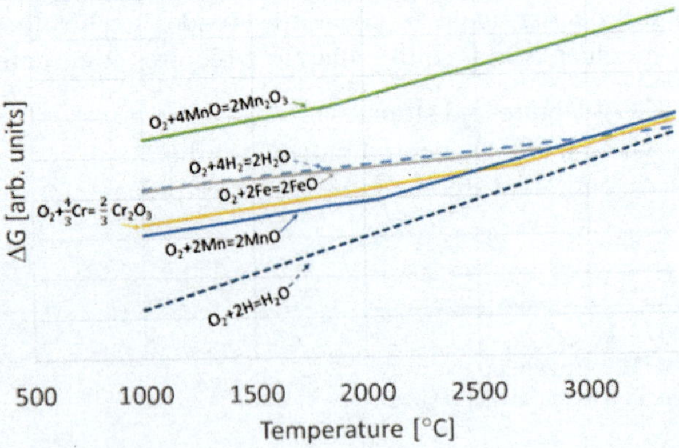


**Fig. 1** Ellingham diagram for selected metal oxides, molecular hydrogen, and monoatomic hydrogen

of chromium during treatment of chromite ore with molecular hydrogen has been reported in the literature, albeit at very low levels [7].

In Fig. 1, the very lowest curve included is that for the reaction between oxygen and monoatomic hydrogen. It lies below all the other included curves for all temperatures below 3500 °C (At even higher temperatures, $H_2O$ and $H_2$ both start to become unstable, so the study of reactions involving either specie as a reactant or product becomes less relevant). Monoatomic hydrogen then, should be able to reduce both chromium and manganese oxides to the metallic state. The plasma would also include charged species like $H^+$ that are even more reactive [3], so the reactivity of the hydrogen plasma would be even greater than indicated by Fig. 1 that only considers H.

The situation of iron is slightly different since iron oxides can be reduced to metallic iron by molecular hydrogen. Indeed, hydrogen-based steelmaking is being explored by several players, including large steel producers [8, 9]. The argument for the study of hydrogen plasma in iron production is thus not based on *enabling* hydrogen use, but rather on *improvement*: it has been reported that the reduction rate of iron oxide using hydrogen in the plasma state is greater than that of the molecular state [10].

SINTEF have launched the HyPla project to investigate these opportunities [11]. The challenging physical conditions involved means that experimental investigations of plasma metallurgy are not straightforward, and a dedicated experimental rig is being built.

Filho et al. [12] reported on a method employing a plasma arc melter/suction caster to expose metal oxides to hydrogen plasma and to study the reduction of hematite. In this work we employ a similar method and apparatus to investigate the interaction between hydrogen plasma and $Fe_2O_3$, $Cr_2O_3$, and MnO.

## Experimental Procedure

The oxides, in powder form, are cyclically exposed to hydrogen plasma in an Arc Melter AM200 (Edmund Bühler GmbH). An electric arc is struck between a tungsten electrode and a copper crucible, which generates a plasma plume extending out from the electrode. Prior to ignition, the process chamber is evacuated by a vacuum pump and then flushed with a gas mixture of 15%$H_2$ in Ar. During plasma exposure, the pressure in the processing chamber is held at a slight vacuum gauge pressure of 700 mbar. The position of the electrode is controlled by the operator and a distance between the tip of the electrode and the oxides is approximately $15 \pm 5$ mm. A schematic of the arc melter setup is shown in Fig. 2.

The experiments presented herein represent the early steps of method development. This means that exposure times and applied currents could not always adhere to a predetermined experimental plan but had to be adapted dynamically. In experiments with hematite, the initial aim was to do 60 s exposure times, but gas formation and foaming in some trials led to a violent reaction and failed experiments. Therefore,

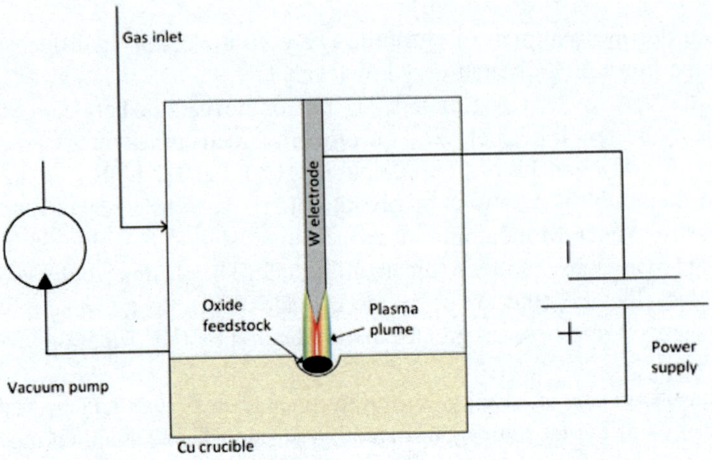


**Fig. 2** Schematic of the experimental setup

in later tests, exposure was discontinued as soon as visual observations indicated gas formation.

The experiments with MnO and $Cr_2O_3$ were done together in a series with four samples, only two of which are reported here. In each cycle, all four samples were sequentially exposed for 30 s each between each evacuation and refill of the furnace chamber. With each cycle, a different sample was the first to be exposed to the plasma arc, to even out how much unreacted hydrogen each sample was exposed to. Also in these experiments, the applied current started out at 30A to avoid higher currents ejecting the powders from the crucible. Once the samples had melted together this was no longer an issue, and the current could gradually be increased over several cycles to 100A. The complete overview of exposure times, current, and gas replenishing intervals are given in Table 1.

The samples were analysed using Electron beam Micro Probe Analyser (EPMA), both for imaging and for use of Electron Diffraction (EDS) for chemical analysis.

## Results and Discussion

### *General Observations*

As mentioned in the experimental description, during experiments with hematite there was considerable gas evolution, which caused the samples to expand. A possible explanation is that this was caused by water vapour forming from the reaction of hematite. However, in all potential reactions, the amount of product $H_2O$ is the same as the reactant $H_2$ (the situation will be the same if 2H is considered instead of $H_2$):

**Table 1** Overview of experiments

| Sample | $Fe_2O_3$ | | $MnO$ | | | $Cr_2O_3$ | | |
|---|---|---|---|---|---|---|---|---|
| Cycle | Current [A] | Time [s] | Current [A] | Time [s] | Exposed as x/4 in cycle | Current [A] | Time [s] | Exposed as x/4 in cycle |
| 1 | 45 | 60 | 30 | 30 | 1 | 30 | 30 | 3 |
| 2 | 45 | 24 | 30 | 30 | 4 | 30 | 30 | 2 |
| 3 | 45 | 16 | 40 | 30 | 3 | 40 | 30 | 1 |
| 4 | 45 | 21 | 50 | 30 | 2 | 50 | 30 | 4 |
| | $H_2$ replenished | | | | | | | |
| 5 | 45 | 25 | 60 | 30 | 1 | 60 | 30 | 3 |
| 6 | 45 | 13 | 70 | 30 | 4 | 70 | 30 | 2 |
| 7 | 45 | 30 | 70 | 30 | 3 | 70 | 30 | 1 |
| | $H_2$ replenished | | | | | | | |
| 8 | 45 | 19 | 80 | 30 | 2 | 80 | 30 | 4 |
| 9 | 45 | 30 | 80 | 30 | 1 | 80 | 30 | 3 |
| 10 | 45 | 30 | 100 | 30 | 4 | 100 | 30 | 2 |
| 11 | – | – | 100 | 30 | 3 | 100 | 30 | 1 |
| 12 | – | – | 100 | 30 | 2 | 100 | 30 | 4 |

During experiments with MnO and $Cr_2O_3$, the hydrogen was replenished after each cycle. For experiments with $Fe_2O_3$, hydrogen was replenished after cycle 4 and cycle 7, as indicated in the table

$$3Fe_2O_3 + H_2 \rightarrow 2Fe_3O_4 + H_2O$$

$$Fe_3O_4 + H_2 \rightarrow 3FeO + H_2O$$

$$FeO + H_2 \rightarrow Fe + H_2O$$

In other words, gas volume expansion as a result of the chemical reaction does not appear to be a viable explanation.

The cooling effect of the water-cooled copper substrate is substantial compared to the thermal mass of the small samples. It is therefore possible that evolved $H_2O$ gas trapped inside the sample could condense once the arc is turned off. During re-heat, these water droplets could boil and expand, but the issue remains unresolved.

When the chamber is vented after each cycle, the pressure drops from 700 mbar to 0.05 mbar. In experiments with hematite, at this point a cloud of mist or dust becomes visible, and the viewport of the instrument fogs over. This could perhaps be water vapour resulting from a reaction between hydrogen and oxide, or solid dust being blowed away from surfaces by the moving gases. In experiments with $Cr_2O_3$ and MnO, the same phenomena appears but to a much lower degree. During exposure,

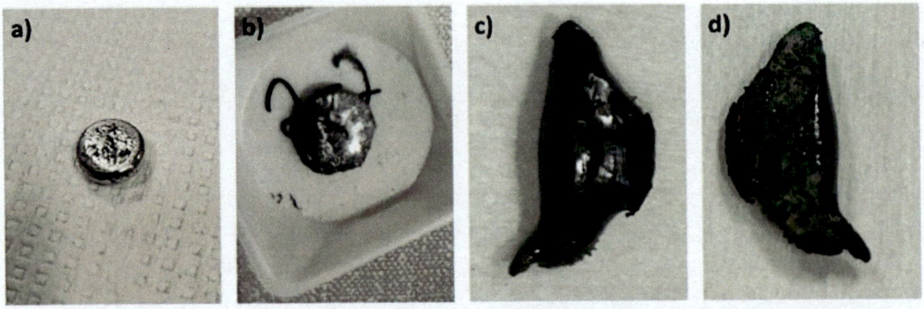


**Fig. 3** Photographs of samples after exposure to hydrogen plasma. The initial material in each figure is respectively: **a** $Fe_2O_3$, **b** $Cr_2O_3$, **c** and **d** MnO

when in contact with the $Cr_2O_3$ sample the plasma plume had hues of green, yellow, and orange, while when in contact with the MnO sample the arc was bluish.

For experiments with both $Cr_2O_3$ and MnO, there was considerable loss of material. This may be because of the volatility of both Cr and Mn at the high temperatures. Dust/condensate was found around the samples at the end of experiments, but these were not analysed at this stage.

Figure 3 shows images of the samples after hydrogen plasma exposure of: $Fe_2O_3$ (a), $Cr_2O_3$ (b), and MnO (c and d). The $Fe_2O_3$ sample appeared metallic to the eye. The bottom face, towards the copper substrate during experiments, was quite shiny, while the top surface that faced out towards the interior of the arc melter, was duller. The sample was magnetic: the bottom surface strongly magnetic, and the top surface weakly magnetic.

The $Cr_2O_3$ and MnO samples showed an even greater contrast between the top and bottom faces. Figure 3c shows the top surface of the MnO sample as being shiny while the bottom (3d) is rock-like.

## *EPMA Analysis and Thermodynamic Considerations*

In this section, the EPMA images of the samples are presented, and the observed phases are discussed. In terms of thermodynamics, FactSage has been used with the FactOxide and FSstel databases [4].

Figure 4a shows the EPMA-image of a part of the $Fe_2O_3$-sample after exposure to hydrogen plasma. EDS analyses were taken in several spots in the sample. The bright phase was determined to be metallic iron by EDS. While the quantification of chemical composition by EDS is somewhat uncertain, an average of 5 measurements in the oxide phase gave an O:Fe mass ratio of 2.63, corresponding to a molar ratio of 1.33. This is consistent with magnetite, $Fe_3O_4$, which matches the observation of magnetism of the sample. The light phase contains only iron, as confirmed by EDS. No difference in the chemical composition of the oxide phase could be detected

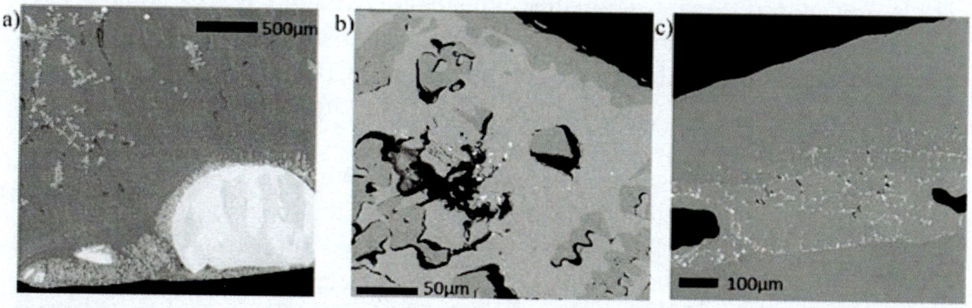


**Fig. 4** EPMA images of samples after exposure to hydrogen plasma. The initial material in each figure is respectively: **a** $Fe_2O_3$, **b** $Cr_2O_3$, **c** MnO. Note the differences in magnification

between the top and the bottom of the sample, and the sample thus appears to consist mainly of magnetite, with droplets of metallic iron inclusions. The observation that the bottom of the sample is more magnetic than the top is not due to a difference in oxide phases, but rather because more of the metallic phase is found at the bottom of the sample. It would be expected that iron would more easily be formed at the top of the sample, where it is exposed to hydrogen plasma and high temperatures, but gravitational separation could cause the denser metal to collect at the bottom of the sample.

Figure 4b shows the EPMA image of the $Cr_2O_3$ sample. Two different oxide phases and a metallic phase are identified using EDS. The darkest phase is consistent with $Cr_2O_3$, while the lighter oxide phase has a Cr:O ratio that corresponds to CrO. The metallic phase is measured at more than 90 wt.% Cr, with an average of 2.5 wt.% Fe 2.3 wt.% Cu and 1.3 wt.% O. It must again be stressed that the measured quantities should not be taken as more than an indication, but this signal is consistent with a metallic phase that is mostly chromium. The oxygen signal can be due to surrounding oxides, the copper can come from the substrate, while the iron is a plausible impurity in the $Cr_2O_3$ starting material (99.6% pure, no other information on impurities given). In other words, metallic chrome has been formed.

CrO is not thermodynamically stable as a solid, but it exists in liquid form at high temperatures. When considering only pure substances, liquid $Cr_2O_3$ decomposes into liquid CrO and solid $CrO_2$ at temperatures above 2690 °C. If liquid CrO cools without interacting with $CrO_2$ (for example due to rapid quenching), it should solidify as equal amounts of Cr and $Cr_2O_3$ at equilibrium.

Some possible explanations for the existence of metallic chromium, the merits of which will be discussed in the following, are:

(A) Cr forms as CrO cools and separates into $Cr_2O_3$ and Cr, the CrO having been formed from thermal decomposition of $Cr_2O_3$ into CrO and $CrO_2$.
(B) Cr forms as CrO cools and separates into $Cr_2O_3$ and Cr, the CrO having been formed CrO forms from the reduction of $Cr_2O_3$ by $H_2$ or H.
(C) Cr is formed by the reduction of $Cr_2O_3$ by $H_2$ or H. First to CrO, and then some CrO is further reduced into Cr.

The lack of any $CrO_2$ is a strike against explanation A, although $CrO_2$ could behave differently to CrO during cooling and reach equilibrium during cooling. Further, the formation of Cr and $Cr_2O_3$ during the solidification of CrO would give Cr as a solid (even with a few % of Fe), meaning that A and B fail to explain how the metallic phase was able to coalesce as the pearls seen in Fig. 4b.

Of the three explanations then, C remains viable. While reaction to metallic Cr by $H_2$ is not particularly favourable thermodynamically speaking ($\Delta G > 0$ for T <3173 °C), it cannot be ruled out. Also, it remains undetermined whether reactions involving monoatomic hydrogen or other exited H-species have played a role.

Figure 4c shows the MnO-sample after exposure to the plasma. The bright phase was confirmed to be metallic manganese. Interestingly, only a single oxide phase was detected, one that corresponds to MnO. In other words, the strikingly different surfaces seen in Fig. 3c, d are the same material. The shiny phase is only quite thin, maybe 100 μm, below which is a cellular structure with metallic manganese on the grain boundaries. The rest of the sample maintains its original chemical and visual characteristics.

If the metallic manganese existed as a liquid at high temperature, it can be imagined that the pattern has been caused by liquid manganese being pushed by the solidifying oxide until it collects in the gaps between the oxide zones.

The distribution of the manganese is reminiscent of precipitates formed at the grain boundaries of a cooling sample. In which case the high-temperature starting phase may have been an oxygen-depleted MnO-phase:

$$MnO_{(l)} + xH_2/2xH = MnO_{1-x(l)}$$

As this phase solidifies, metallic manganese could be expulsed

$$MnO_{1-x(l)} = (1-x)MnO_{(s)} + xMn_{(l)}$$

However, we have found no mention in the literature of oxygen deficient liquid MnO.

Another explanation is that during exposure, liquid Mn exists on the surface of the sample, but as the plasma is turned off and the temperature starts dropping it will to some extent react with $H_2O$ to re-form MnO at the top of the sample most exposed to the $H_2O$-containing atmosphere. Some metallic manganese below the surface could remain unreacted between a protective layer of MnO, but this does not adequately explain the formation of the observed microstructure. Thus, the exact mechanism remains unconfirmed, but it seems most likely that metallic Mn has been pushed by solidifying MnO, perhaps in a combination of reactions between $H_2O$ and Mn.

Filho et al. [12] presented images of partially converted samples that showed that over the course of 30 1-min cycles the hematite sample transformed first into magnetite and then into metallic iron. Compared to those results, our $Fe_3O_4$ sample shows less conversion than the one they present as having been exposed to five cycles. This indicates that with an increased number of cycles, eventual full conversion to metallic iron could be possible in our setup, although electrical parameters also differ (800A vs. 45A). This in turn suggests that increased amounts of metallic manganese and chromium could also be obtained with increased numbers of cycles and/or increased currents. In our setup we have experienced considerable electrode wear at higher currents, so substantial changes to the setup might be necessary. It must also be noted that there is a limit to how many cycles the samples can effectively be exposed to, as the material is lost as vapour and/or dust during exposure to plasma. Furthermore, as was already mentioned, back-reactions with $H_2O$ may also occur.

However, it is important to note that these challenges are mostly related to the experimental method. The main message of the current work is that $Cr_2O_3$ and MnO have been exposed to hydrogen plasma at high temperatures, successfully obtaining metallic products completely without carbonaceous reductants. Of course, these are early days and a lot of work remains, for example in avoiding the produced metals to be re-oxidised by the water vapour. But if these results can eventually be reproduced and improved in an industrial setting, it can be a significant step forward for sustainable metal production. If this happens it will likely be in conditions not directly comparable to those obtained in a plasma arc melter. Other experimental setups will likely be necessary for the process development.

## Conclusions

Using a plasma arc melter, samples of $Fe_2O_3$, MnO and $Cr_2O_3$ have been exposed to hydrogen plasma, producing metallic Fe, Mn, and Cr.

The $Fe_2O_3$ sample was converted to $Fe_3O_4$ with significant amounts of metal droplets. In the $Cr_2O_3$ and MnO samples much less metal was found than in the $Fe_2O_3$ sample, and in the $Cr_2O_3$ sample the metastable phase CrO was found in significant amounts.

The mechanisms for Cr and Mn formation remain unclear, even though it is suspected that plasma radicals such as monoatomic hydrogen have played a role.

Comparisons with similar work suggest that iron conversion could be increased by increased current and/or exposure time, while it is less certain how well this approach would work for $Cr_2O_3$ and MnO due to increased material loss.

## References

1. Kero IT, Dalaker Halvor, Osen KS, Ringdalen E (2021) Some carbon-free technologies for manganese ferroalloy production. In: Proceedings of the 16th international ferro-alloys congress (INFACON XVI). Available at SSRN: https://ssrn.com/abstract=3926069 or https://doi.org/10.2139/ssrn.3926069
2. European Commission (2020) Communication from the commission to the European parliament. The Council, The European economic and social committee and the committee of the regions—a hydrogen strategy for a climate-neutral Europe. https://ec.europa.eu/energy/sites/ener/files/hydrogen_strategy.pdf

3. Sabat KC, Rajput P, Paramguru RK, Bhoi B, Mishra BK (2014) Reduction of oxide minerals by hydrogen plasma: an overview. Plasma Chem Plasma Process 34:1–23. https://doi.org/10.1007/s11090-013-9484-2
4. Bale CW, Bélisle E, Chartrand P, Decterov SA, Eriksson G, Gheribi AE, Hack K, Jung IH, Kang YB, Melançon J, Pelton AD, Petersen S, Robelin C. Sangster J, Van Ende MA (2016) FactSage thermochemical software and databases, 2010–2016. Calphad 54:35–53. www.factsage.com
5. Barner HE, Mantell CL (1968) Kinetics of hydrogen reduction of manganese dioxide. Ind Eng Chem Process Des Dev 7(2):285–294. https://doi.org/10.1021/i260026a023
6. Safarian J (2022) A sustainable process to produce manganese and its alloys through hydrogen and aluminothermic reduction. Processes 10(1):27. https://doi.org/10.3390/pr10010027
7. Davies J, Paktunc D, Ramos-Hernandez J, Tangstad M, Ringdalen E, Beukes J, Bessarabov D, du Preez S (2022) The use of hydrogen as a potential reductant in the chromite smelting industry. Minerals 12:534. https://doi.org/10.3390/min12050534
8. SSAB (2021) The world's first fossil-free steel ready for delivery. https://www.ssab.com/en/news/2021/08/the-worlds-first-fossilfree-steel-ready-for-delivery [Accessed 03 Aug 2022]
9. Thyssenkrupp (2021) Green hydrogen for the decarbonization of the steel industry: Thyssenkrupp Steel and STEAG agree on hydrogen supply. https://www.thyssenkrupp-steel.com/en/newsroom/press-releases/thyssenkrupp-steel-and-steag-agree-on-hydrogen-supply.html [Accessed 03 Aug 2022]
10. Seftejani MN, Schenk J (2018) Thermodynamic of liquid iron ore reduction by hydrogen thermal plasma. Metals 8(12):1051. https://doi.org/10.3390/met8121051
11. SINTEF (2021) HyPla-Hydrogen plasma for $CO_2$-free metal production. https://www.sintef.no/en/projects/2020/hypla-hydrogen-plasma-for-co2-free-metal-production/ [Accessed 03 Aug 2022]
12. Souza Filho IR, Ma Y, Kulse M, Ponge D, Gault B, Springer H, Raabe D (2021) Sustainable steel through hydrogen plasma reduction of iron ore: process, kinetics, microstructure, chemistry. Acta Mater 213:116971. https://doi.org/10.1016/j.actamat.2021.116971

# Effect of Hydrogen-Rich Atmosphere on Softening and Melting Behaviors of Ferrous Burden in Blast Furnace Cohesive Zone

**Binbin Lyu, Fan Yang, Guang Wang, Haibin Zuo, Qingguo Xue, and Jingsong Wang**

**Abstract** To evaluate the effect of hydrogen-rich smelting on the softening and melting behaviors of the ferrous burden in a blast furnace, softening and melt-dripping performance tests were performed on the burden under different atmospheric ratios. The results showed that the softening start temperature decreased as the $H_2$ ratio increased, giving the blast furnace a broader softening zone. As less FeO-containing low-melting-point phase was generated, the melting start temperature increased, the temperature ranges of the melting zone narrowed and moved to the high-temperature zone, and the permeability of the burden significantly improved. Comparing the slag-iron-coke interface characteristics demonstrated that the carbon content in its metallic iron decreased as the ratio of $H_2$ increased.

**Keywords** Ferrous burden · Hydrogen · Softening and melting behaviors · Cohesive zone · Slag-iron-coke interface

## Introduction

Rapid industry development has increased concern about the greenhouse effect and environmental problems caused by carbon emissions [1]. The steel industry, an important basic support industry, accounts for 6.7% of global $CO_2$ emissions annually, and the energy consumption and $CO_2$ emissions of ironmaking account for approximately 70% of the total for the entire steel process [2]. Therefore, improvements in blast furnaces, the main ironmaking technology currently in use, have great potential for reducing $CO_2$ emissions. Hydrogen-rich blast furnaces are an effective way to realize this potential [3]. Hydrogen has been shown to decrease carbon emissions at the source by acting as both a reducing agent and an energy carrier that

B. Lyu · F. Yang · G. Wang (✉) · H. Zuo · Q. Xue · J. Wang (✉)
State Key Laboratory of Advanced Metallurgy, University of Science and Technology Beijing, Beijing, People's Republic of China
e-mail: wangguang@ustb.edu.cn

J. Wang
e-mail: wangjingsong@ustb.edu.cn



C. Fleuriault et al. (eds.), *Advances in Pyrometallurgy*, The Minerals, Metals & Materials Series, https://doi.org/10.1007/978-3-031-22634-2_9

can replace carbon. As a new technology of green metallurgy, hydrogen metallurgy is an important direction for low-carbon development and achieving low carbon green development in the field of metallurgy, which is a key concern of researchers worldwide.

The key to fundamentally solving the $CO_2$ emission problem of the ironmaking process is to decrease the amount of carbon fed into the blast furnace, which results in low-coke-ratio smelting in the blast furnace. This has some disadvantage: the function of coke as the "skeleton" of the column is decreased, and permeability deteriorates in the blast furnace. In the blast furnace smelting process, the softening and melting behaviors of the ferrous burden determine the location and thickness of the cohesive zone and play an important role in the secondary distribution of airflow, energy savings, and consumption of resources. The effect of hydrogen as a reducing agent in blast furnaces on the softening-melting-dripping behaviors of the burden under hydrogen-rich conditions is not yet fully understood. In this study, the softening-melting-dripping behaviors of the ferrous burden under hydrogen-rich conditions were systematically investigated under laboratory conditions. The aim was to provide a theoretical basis and process guidance for current and future low-carbon ironmaking processes in blast furnaces.

## Experimental

### *Experimental Samples*

The pellets and coke used in this study were obtained from a commercial steel company. These burdens were broken down into some small particles with a diameter of 10–12.5 mm. Chemical analyses of the pellets and coke are presented in Tables 1 and 2, respectively.

**Table 1** Chemical composition of pellets used in experiments (wt.%)

| Burden | TFe | FeO | $Al_2O_3$ | CaO | MgO | $SiO_2$ | Basicity |
|---|---|---|---|---|---|---|---|
| Pellet | 63.49 | 1.02 | 0.87 | 0.94 | 0.67 | 3.53 | 0.27 |

**Table 2** Chemical composition of coke used in experiments (wt.%)

| Fixed carbon | Volatile matter | Ash content | Moisture |
|---|---|---|---|
| 86.85 | 0.82 | 12.31 | 0.021 |

## Experimental Apparatus and Conditions

The softening and melting properties of the burden were determined using a softening-melting experimental apparatus, as shown in Fig. 1. A 500 g pellet was loaded into a graphite crucible (94 mm inner diameter and 210 mm depth) with 40 and 80 g of coke above and below it, respectively. The reducing gas was introduced from the lower part during the experiment, and the graphite crucible had 12 holes at the bottom to facilitate full contact with the reducing gas and solid coke. The heating rate was 10 °C/min in the temperature range below 900 °C, 3 °C/min in the range of 900–1020 °C, and 5 °C/min from 1020 °C to the end of the experiment. During the temperature increase from room temperature to 900 °C, 5 L/min of $N_2$ was passed, and 12 L/min of reducing gas was passed after exceeding 900 °C. The composition of the reducing gas is presented in Table 3. A constant load of 9.8 N/cm$^2$ was applied to the burden using a graphite pusher. The heating was stopped when the first drop was produced, and the droplets were collected by the lower graphite crucible and cooled to room temperature by passing $N_2$. The index and physical significance of the softening, melting, and dripping of the ferrous burden are listed in Table 4. During the test, the softening and melt-dripping parameters of the ferrous burden are automatically recorded using a computer. The microstructure of the residual burden was analyzed by using scanning electron microscopy-energy scattering spectroscopy (SEM–EDS).

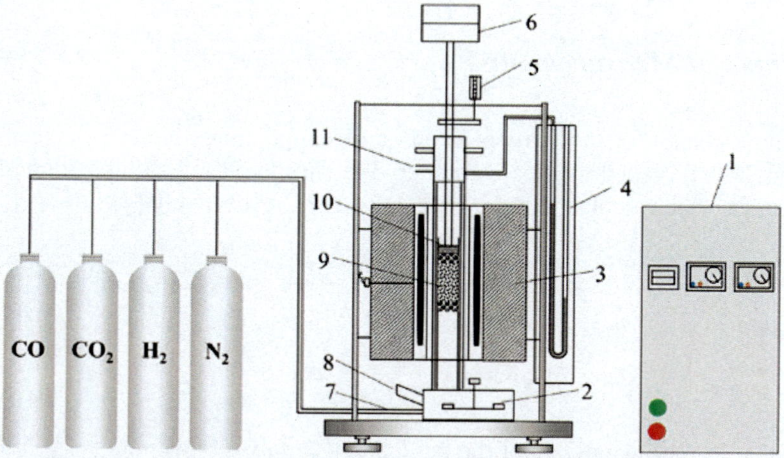


1.Temperature control cabinet; 2.Sampling dish; 3.Electric furnace; 4.Differential pressure gauge; 5.Displacement meter; 6.Load (1 Kg); 7. Gas inlet; 8.Sight glass; 9.Burden; 10. Graphite pusher; 11.Gas outlet

**Fig. 1** Schematic diagram of the softening-melting experimental apparatus

**Table 3** Composition of reducing gases (vol.%)

| No. | CO | $H_2$ | $N_2$ |
|---|---|---|---|
| 1# | 40 | 0 | 60 |
| 2# | 30 | 10 | 60 |
| 3# | 20 | 20 | 60 |
| 4# | 10 | 30 | 60 |
| 5# | 0 | 40 | 60 |

**Table 4** Softening, melting, and dripping index and physical significance

| Parameters | Physical significance |
|---|---|
| $T_{10\%}$/°C | Softening start temperature, the burden lays shrinkage of 10% |
| $T_{40\%}$/°C | Softening end temperature, the burden lays shrinkage of 40% |
| $T_s$/°C | Melting start temperature, the temperature in which the differential pressure rises rapidly |
| $T_d$/°C | Dripping temperature, the temperature at which dripping starts |
| $\Delta T_S$/°C | Softening zone ($T40$–$T10\%$) |
| $\Delta T_M$/°C | Melting zone ($T_d$–$T_s$) |
| $\Delta T_{SM}$/°C | Cohesive zone ($T_d$–$T_{10\%}$) |
| $\Delta P_{max}$/Kpa | Maximum differential pressure, KPa |
| S/Kpa·°C | Permeabllity index, $S = \int_{T_s}^{T_d} (\Delta P_T - \Delta P_S)\mathrm{d}T$ |

## Experimental Measurements

The volume change of the burden at different temperatures under hydrogen-rich conditions was recorded using a computer. The shrinkage and shrinkage rate (SR) of the burden were calculated using Eqs. (1) and (2), respectively [4]:

$$\text{Shrinkage} = \frac{H_0 - H_t}{H_0} \tag{1}$$

$$\text{SR} = \Delta \frac{H_0 - H_t}{H_0} / \Delta T \tag{2}$$

where SR is the shrinkage rate of the burden at a certain temperature (%/°C), $H_0$ is the initial height of the burden (mm), and $H_t$ is the transient height of the burden (mm).

## Results and Discussion

### *Effect of Hydrogen-Rich Atmosphere on the Shrinkage Behavior of Ferrous Burden*

Figure 2 shows the effect of burden shrinkage during the softening-melting-dripping process under different atmospheres. The shrinkage trend can be divided into three stages. In the first stage (900–1000 °C), the burden shrinkage was negative, indicating that the burden was swelling. Previous studies have shown that the swelling that occurs during the reduction of pellets in the temperature range of 900–1100 °C is mainly caused by changes in the lattice ($Fe_2O_3 \rightarrow Fe_3O_4$) and thermal expansion during the reduction process [5, 6]. With an increase in the ratio of $H_2$ in the gas, the maximum negative shrinkage rate of the pellets decreased by 0.58%, from 7.33% to 6.75%, and its maximum negative shrinkage temperature decreased by 31 °C, from 997 °C to 966 °C. $H_2$ has a smaller volume and lower viscosity than other components of the gas, which makes it easier to enter the burden for reduction; thus, the growth of iron whiskers is restrained, and spherical and very fine iron whiskers are formed [7]. This decreases the swelling of the pellets. From the gas-phase equilibrium of iron oxide reduction by CO and $H_2$ at different temperatures, it can be concluded that at temperatures greater than 900 °C, the reduction of iron oxide by $H_2$ is greater than that by CO, which leads to a decrease in its reduction start temperature and, thus, a decrease in its maximum negative shrinkage temperature. In the second stage (1000–1150 °C), shrinkage of the burden occurred. As the ratio of $H_2$ in the gas increased, the SR of the pellets decreased from 0.29%/°C to 0.22%/°C, which is a decrease of 0.06%/°C. In the third stage (1150–1400 °C), the SR of the burden further increased. As the ratio of $H_2$ in the gas increased, the SR of the pellets decreased from 0.46%/°C to 0.19%/°C, which is a decrease of 0.27%/°C. Qie et al. [8] showed that with an increase in the $H_2$ content in the reducing gas, the reduction potential of the gas was promoted and the surface of the burden was rapidly reduced to a thick metallic iron shell, which had a higher resistance to deformation and retarded the shrinkage of the burden. This was beneficial for improving the permeability of the burden layer.

### *Effect of Hydrogen-Rich Atmosphere on the Softening-Melting Behavior of Ferrous Burden*

The experimental results for the burden under different atmospheres are listed in Table 5. The data in Table 5 were plotted to obtain the characteristics of the cohesive zone (Fig. 3). As shown in Fig. 3, as the ratio of $H_2$ in the gas increased, $T_{10\%}$ decreased from 1103 to 1110 °C, $T_{40\%}$ increased from 1150 to 1167 °C, and the temperature

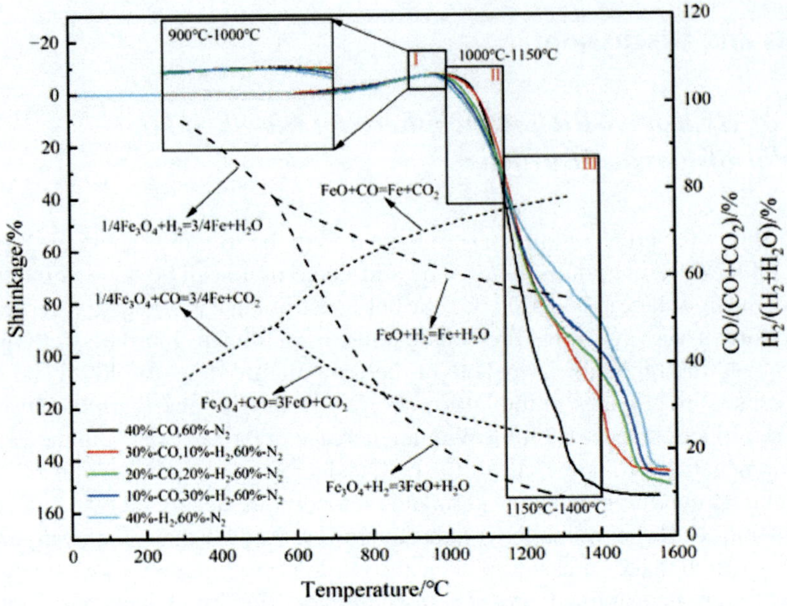


**Fig. 2** Effect of different atmospheres on burden shrinkage

range of the softening zone increased from 47 °C (1103–1150 °C) to 86 °C (1080–1166 °C), which is favorable for the gas–solid reduction of the upper burden of the blast furnace. The hydrogen-rich blast furnace shortens the reduction process from $Fe_2O_3$ to $Fe_3O_4$ and produces large amounts of FeO and metallic iron at lower temperatures, which promotes the formation of low-melting-point phases and thus lowers the softening start temperature. Therefore, a low softening start temperature and a wide softening zone are observed under hydrogen-rich conditions.

As the ratio of $H_2$ in the gas increased, $T_s$ increased from 1230 to 1515 °C, $T_d$ increased from 1488 to 1551 °C, and the temperature range of the melting zone decreased from 258°C (1230–1488 °C) to 3 °C (1515–1518 °C). As the melting zone becomes narrower, it helps improve the permeability of the burden. It is generally believed that the cohesive zone of a blast furnace plays an important role in determining the permeability of the burden layer [9]. From Fig. 3, it can be clearly seen

**Table 5** Experimental measurements of burden under different atmospheric conditions

| | $T_{10\%}$/°C | $T_{40\%}$/°C | $T_S$/°C | $T_d$/°C | ΔTS/°C | $\Delta T_M$/°C | $\Delta T_{SM}$/°C | $\Delta P_{max}$/KPa | S/(kPa·°C) |
|---|---|---|---|---|---|---|---|---|---|
| 1# | 1103 | 1150 | 1230 | 1488 | 47 | 258 | 385 | 33 | 2100.3 |
| 2# | 1110 | 1167 | 1294 | 1544 | 57 | 250 | 434 | 29.9 | 1670 |
| 3# | 1094 | 1161 | 1351 | 1551 | 67 | 200 | 457 | 13.9 | 901.3 |
| 4# | 1082 | 1156 | 1402 | 1527 | 74 | 125 | 445 | 3.7 | 211.9 |
| 5# | 1080 | 1166 | 1515 | 1518 | 86 | 3 | 438 | 1 | 1.7 |

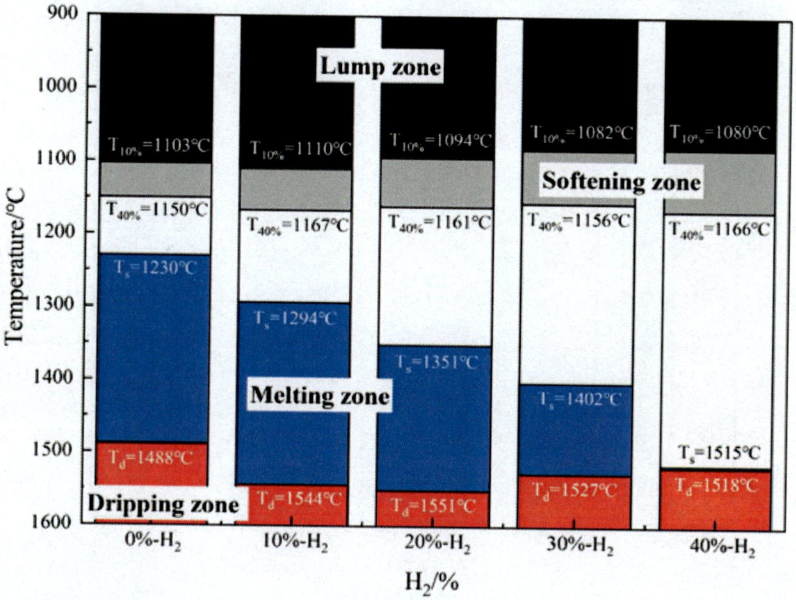


**Fig. 3** Effect of different atmospheres on cohesive zone of burden

that with an increase in the $H_2$ ratio in the gas, the softening zone of the cohesive zone became significantly wider, and the melting zone became narrower, which is beneficial for blast furnace smelting.

The permeability of the burden layer depends primarily on its permeability in the cohesive zone. Decreasing the width of the melting zone and the pressure difference in the cohesive zone can improve the permeability of the burden layer. Figure 4 shows the effect of different atmospheres on the maximum differential pressure and permeability index of the burden. As shown in Fig. 4, the volume fraction of $H_2$ in the reducing gas had a significant effect on both the maximum pressure difference and permeability index. As the ratio of $H_2$ in the gas increased, $\Delta P_{max}$ decreased 32 kPa, from 33 to 1 kPa, and S decreased 2098.6 kPa·°C from 2100.3 to 1.7 kPa·°C. The increase in $H_2$ content promotes the permeability of the burden layer because $H_2$ has a smaller volume and lower viscosity, thus reducing the density and viscosity of the gas.

## *Effect of Hydrogen-Rich Atmosphere on the Melt-Dripping Behavior of Ferrous Burden*

The morphologies of droplets formed under different atmospheric conditions are shown in Fig. 5. When the ratio of $H_2$ in the atmosphere was less than 20%, a slag phase existed on the surface of the droplets. When the ratio of $H_2$ in the atmosphere

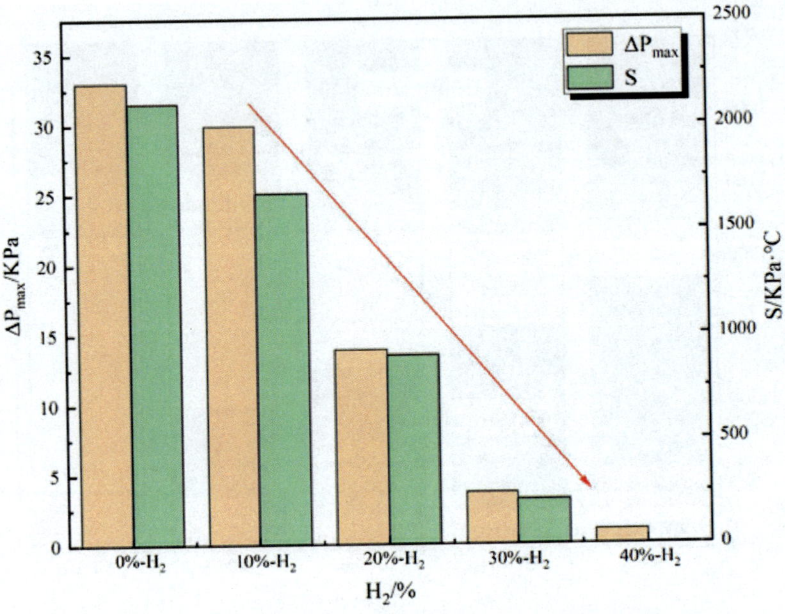


**Fig. 4** Effect of different atmospheres on permeability index and maximum differential pressure of burden

was greater than 20%, no slag phase was observed on the surface of the droplets. The weight of the droplets and the carbon content of the iron in the droplets under different atmospheric conditions are shown in Fig. 6. As the ratio of $H_2$ in the gas increased, both the droplet weight and carbon content of the iron decreased. This implies that more metallic iron could not be separated from the slag and remained in the coke bed. The higher reduction potential in the gas decreased the FeO content of the slag and increased the melting point and viscosity of the slag, resulting in less metallic iron being able to pass through the coke layer to produce drops. Pan et al. [10] showed that increasing the degree of reduction in the burden leads to more iron remaining in the slag, which makes dripping difficult.

The chemical compositions of the residual slag and dripping slag are listed in Table 6. Under a CO atmosphere, the FeO content in the residual slag was 1.71%

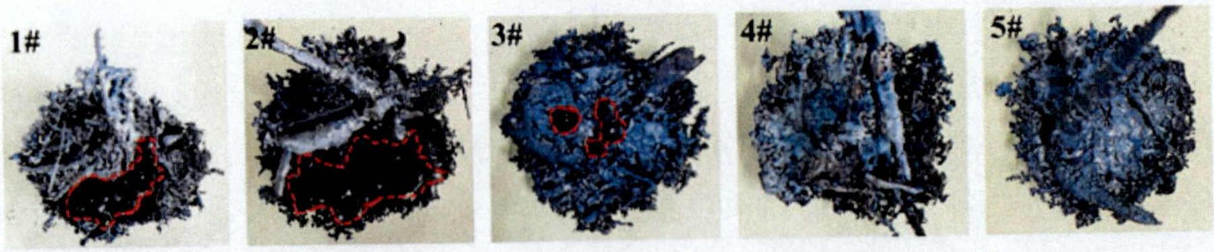


**Fig. 5** Morphology of droplets formed under different atmospheres

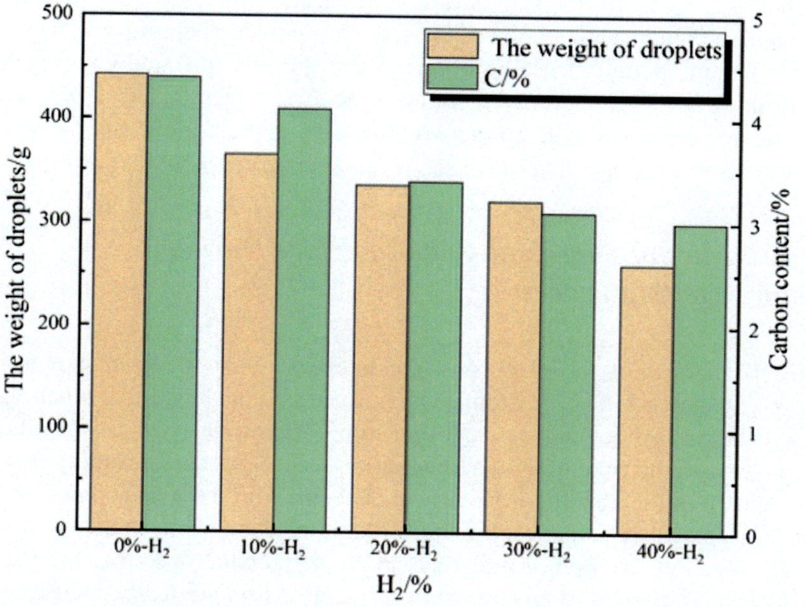


**Fig. 6** Effect of different atmospheres on the weight of droplets and carbon content in the iron in droplets

and that in the dripping slag was 0.12%. However, with an increase in the $H_2$ content of the gas, the FeO content in the residual slag gradually decreased.

The droplet weight was positively correlated with the carbon content of the metallic iron. The amount of carburization in iron determines its melting temperature, and thus affects the separation of slag and iron. In a CO atmosphere, the high FeO content in the residual slag leads to an increase in the low-melting phase, and the FeO-containing slag is fully in contact with the coke for direct reduction reactions. The iron in the slag undergoes carburization and melting, forming more molten Fe–C alloys, thus lowering the melting temperature of the iron. As the ratio of $H_2$ in the

**Table 6** Chemical composition of dripping slag and residual slag formed under different atmospheres

| Sample | Slag | CaO | $SiO_2$ | MgO | $Al_2O_3$ | FeO |
|---|---|---|---|---|---|---|
| 1# | Residual slag | 18.2 | 26.2 | 1.26 | 8.33 | 1.71 |
| | Dripping slag | 17.5 | 26.5 | 1.43 | 8.56 | 0.12 |
| 2# | Residual slag | 20.3 | 26.7 | 1.79 | 8.82 | 1.68 |
| | Dripping slag | 20.5 | 26.9 | 1.53 | 8.9 | <0.1 |
| 3# | Residual slag | 19.2 | 26.1 | 1.34 | 8.39 | 1.61 |
| 4# | Residual slag | 18.5 | 26.5 | 1.25 | 8.43 | 1.43 |
| 5# | Residual slag | 19.9 | 26.4 | 1.41 | 8.64 | 1.42 |

gas increases, the FeO in the slag is rapidly reduced to metallic iron, which has a higher melting point than that of FeO. Thus, the dripping temperature was increased. When the melting point of metallic iron is reached, iron drips rapidly into the coke layer, resulting in a very short carburization time. Consequently, the carbon content in metallic iron decreased with an increase in the $H_2$ content in the gas.

## *Characteristics of Slag-Iron-Coke Interface During Softening-Melting Process*

SEM–EDS analysis of the slag-iron-coke interface under different atmospheric conditions is shown in Fig. 7. From Fig. 7, it can be concluded that when the $H_2$ content in the gas is 0%, there is a clear slag-iron-coke interface, whose metallic iron is in contact with the coke; thus, a carburization reaction occurs, resulting in a high carbon content in the metallic iron. As the $H_2$ content in the gas increases, a large amount of slag gathers at the reaction interface of the coke, which hinders the direct contact between the iron phase and coke and is not conducive to the carburization reaction. Therefore, the carbon content in its metallic iron gradually decreases.

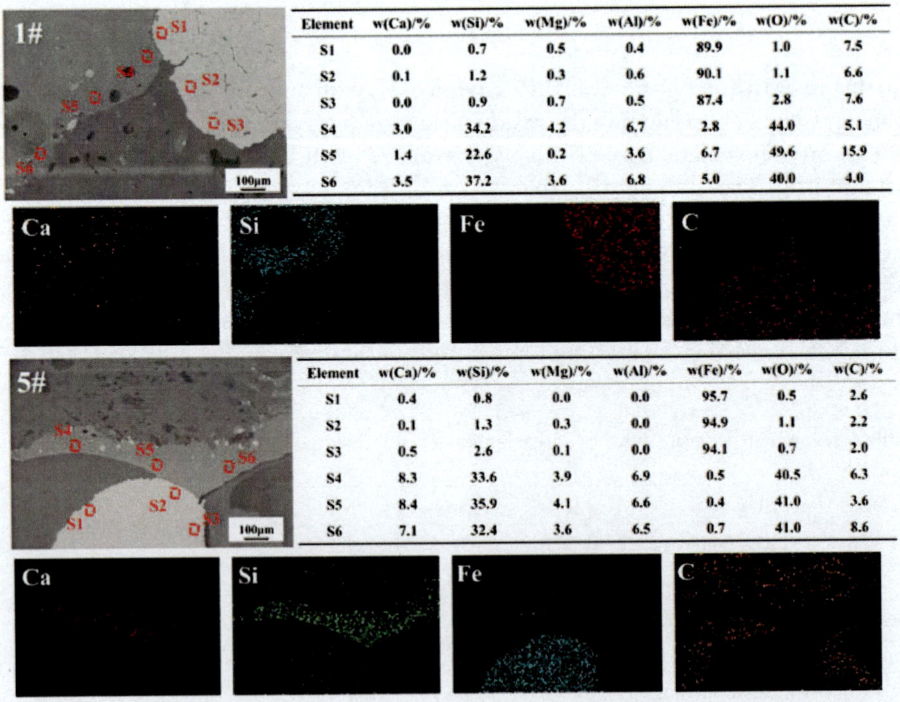


1#

| Element | w(Ca)/% | w(Si)/% | w(Mg)/% | w(Al)/% | w(Fe)/% | w(O)/% | w(C)/% |
|---|---|---|---|---|---|---|---|
| S1 | 0.0 | 0.7 | 0.5 | 0.4 | 89.9 | 1.0 | 7.5 |
| S2 | 0.1 | 1.2 | 0.3 | 0.6 | 90.1 | 1.1 | 6.6 |
| S3 | 0.0 | 0.9 | 0.7 | 0.5 | 87.4 | 2.8 | 7.6 |
| S4 | 3.0 | 34.2 | 4.2 | 6.7 | 2.8 | 44.0 | 5.1 |
| S5 | 1.4 | 22.6 | 0.2 | 3.6 | 6.7 | 49.6 | 15.9 |
| S6 | 3.5 | 37.2 | 3.6 | 6.8 | 5.0 | 40.0 | 4.0 |

5#

| Element | w(Ca)/% | w(Si)/% | w(Mg)/% | w(Al)/% | w(Fe)/% | w(O)/% | w(C)/% |
|---|---|---|---|---|---|---|---|
| S1 | 0.4 | 0.8 | 0.0 | 0.0 | 95.7 | 0.5 | 2.6 |
| S2 | 0.1 | 1.3 | 0.3 | 0.0 | 94.9 | 1.1 | 2.2 |
| S3 | 0.5 | 2.6 | 0.1 | 0.0 | 94.1 | 0.7 | 2.0 |
| S4 | 8.3 | 33.6 | 3.9 | 6.9 | 0.5 | 40.5 | 6.3 |
| S5 | 8.4 | 35.9 | 4.1 | 6.6 | 0.4 | 41.0 | 3.6 |
| S6 | 7.1 | 32.4 | 3.6 | 6.5 | 0.7 | 41.0 | 8.6 |

**Fig. 7** SEM and EDS analysis of the interaction between coke and slag-iron

## Conclusions

This study systematically investigated the softening-melting-dripping behaviors of the ferrous burden in a blast furnace under different ratios of atmospheric hydrogen. The key findings can be summarized as follows.

(1) Hydrogen-rich smelting in a blast furnace decreased the maximum negative shrinkage rate and maximum negative shrinkage temperature of its burden. The rapid formation of a metallic iron shell on the burden led to a decrease in its SR and improved the permeability of the burden layer.
(2) With an increase in the $H_2$ content of the atmosphere, the softening start temperature gradually decreased and the softening end temperature increased, resulting in a wider softening zone. The melting start temperature and dripping temperature increased, but the melting zone narrowed, which is beneficial for blast furnace smelting.
(3) The carbon content in the metallic iron in the slag-iron-coke interface decreased with an increase in the $H_2$ content in the gas.

**Acknowledgements** The authors gratefully acknowledge the financial support of the National Natural Science Foundation of China (No. U1960205, No. 51804024).

**Conflict of Interest** No potential conflict of interest was reported by the author(s).

## References

1. Ariyama T, Takahashi K, Kawashiri Y, Nouchi T (2019) Diversification of the ironmaking process toward the long-term global goal for carbon dioxide mitigation. J Sustain Metall 5(3):276–294
2. Ma K, Deng J, Wang G, Zhou Q, Xu J (2021) Utilization and impacts of hydrogen in the ironmaking processes: a review from lab-scale basics to industrial practices. Int J Hydrogen Energ 46(52):26646–26664
3. Yilmaz C, Wendelstorf J, Turek T (2017) Modeling and simulation of hydrogen injection into a blast furnace to reduce carbon dioxide emissions. J Clean Prod 154:488–501
4. Ma K, Xu J, Deng J, Wang D, Xu Y, Liao Z (2018) Transient interaction between reduction and slagging reactions of wustite in simulated cohesive zone of blast furnace. Metall Mater Trans B 49(5):2308–2321
5. Hayashi S, Iguchi Y (2005) Influence of several conditions on abnormal swelling of hematite pellets during reduction with $H_2$-CO gas mixtures. Ironmak Steelmak 32(4):353–358
6. Wang H, Sohn H (2011) Effect of CaO and $SiO_2$ on swelling and iron whisker formation during reduction of iron oxide compact. Ironmak Steelmak 38(6):447–452
7. Li F, Chu M, Tang J, Liu Z, Feng C, Tang Y (2017) Swelling behavior of high-chromium, vanadium-bearing titanomagnetite pellets in $H_2$-CO-$CO_2$ gas mixtures. JOM 69(10):1751–1758
8. Qie Y, Lyu Q, Liu X, Li J, Lan C, Zhang S (2018) Effect of hydrogen addition on softening and melting reduction behaviors of ferrous burden in gas-injection blast furnace. Metall Mater Trans B 49(5):2622–2632

9. Yang W, Zhou Z, Yu A, Pinson D (2015) Particle scale simulation of softening-melting behaviour of multiple layers of particles in a blast furnace cohesive zone. Powder Technol 279:134–145
10. Pan Y, Zuo H, Wang B, Wang J, Wang G, Liu Y (2020) Effect of reduction degree on cohesive zone and permeability of mixed burden. Ironmak Steelmak 47(3):322–327

# Hydrogen-Based Direct Reduction of Iron Oxides

**Dierk Raabe, Hauke Springer, Isnaldi Souza Filho, and Yan Ma**

**Abstract** The lecture presents some recent progress in understanding the key mechanisms of hydrogen-based direct reduction. The kinetics of the solid-state reduction reactions strongly depend on mass transport kinetics, nucleation during the multiple phase transformations, the oxide's chemistry and microstructure, and on plasticity, damage, and fracture associated with the phase transformation and mass transport phenomena occurring during reduction. Understanding these effects is key to produce hydrogen-based green steel and design corresponding direct reduction shaft or fluidized bed reactors, enabling massive $CO_2$ reductions.

**Keywords** Sustainability · Iron and steel · Pyrometallurgy

## Extended Abstract

Iron- and steelmaking stand for about 7–8% of all global greenhouse gas emissions, accelerating global warming. This staggering environmental damage is caused by the use of fossil carbon carriers as precursor materials for the reduction of iron oxides. Carbon is turned in blast furnaces into CO and—through the redox processes behind iron making—into $CO_2$, producing about 2 tons $CO_2$ for each ton of steel produced.

One mitigation strategy consists in the replacement of fossil carbon carriers by hydrogen as an alternative reductant, to massively cut these $CO_2$ emissions, thereby lying the foundation for transforming a 3000 years old industry within a few years [1].

As the sustainable production of hydrogen using renewable energy is currently expensive, thus acting as a severe bottleneck in green steel making, at least during the next decade, the gigantic annual steel production of 1.85 billion tons requires strategies to use hydrogen very efficiently and to yield high metallization at fast reduction kinetic.

D. Raabe (✉) · H. Springer · I. S. Filho · Y. Ma
Max-Planck-Institut Für Eisenforschung, Max-Planck-Str. 1, 40237 Düsseldorf, Germany
e-mail: d.raabe@mpie.de



C. Fleuriault et al. (eds.), *Advances in Pyrometallurgy*, The Minerals, Metals & Materials Series, https://doi.org/10.1007/978-3-031-22634-2_10

In this presentation we therefore discuss some recent progress in understanding the governing mechanisms of hydrogen-based direct reduction of iron oxides [2–5].

Metallization degree, reduction kinetics, and their dependence on the underlying solid-state redox reactions in hydrogen-containing direct reduction strongly depend on mass transport kinetics, Kirkendall effects, nucleation phenomena during the multiple phase transformations, chemical and stress partitioning, the oxide's chemistry and microstructure, the acquired (from sintering) and evolving (from oxygen loss) porosity, crystal plasticity, damage and fracture effects associated with the phase transformation phenomena occurring during reduction. Understanding these effects—together with external boundary conditions such as reductant gas mixtures, oxide feedstock composition, pressure, and temperature—is key to produce hydrogen-based green steel and design corresponding direct reduction shaft or fluidized bed reactors, enabling the required massive $C0_2$ reductions at affordable costs. Possible simulation approaches that are capable of capturing some of these phenomena and their interplay are also discussed [6].

## References


1. Raabe D, Tasan CC, Olivetti EA (2019) Strategies for improving the sustainability of structural metals. Nature 575:64–74
2. Kim SH et al (2021) Influence of microstructure and atomic-scale chemistry on the direct reduction of iron ore with hydrogen at 700°C. Acta Mater 212:116933
3. Filho IRS et al (2022) Green steel at its crossroads: Hybrid hydrogen-based reduction of iron ores. J Clean Prod 340:130805
4. Ma Y et al (2022) Hierarchical nature of hydrogen-based direct reduction of iron oxides. Scr Mater 114571
5. Ma Y et al (2022) Hydrogen-based direct reduction of iron oxide at 700°C: Heterogeneity at pellet and microstructure scales. Int J Miner Metall Mater 29:1901–1907
6. Bai Y et al (2021) Chemo-mechanical phase-field modeling of iron oxide reduction with hydrogen. Acta Mater 231:117899

# Part III
# Biocarbon and Alternative Reduction Methods

# Ferroalloy Production Without Use of Fossil Carbon—Some Alternatives

**Eli Ringdalen and Trine A. Larsen**

**Abstract** To reach the goal of fossil free ferroalloy production within 2050, several alternative materials and processes are being investigated. Fast transition requires that the more mature technologies are implemented first. In parallel technologies with lower TRL levels must be further developed. An overview of different options is given and discussed. Substituting fossil carbon with biocarbon seems to be among the first to be implemented. How the differences in key properties between these two reductants may affect the operation of Mn-alloy and Si/FeSi furnaces will be discussed. Pretreatment and prereduction by sustainable energy sources are another way for reducing fossil $CO_2$ emission. Pretreatment of Mn-ores with CO-rich off-gases and with thermal solar heated air have been investigated and tested on a pilot-scale within the PreMa project. The effect of different temperatures and gas atmospheres has been studied. Effect on $CO_2$ emissions on integrated pre-treatment units and furnaces are evaluated by comparing various scenarios in an HSC Sim model adapted for this purpose.

**Keyword** Ferroalloys · $CO_2$ emissions · Biocarbon

## $CO_2$ Emissions from Ferroalloy Production

Ferroalloys are mainly used in and required for steel production, both as functional element in production and as alloying element to improve their properties. Steel is crucial for the modern society and an important part of our daily life. At the same time production of iron and steel is a source of $CO_2$ emissions that have a negative impact on global climate. Global $CO_2$ emission was in 2020 around 34 billion tons [1]. 24% of this were $CO_2$ emission from energy use in industry and around 7.2% were from energy use in iron and steel production. Around 2.6 billion tons of $CO_2$ [2] were generated from the 1800 million tons of steel produced in 2020. Ferroalloys and Silicon are in addition to their use in the steel production additionally used in

E. Ringdalen (✉) · T. A. Larsen
SINTEF, Trondheim, Norway
e-mail: eli.ringdalen@sintef.no



C. Fleuriault et al. (eds.), *Advances in Pyrometallurgy*, The Minerals, Metals & Materials Series, https://doi.org/10.1007/978-3-031-22634-2_11

**Table 1** $CO_2$ emissions from ferroalloy and silicon production (ex US) in 2016. Based on [3]

| | FeCr | FeMn | SiMn | FeSi | Si |
|---|---|---|---|---|---|
| Global production million of tons/year | 11.900 | 4.655 | 12.500 | 6.870 | 2.850 |
| Ton $CO_2$/ton liquid metal | 1.31 | 1.3 | 1.4 | 3.6 | 5.0 |
| Million tons $CO_2$/year | 15.5 | 6.1 | 17.5 | 24.7 | 14.3 |

several other products as silicones and electronics. They are important constituents in products needed to reduce the $CO_2$ emission. Examples are products for sustainable energy production as PV solar and windmills, as well as, alloys for lighter and thus more energy efficient vehicles as TWIN/TRIP steels.

Production of ferroalloys and silicon generates $CO_2$ emission. The emission depends both on a produced ton of alloy, as well as, on $CO_2$ intensity given as $CO_2$ emission pr ton of alloy. It is as shown in Table 1 around 5–25 million tons of $CO_2$ pr year (2016) [3]. Although silicon has a higher $CO_2$ intensity than SiMn, yearly $CO_2$ emissions where higher for SiMn due to its higher production.

Since ferroalloys and silicon are required both for our daily life and as mean to reduce our climate footprint, alternatives to current production are needed to ensure the future supply of these materials without $CO_2$ emission.

## Alternatives to Reduce $CO_2$ Emissions

### *Current Processes*

The ferroalloys discussed here, HCFeMn, SiMn, FeSi, and metallurgical silicon are all mainly produced in submerged arc furnaces (SAF), as illustrated in Fig. 1. The furnaces can be fully open at the top, semi-open as is usual in Si and FeSi production, or sealed as shown here. Solid raw materials are fed at the top of the furnace and liquid alloy tapped out near the bottom. Electrical energy is supplied through the electrodes inserted from the top of the furnace: They generate heat near their tip in the lower part of the furnace where most of the reduction takes place.

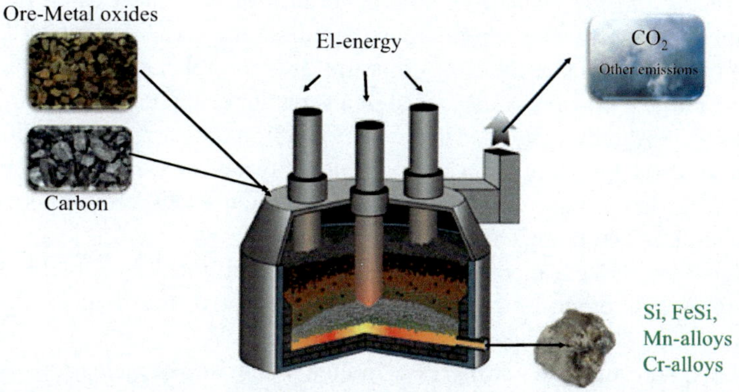


**Fig. 1** Illustration of ferroalloys production in submerged arc furnaces (SAF) The figure is based on Mn-alloy production in a sealed furnace. Adapted from [12]

Metal oxides are reduced to metal by carbon reacting in the furnaces to CO by the overall reactions (1) to (3). CO produced by these reactions will further either reduce higher Mn-oxides or Fe-oxides, but mainly react above charge top with air to $CO_2$. Iron oxides in the ore can be reduced directly with CO gas to $CO_2$ as in reaction (4)

$$MnO + C = Mn + CO \tag{1}$$

$$FeCr_2O_4 + 4C = Fe + 2Cr + 4CO \tag{2}$$

$$SiO_2 + 2C = Si + 2CO \tag{3}$$

$$Fe_2O_3 + 3CO = 2Fe + 3CO_2 \tag{4}$$

In current processes for the production of alloys in SAF, carbon is required as a reductant to remove oxygen from the ores. It also acts as an electrical conductor needed for the dissipation of energy. When carbon is removed from the process to avoid $CO_2$ emissions, alternatives to carbon both as reductant and as electrical conductors must be found.

## *Alternative Processes*

Several alternative processes have been proposed and are currently being investigated [4, 5]. They can be broadly divided into the following groups:

*Biocarbon*: This can be a substitute for fossil carbon as discussed in the review given by Sommerfeldt et al. [3]. It is described more in detail in a separate chapter.

*Pre-treatment and pre-reduction*: The materials are then pre-treated before the SAF to reduce the energy and carbon consumption in the furnace. This will reduce, but not completely eliminate the $CO_2$ emission. It is not relevant for Si production. The pretreatment of Mn-ores is discussed in a separate chapter.

*Hydrogen*: Hydrogen can be used in metal production both for pre-reduction [6], as hydrogen plasma for final reduction and in completely new process routes. Hydrogen is currently regarded to have a high potential for reducing $CO_2$ emissions and are a topic for several investigations and publications.

*Methallothermi*: Here a metal with higher reduction potential substitute carbon as reductant. Aluminothermic reduction is one alternative that is possible for the production of both Si [7] and Mn [8].

*Electrolysis*: Alternative electrolytes as molten oxides and new electrode materials give a potential for producing Si and Mn by electrolysis.

*Natural gas*: Substituting coke and coal with natural gas will reduce, but not eliminate $CO_2$ emissions. Various methods have been proposed as reduction in the solid state with natural gas [9, 10] and deposition of carbon on raw materials with $H_2$ as a by-product [11] followed by a final reduction in SAF.

*Other alternatives and methods*: Other new processes, various combinations of process routes as well as Carbon capture and storage (CCS) and Carbon capture and utilisation (CCU) present possibilities to reduce $CO_2$ emissions.

## *Important Considerations*

When alternative technologies are evaluated, their total $CO_2$ emissions must be considered. In addition to the direct $CO_2$ emission from the process, Scope 1, and Scope 2, the indirect $CO_2$ emission as the emissions from energy use, as well as Scope 3, $CO_2$ emissions from value chain outside the producer, must be considered. In power intensive processes, as metal production, $CO_2$ emission from production of energy used in the process is especially important. This depends both on the energy source and the specific energy consumption for the actual alloy as shown in Tables 2 and 3.

In evaluation of alternative processes, both $CO_2$ emissions from the new process, as well as $CO_2$ emissions from energy consumption in the new process must be compared with similar values for the current processes. The most important sources for $CO_2$ emissions are summarised in Table 4 for selected processes.

**Table 2** Typical values for specific energy consumption [12, 13]

| | MWh/ton |
|---|---|
| FeMn [12] | 2.4 |
| SiMn [12] | 3.5 |
| Si [13] | 12 |
| FeSi [13] | 8.2 |

**Table 3** Typical $CO_2$ emissions from different energy sources [14]

| | Kg $CO_2$/kWh |
|---|---|
| Hydropower | 0 |
| Coal [14] | ≈1 |
| Natural gas | 0–1 |

**Table 4** Main sources for $CO_2$ emissions in some alternative materials

| Reductant | Direct CO2 emissions | $CO_2$ from electrical energy |
|---|---|---|
| Bio carbon | Volatiles in bio carbon | |
| Hydrogen from water | | Energy for electrolysis |
| Hydrogen from natural gas | $CO_2$ in gas, $CH_4$ leakage | |
| Other alloys | From production of alloys | Energy for alloy production |

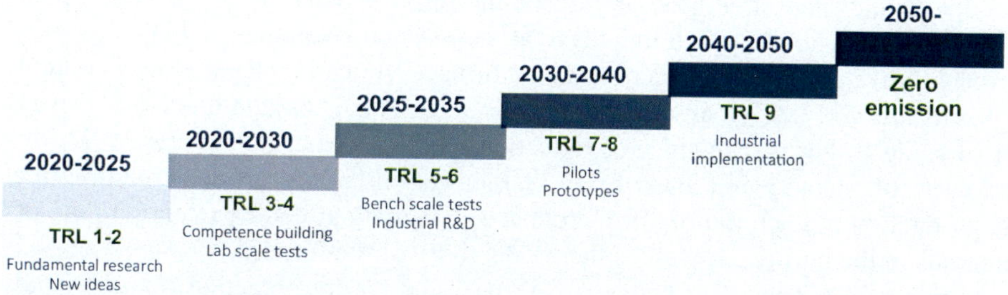


**Fig. 2** Timescale for development of new processes for metal production. Adapted and based on [15]

Evaluation of alternative processes must also take into account expected time for industrialisation. This will depend on Technology Readiness level (TRL) for the new technology Development of new processes, will as illustrated in Fig. 2, take a long time. A process that needs to be industrialised in 2030, must thus be at TRL 7–8 today. The figure also illustrates that development of new break-through processes must start today if they are to be industrialised within and contribute to reduced $CO_2$ emissions in 2050.

## Biocarbon

Substituting fossil carbon with biocarbon seems to be the easiest methods to reduce $CO_2$ emissions in production of Si and ferroalloy. It is already used in both FeSi and Mn-alloy production in Brazil. But there are major concerns related both to the availability of enough biocarbon for all that plans to use it and the total environmental impact of its use [3, 16], as well as, its economic feasibility. In addition, biocarbon

**Table 5** Typical properties of carbon reductants [17]

| | Coal | Metallurgical Coke | Charcoal | Woodchips |
|---|---|---|---|---|
| Fixed carbon (%) | 60 | 87 | 65–87 | 12 |
| Density (kg/m$^3$) | 800 | 500 | 180–300 | 250 |
| Volatiles (%) | 40 | 2 | 10–30 | 35 |

has different properties than the fossil reductants, carbon and coke used today (see Table 5).

These properties will affect the alloy production in various ways as illustrated in Fig. 3

Some properties will affect both Si/FeSi and Mn-alloy furnaces in the same way although their relative impact on the furnaces may vary between the alloys. Low cold strength of biocarbon will increase dust formation during transport and handling of biocarbon outside the furnace while low warm strength will led to an increased amount of fines and affect gas flow inside the furnace. Volatiles and moisture will increase the gas flow through the gas treatment system. Volatiles and $H_2$ may also have a positive effect on the reactions in the furnace. Biocarbon have a lower content of fixed carbon (Fix C) than other carbon sources. This is carbon that is not driven off in volatiles but is available for reduction work at higher temperatures. Due to its lower bulk density and lower FixC a larger volume of biocarbon is required for the production of each ton of alloy produced. This may affect the retention time of materials in the burden, as well as the productivity of the furnace.

A high rate for reaction between SiO gas, and carbon, reaction (5) is crucial to obtain a high yield in silicon and ferrosilicon production. The higher SiO reactivity[18] of biocarbon compared to coal is thus an advantage. In Mn-alloy production, $CO_2$ is generated in the burden. It will react at temperatures above around

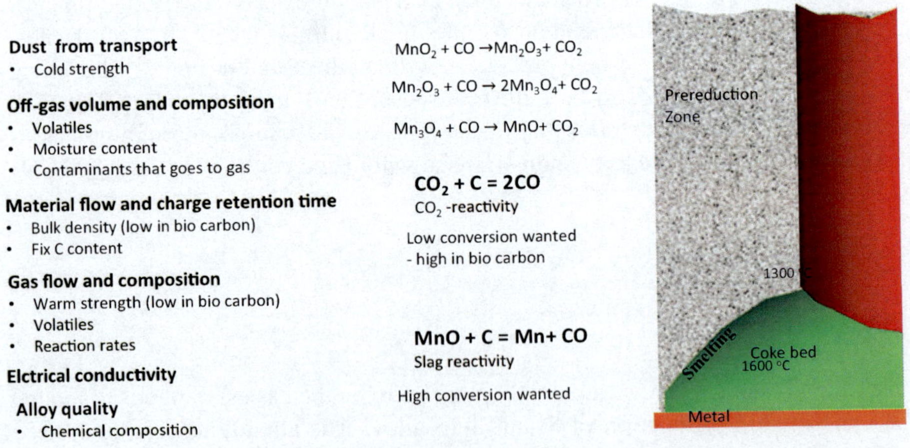


**Fig. 3** Effect of different biocarbon properties on Mn-alloy production

800 °C, with carbon according to Boudouard reaction (6), and by this increase energy consumption and $CO_2$ emissions for the process. The higher rate for reaction between carbon and $CO_2$, the $CO_2$, reactivity, for biocarbon compared to coke can thus give a negative impact on furnace performance when biocarbon is used. Final reduction of MnO in slag to Mn-alloy, reaction (7) is also affected by carbon properties and is often referred to as the slag reactivity of the carbon material. Its impact on furnace performance is not clear

$$\mathrm{SiO(g) + 2C = SiC + CO(g)} \tag{5}$$

$$\mathrm{CO_2(g) + C = 2CO(g)} \tag{6}$$

$$\mathrm{MnO(l) + C = Mn(l) + CO(g)} \tag{7}$$

To avoid major differences in properties between biocarbon and traditional carbon materials, biocarbon with designated properties can be produced. Another alternative is to adopt operational strategies and equipment design to the new carbon properties.

## Pretreatment of Mn-Ores: PreMa Project

A process for pre-treatment of Mn-ores is developed as a part of the PreMa project, an EU Horizon 2020 project that started in 2018. The main idea behind the project is, as shown in Fig. 4, to pre-treat manganese ores with sustainable energy sources in a separate unit before the ore is reduced to alloy in a traditional SAF. This is expected to reduce the electrical power consumption in the order of 20% and the $CO_2$ emissions from the process in the order of 15%. Removal and water in the pre-treatment unit will in addition give a more stable operation of the submerged arc furnace.

The manganese ores are fed into the pre-treatment unit with a temperature of 25 °C and with amount of higher Mn-oxides given by the actual ore used. In the pre-treatment unit the ores are heated to 600 °C with sustainable energy sources, either CO-rich off-gas, biocarbon or thermal solar heated air. During pre-treatment

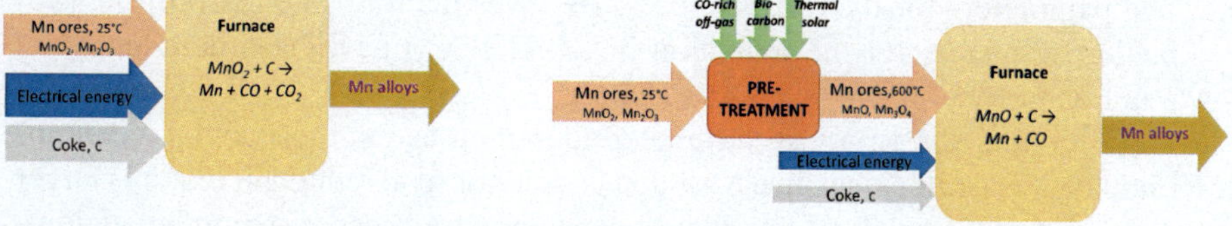


**Fig. 4** Schematic overview of the PreMa process

higher Mn-oxides in the ore are reduced, carbonates are calcinated, and moisture is removed. The ores are transferred warm, with a temperature of 600 °C from the pre-treatment unit to the submerged arc furnace where they are reduced to alloy.

In the pre-reduction zone of the Mn-alloy furnace descending raw materials are heated and reacted with ascending CO-gas generated during final reduction of the coke bed in the lower part of the furnace. Higher Mn-oxides, as $MnO_2$, $Mn_2O_3$ and $Mn_3O_4$ are then reduced with CO -gas to MnO according to reactions (8), (9), and (10). These reactions are exothermic and contribute with energy to the process. If they are moved to a separate process step, it is crucial for the energy balance that this energy is not lost to the surroundings. Thus, the material must be transferred from the pre-treatment unit to the furnace with minimum energy losses. In the furnace, reaction (10) will not be completed before 800 °C is reached and reaction (11) take place. Overall reduction of $Mn_2O_3$ will then take place by carbon instead of CO-gas with increased energy and carbon consumption as a result. In industrial Mn-alloy furnaces 60–90% of $Mn_2O_3$ is reduced with C instead of CO. By reducing the higher Mn-oxides to MnO in the pre-treatment unit before they meet carbon, this extra C and energy consumption can be avoided.

$$MnO_2 + \tfrac{1}{2}CO(g) = \tfrac{1}{2}Mn_2O_3 + \tfrac{1}{2}CO_2(g)\ \Delta H^0_{298} = -100.9\,\text{kJ} \tag{8}$$

$$Mn_2O_3 + \tfrac{1}{3}CO(g) = \tfrac{1}{3}Mn_3O_4 + \tfrac{1}{3}CO_2\ \Delta H^0_{298} = -57.6\,\text{kJ} \tag{9}$$

$$Mn_3O_4 + CO(g) = 3MnO + CO_2(g)\ \Delta H^0_{298} = -55.2\,\text{kJ} \tag{10}$$

$$CO_2 + C = 2CO(g)\quad \Delta H^0_{298} = 171.9\,\text{kJ} \tag{11}$$

In the PreMa project four different manganese ores have been investigated: Comilog, Assmang-Nchwaneng, UMK, and Kudumane. Pretreatment with different gases in experiments in kg scale [19] have shown that when heated in Ar, $CO_2$, or air, Comilog ore is reduced from $MnO_2$ to $Mn_2O_3$ while there are no changes in Nchwaneng and UMK ore that originally are $Mn_2O_3$ ores. When heated in a mixture of 70% CO/30% $CO_2$ all the ores are, as illustrated in Fig. 5, reduced to MnO at 800 °C. Comilog ore with the highest initial oxygen content is reduced fastest nearly to MnO at 400 °C. In Comilog ore, the fast exothermic reduction of $MnO_2$ resulted in a super heating of the particles presenting a risk for hotspots in a rector.

The reduction of Mn-oxides is shown to be governed by kinetics with different rates and parameters for the different ores. By investigations in other projects [20] the kinetic parameters for the reduction in $CO/CO_2$ gas have been determined as shown in Fig. 6.

Rotary kiln and shaft furnace have been identified as the most actual technologies. Fluid bed was ruled out since SAF that is used for final reduction requires larger particle so material from fluid bed would need an extra process step to be agglomerated. Pre-treatment in rotary kiln has been investigated in pilot experiments [21]

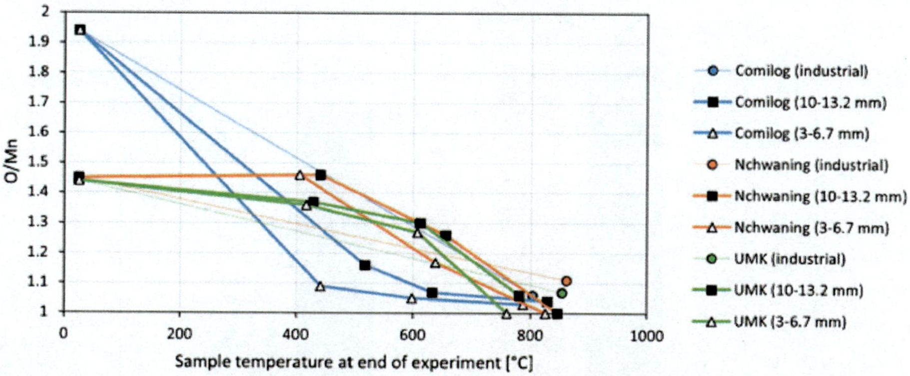


**Fig. 5** Pre-treatment of different ores in a mixture of 70% CO–30% $CO_2$ [19]

**Fig. 6** Kinetics for prereduction of different ores in a mixture of CO and $CO_2$

| | Comilog | Nchwaning |
|---|---|---|
| Ea [kJ/mol] | 17 | 63 |
| $k_0$ [cm/min·atm] | 0.53 | 60.4 |
| m (in $p_{CO}^m$) | 0.7 | 1.5 |

with similar results as in lab-scale experiments described above. Use of pre-treated materials in SAF is investigated in 10 submerge arc pilot experiments. These showed that pre-reduction did not affect final reduction to alloy [22]. Effect of pre-reduction on energy consumption will be derived based on generated data. Production of air at 900 °C by thermal solar technology has been demonstrated, as well as the heating of Mn ores to more than 700 °C with warm air flowing through a stationary bed in a shaft.

To evaluate and compare the effect on $CO_2$ emissions with different scenarios for preheating a model in HSC Sim has been developed. To be able to compare different ores, kinetics for their prereduction must be included. This is done by dividing the furnace into temperature zones and in each zone defining extent of the reaction for all the actual reactions separately for each of the ores. The defined extent of reaction is based on experimental data from PreMa and from the literature. Several scenarios have been simulated by the model. Resulting total power consumption and $CO_2$ emission vary considerably between the different ore mixtures and are generally lowest in ore mixtures with a high content of Comilog. Power consumption, both for SAF and in total, as well as $CO_2$ emissions depend on whether the gas is burned and used only for heating or if CO is used as a reductant, and vary in addition with the temperature in the kiln. Use of CO in the off gas as reductant will give lower $CO_2$ emissions than if the CO-gas is burned as seen Fig. 7. This will also give lower power consumptions in the SAF. But also heating alone will reduce the $CO_2$ emissions. An increase of the temperature in the pre-treatment unit from 600 °C to 1000 °C, will give a major reduction in $CO_2$ emissions.

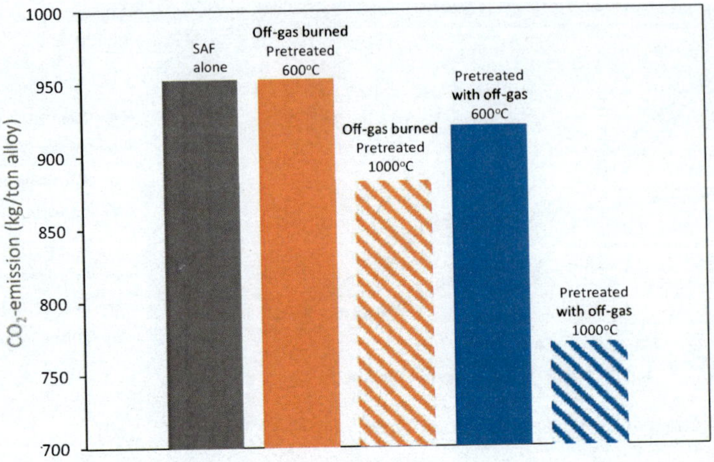


**Fig. 7** $CO_2$ emission for a SAF without pre-treatment unit compared with CO2 emission with a pre-treatment with different pre-treatment options. The off gas is either burned and the heat used for pre-treatment, or the off-gas is fed directly to the pre-treatment unit. Temperature in the pre-treatment unit is either 600 °C or 1000 °C

## Conclusions

$CO_2$ emission from metal production can be reduced by several different methods. Evaluation of best options for each case require consideration of $CO_2$ emission prom the process itself and for the other steps along the value chain. Specific energy consumption and method for energy production are important factors. Use of bio carbon is an actual method in many cases. Due to the difference between traditional carbon reductant and bio carbon it may require either production of bio carbon with designated properties or adaption of operational strategies and equipment to materials with other properties.

Pre-treatment of Mn-ores in a separate unit before SAF with CO-rich off-gas will reduce $CO_2$ emissions for overall production. Use of the gas as a reductant will have a larger effect than to use it for heating. An increase of temperature in the pre-treatment unit from 600 °C to 1000 °C will give a major reduction in $CO_2$ emissions.

**Acknowledgements** This work has been funded by the Norwegian Ferroalloy Research Association (FFF) and the Research Council of Norway through the RCN projects 280968 KPN Reduced $CO_2$ and through the PreMa project financed by the European Union's Horizon 2020 Research and Innovation Programme under grant Agreement No 820561.

**Conflict of Interest** On behalf of all authors, the corresponding author states that there are no conflicts of interest.

## References

1. Our World in data org. Source: Climate watch, The world Resource institute
2. Worldsteel.org/publications/policy papers
3. Sommerfeldt and Friedrich Replacing fossil carbon in the production of Ferroalloys. Minerals 11/2021 1286
4. Kero et al Some carbon-free technologies for manganese ferroalloy production. INFACON XVI, SSRN id 3936069
5. Kero et al (2022) Technologies with potential for climate neutral silicon production: Silicon for the chemical industry XVI
6. Schanche T, Tangstad M (2021) Prereduction of Nchwaning Ore in CO/CO2/H2 gas mixtures. Minerals 11(10):1097–1114
7. Camarasa B (2022) New metallurgical way for the solar silicon production. The SisAl project. t Silicon for the chemical and solar industy XVI, Trondeheim, Norway
8. Safarian (2022) A sustainable process to produce manganese and its alloys through hydrogen and aluminothermic reduction. Processes, MDPI 10:27. https://doi.org/10.3390/pr10010027
9. Elliott R, Coley K, Mostaghel S, Barati M (2018) Review of manganese processing for production of TRIP/TWIP steels, Part 2: reduction studies. JOM 70:680–690
10. Ostrovski O, Zhang G (2006) Reduction and carburization of metal oxides by methane-containing gas. AIChE J 52:300–310
11. Dalaker et al Deposition of carbon from Methane on manganese sources. Scientific reports. Submitted
12. Olsen S, Tangstad M, Lindstad T (2007) Production of manganese ferroalloys. Tapir Academic Press, Trondheim, Norway, ISBN 978-82-519-2191-6
13. Tangstad M (2013) Introduction to metal production. Metal Production in Norway, Akademika Publishing ISBN 987-82-321-0241-9
14. CoalTrans conferences: Carbon intensity of coal per KWh.15.08.2021 by Sergey Belozerov, CCO, EcoCarbon
15. Process 21 report: Ny prosessteknologi med redusert karbon avtrykk inkludert CCU, Industriell strategi for prosessindustrien (prosess21.no)
16. Surup et al (2021) Life cycle assessment of renewable reductants in the ferromanganes alloy production: a review. Processes
17. Hanssen S et al Biocarbon materials in metallurgical processes-investigations of critical properties: to be presented at TMS 2023
18. Jayakumari S, Ringdalen E (2022) Effect of varing SiO contents on Si and FeSi production. Silicon for the chemical and solar industy XVI, Trondeheim, Norway
19. Ringdalen et al (2021) Pretreatment of manganese ores in different gas atmospheres- a method to reduce energy consumption and $CO_2$ emissions in Mn-alloy production.: INFACON XVI. Available at SSRN: https://ssrn.com/abstract=3930059 or https://doi.org/10.2139/ssrn.3930059
20. Larssen TA, Senk D, Tangstad M (2021) Reaction rate analysis of manganese ore prereduction in CO-$CO_2$ atmosphere. Metall Mater Trans B 52:2087–2100. https://doi.org/10.1007/s11663-021-02162-1
21. Julia et al (2021) Preheating of manganese ore in a pilot scale rotary kiln: Proceedings of the 16th International Ferro-Alloys Congress (INFACON XVI). Available at SSRN: https://ssrn.com/abstract=3930059 or https://doi.org/10.2139/ssrn.3926242
22. Mukono T et al (2022) Utilisation of pretreated mn-ore in a pilot-scale ferromanganes furnace: effect of ore pretreatment on carbon an energy consumption. Presented at COM, Montreal, Canada, To be printed in proceedings

# The Path to Zero Carbon Dioxide Emissions in Silicon Production

Gudrun Saevarsdottir, Halvor Kvande, and Thordur Magnusson

**Abstract** The global community has set a path towards carbon neutrality by 2050. Almost one-fourth of the global emissions is attributed to direct emissions from industrial processes. Therefore, a zero-carbon alternative must be developed for each process, including the production of silicon. The silicon industry is exploring ways to efficiently capture $CO_2$ from the flue gases from the submerged arc furnaces, for example by increasing the $CO_2$ concentration. Replacing fossil reducing agents with biofuel is a carbon neutral alternative, while recycling waste streams from aluminium production as a reducing agent for silicon, is a more recent development. Electrochemical methods have also been explored at a laboratory scale. This paper gives a review of the efforts to date, from industry and academia, to decarbonize the production of silicon. The development of the largest part of the carbon footprint, arising from the production of the electrical energy used, is also discussed.

**Keywords** Pyrometallurgy · Sustainability · Silicon · Decarbonization · Process emissions · Indirect emissions

## Introduction

As a growing part of the global population has been gaining economic prosperity, the demand for various materials has been increasing. Silicon products are no exception to this, and the basis for most silicon products is Metallurgical Grade Silicon (MG-Si), which has seen a rapid increase in demand, with annual global production in 2021 having increased by a factor of 2.6 since 2000. The scope 1 emissions of greenhouse gases from the silicon production process have increased linearly with the production

G. Saevarsdottir (✉)
Reykjavik University, Reykjavik, Iceland
e-mail: gudrunsa@ru.is

H. Kvande
Formerly NTNU, Trondheim, Norway

T. Magnusson
Normi ehf, Vogar, Iceland



C. Fleuriault et al. (eds.), *Advances in Pyrometallurgy*, The Minerals, Metals & Materials Series, https://doi.org/10.1007/978-3-031-22634-2_12

volume, but in addition comes the scope 2 emissions from the electric energy used to power the process. Due to limited access to electrical energy generated from low carbon energy resources such as renewables or nuclear energy, a larger share of the global energy mix for MG-Si production is now derived from fossil fuels. In 2021, 50% of the MG-silicon production was based on electricity from fossil fuels, mainly coal (46%) and natural gas (3%).

According to the most recent report from IPCC [1], limiting the rise of global temperature to 1.5 °C from pre-industrial levels requires a reduction in greenhouse gas emissions by 45% from 2010 levels by 2030, and "net zero" emissions by around 2050. The silicon industry must find a way to cut emissions, both from the process itself and the electric energy used, in order to reach this ambitious goal. The various strategies that are currently being explored by the industry are discussed in this paper, but it is clear that the de-carbonizing of the source of electrical energy used is the most important factor, in particular avoiding the use of electricity produced by coal thermal power plants. This can be obtained by increasing the silicon production in regions where $CO_2$-emission free electricity is available or changing the energy mix for existing production facilities.

## Sources of Greenhouse Gas Emissions from MG-Silicon Production

From the mining of raw materials to cast MG-silicon, the total global average emissions are estimated in this paper to be around 12 metric tons of $CO_2$-equivalents per ton of MG-silicon produced (t $CO_2$e/t Si). This number includes both the scope 1 direct process emissions and scope 2 indirect emissions (mainly from energy sources). These emissions are significant, and the industry is facing pressure to reduce them in the near future.

There are three categories of contributions to greenhouse gas emissions in MG-Si production [2, 3]:

- The emission range for the carbothermic reduction of silicon from quartz in the submerged arc furnace is estimated in the literature [2, 3] to be 4.7–5 t $CO_2$e/t Si, including the contribution from the carbon electrodes. This number can be reduced by increasing the share of biocarbon among reductants, as this material is considered carbon neutral.
- The specific energy consumption in MG-silicon production, in terms of the electric energy fed to the submerged arc furnace, is in the range of 10.5–12 kWh/kg Si [4]. In addition to this comes around 1 kWh/kg Si of electric energy use to power auxiliaries, such as the operation of the bag-house filter, fans, cooling systems, conveyor belts, etc. The emissions that occur in the production of the electric energy used are termed *indirect* emissions, with a global average of about 7 t $CO_2$e/t Si in 2021. Globally the emissions from electric energy production range from practically zero when using nuclear or renewable power sources such as

hydro, geothermal, wind or solar, up to 12–17 t $CO_2$e/t Si, when electric energy from coal is used for silicon production, dependent on the type of coal, technology, and the actual specific energy consumption. The estimates in this paper are based on source emission estimates published by IPCC [5] and partly based on the International Aluminium Institutes regional estimates for Chinese smelters that do not have hydropower [6]. The numbers for energy use and regional-specific energy consumption were shared by CRU.

- The mining and transport of raw materials, including electrode materials are low compared to the other factors, and they are included by using a relatively high estimate for process emissions.

This paper addresses the pathways pursued by the industry to reduce process emissions, and also addresses how the development of the energy mix for the industry affects the indirect emissions.

## Direct Emissions from the Silicon Metal Production Process

Metallurgical grade silicon is produced by carbothermic reduction in submerged arc furnaces. The silicon containing raw material, $SiO_2$ in the form of quartz, and the carbon containing raw materials that are a mixture of coal, coke, charcoal and woodchips, are fed into a furnace. Electrodes penetrate the raw material mix, or charge, from above. In most cases there are three electrodes carrying three-phase electric current, bringing electric energy to the process. The current passes between the electrodes and cancels in a star point in the furnace, through the intermediate reaction products filling the furnace, partly passing through an electric arc. The stoichiometry of the overall reaction is ideally written as [4]

$$SiO_2 + 2C = Si + 2CO \tag{1}$$

This overall reaction is a combination of sub-reactions occurring at different locations and temperatures within the furnace. In the upper part of the furnace, the carbon materials react with SiO gas formed in the hotter zones further down in the furnace, to form SiC (Eq. 2) at temperatures above 1500 °C. SiO gas also condenses to form a condensate made of a mixture of $SiO_2$ and Si (Eq. 3) at lower temperatures.

$$SiO + 2C = SiC + CO \tag{2}$$

$$2SiO = SiO_2 + Si \tag{3}$$

As SiC reaches the hotter, lower regions of the furnace it reacts with SiO gas to form Si metal as shown in Reaction 4, while the SiO is formed by the reversed Reaction 3 which runs strongly to the left at high temperatures. Silicon formation by these reactions requires temperatures higher than 1811 °C.

$$SiO + SiC = 2Si + CO \tag{4}$$

For a good silicon yield, which is defined as the fraction of Si entering the furnace in the quarts, which is tapped as metal, the reaction temperature needs to be close to 2000 °C, which is supported by arcing in a cavity surrounding the electrode tip in well operating furnaces. Thus this furnace region is often termed a "crater zone". A part of the SiO gas formed by the reversed Reaction 3 in the crater zone, does not participate in Reactions 2 and 4 or condense as in Reaction 3, but instead leaves through the charge surface, where it reacts with oxygen, forming amorphous $SiO_2$ particles, or silica fumes, that are collected and sold as a by-product. This removal of Si from the furnace reduces the yield. Also, the CO gas reacts to $CO_2$ above the charge surface.

According to the ideal stoichiometry of Reaction 1, around 3.14 kg of $CO_2$ should be formed for each kg of Si metal produced. This cannot be avoided in a carbothermic process, which is based on using carbon to capture the oxygen from the quartz at elevated temperature leaving Si metal behind. There are inevitable inefficiencies and losses in the furnace that increase the carbon consumption per unit metal produced. For instance, reduced yield due to silica fume formation reduces the amount of Si metal in the denominator, increasing the t $CO_2$e/t Si produced. Thus, the greenhouse gas emissions, as well as the energy consumption of the process, are closely linked to the silicon yield for the process. Also, volatiles in the carbon raw materials evaporate from the charge and burn, forming $CO_2$, but do not contribute to the reduction, and as the carbon electrodes are consumed they contribute to the emissions. The electrodes, however, do not efficiently contribute to the reactions, as they partly enter the crater zone as unreacted carbon, which reduces the silicon yield. Unreacted carbon enters Eq. 4, which takes the form:

$$(1 + x)SiO + SiC + x\,C = (2 + x)Si + (1 + x)CO \tag{5}$$

$$(1 + x)/2\ Si + (1 + x)/2SiO_2 = (1 + x)SiO \tag{6}$$

As the presence of unreacted C on the left side of Reaction 5 leads to CO generation without consuming SiC, and more Si metal is consumed to produce more SiO, the result is reduced Si yield.

The carbon raw materials can be petroleum coke, coal, charcoal and woodchips. Charcoal and woodchips have a good influence on the furnace operation, as they contribute to permeable charge, which reduces SiO losses and improves the yield. In addition these are forms of biocarbon that are categorized as carbon neutral and contribute to lowering the carbon footprint, as discussed below.

The IPCC estimate for the process emissions from MG-Si production, or emission factor, is 5 kg $CO_2$e/kg Si [7]. This official emission factor is cited by Lindstad [8] and by Kero et al. [3], but the effect of biocarbon as in woodchips is not considered in that number. Monsen et al. [2] reported an estimate of 4.68 kg $CO_2$/kg Si, assuming that 10% of the carbon content provided was from woodchips.

## The Use of Biocarbon to Reduce Emissions in the Production Process

Woodchips are normally a part of the raw material mix for the MG-silicon production process. The presence of woodchips contributes to improve the furnace operation as it increases the permeability of the charge, enabling the SiO gas to evenly penetrate the charge, increasing its capture by Reactions 2 and 3, reducing the silica fume formation and increasing the silicon yield. The availability and price of woodchips are very location dependent, and the quantity used in the raw material mix inevitably depends on those factors. As woodchips are biocarbon, their carbon content replaces a part of the fossil carbon sources but does not contribute to the carbon footprint. Wood products can also be turned into charcoal, which can with benefit replace coal in the raw material mix. In addition to containing less polluting elements, such as sulfur, ash, and metals, charcoal has a porous structure, which makes it very reactive to SiO gas, making it prone to react efficiently to SiC in Reaction 2, reducing the risk of unreacted carbon entering the crater zone, which harms silicon yield as shown in Reactions 5 and 6.

Many silicon smelters have charcoal as a part of their raw material mix in an effort to reduce their carbon footprint. The handling of biocarbon and production of charcoal inevitably leads to emissions, which can be estimated through life cycle analysis, but it is clear that replacing fossil carbon with woodchips and charcoal is a significant opportunity to reduce the carbon footprint [9]. However, the availability and cost of biocarbon are very location dependent. The carbon content by weight is much lower for woodchips than coal and coke, which increases the contribution of transportation cost for the raw materials, and in most locations charcoal will be more costly than fossil carbon sources. It can be economically feasible to run the furnace almost exclusively on locally sourced biocarbon in South America, for example. It is claimed on Elkem´s webpage [10] that Elkem´s Limpio plant in Paraguay has been running on 100% biocarbon since 2019. The plant runs on hydropower, so the silicon produced there is as close to carbon neutrality as practically possible. Elkem also states that they are currently running their smelters in Norway with 20% biocarbon, and that their goal is to run their smelters, globally, on 50% biocarbon by 2030, including also their ferrosilicon smelters. It is unclear what that goal infers for their ferrosilicon facility in Iceland, where most of the biocarbon currently needs to be imported.

## Carbon Capture and Storage/Sequestration (CCS)

Provided that $CO_2$ can be accessed in sufficiently high concentrations, it is possible to remove it in a reasonably reliable manner. In areas with geology rich in mafic rock, such as basalt or gabbro, the $CO_2$ can be fixed in the form of a mineral. This has been demonstrated in the CarbFix project in the volcanic Iceland. $CO_2$ in the non-condensable gases from a geothermal powerplant was dissolved in pressurized water, pumped underground, where through a series of geochemical reactions, carbonates were formed [11, 12]. Alternatively, in geologies with hydrocarbon resources, there are initiatives focusing on storing the $CO_2$ in depleted oil and gas reservoirs [13]. Then the task is to capture the $CO_2$ from the silicon smelter and increase the concentration to a sufficient level so that such methods can be applied. Research and development in this field have focused on processes that are larger in volume than ferroalloys, such as steel and cement production, but solution developed for those industries are likely to be transferrable to ferroalloy producers, such as silicon smelters [14]. The concentration of $CO_2$ in the flue gas from the carbothermic process can be up to 4% [15], which makes it marginally compatible with existing technologies for carbon capture and up-concentration for either sequestering or reuse. Mathiesen et al. estimated the cost of capture to be in the range of 45–55 €/t $CO_2$ captured [15]. According to the EU Carbon Price Tracker, the EU ETS Futures price has been fluctuating between 60 and 100 €/t $CO_2$ from September 2021 to September 2022 [16]. If we also take into account the cost of permanently disposing of the $CO_2$, which according to CarbFix [17] was around 25 US$/ton, the combined cost can be estimated to be within the same range as the 2022 price level for the EU ETS system. In the case of the CarbFix method, the cost of $CO_2$ capturing can be significantly reduced if the concentration in the flue gas can be brought above 10%, as for $CO_2$ concentrations higher than 10% direct dissolution in water is feasible without further up-concentration using an absorption medium [18]. This would also make it possible to avoid costly impurity removal from the flue gas, as the mineralization process can proceed in the presence of sulfur and other impurities. Current efforts to recycle flue gas back into the furnace, with intermediate cooling, in order to up-concentrate the $CO_2$, and reduce the $O_2$ content to reduce $NO_x$ formation, can possibly reach $CO_2$ fraction of up to 20%, which would enable the direct application of the CarbFix method [19, 20].

New carbon capture-ready iron ore smelting technologies such as HIsarna, currently piloted at Tata Steel in Ijmuiden in the Netherlands, show that CCS in theory can be used in smelting processes, in particular for carbothermic processes, and it should therefore be applicable for the silicon process. If CCS is applied in combination with biocarbon such as woodchips and charcoal as a large part of the carbon raw materials, the production can even be considered as carbon negative. It is likely that CCS will be a part of a range of solutions applied to decarbonizing the industry.

## Alternative Production Processes

There are several on-going initiatives aimed at developing alternative processes for silicon production, that are not based on carbothermic reduction. One option is to use different reduction agents for the quartz. Using methane gas as a reduction agent in a plasma furnace has been proposed as the bases for a new process. In that case the $CO_2$ emissions would be significantly reduced, but not eliminated. The steel industry considers the use of hydrogen as a reducing agent for iron smelting, replacing the role of carbon in the blast furnace, and many companies are investing heavily in that area [21]. In the case of reducing $SiO_2$ to form silicon, hydrogen does not offer a thermodynamically favorable chemistry to produce silicon metal, but it readily produces the intermediate SiO gas by reaction with $SiO_2$ [22]. In the carbothermic process the final metal-producing step involves a reaction between the intermediate compounds SiO and SiC as shown in Reaction 4, but an alternative silicon forming reaction seems elusive when using only hydrogen as a reduction agent. The reduction of SiO to Si by $H_2$ is not favored by thermodynamics. Using methane as a reduction agent can overcome this obstacle, as the carbon in methane will react with the silica to form SiC [23] enabling the metal-forming reaction to occur, but a reactor design to achieve this outcome may prove challenging. Condensation of SiO into a condensate containing $SiO_2$ and Si metal droplets forms metal, but a process based on this is likely to be very inefficient.

Silicon can be produced using aluminium, contained for example in a waste product such as aluminium dross, as a reducing agent through an aluminothermic reaction from a silica-containing slag. There is an on-going Horizon 2020 project, "SisAl Slag Valorization", which aims at upgrading waste products from a range of processes to create value, in the spirit of circular economy [24]. Some promising results have been published from this on-going project, such as Philipson et al. [25], indicating that silicon can indeed be produced from such waste streams with a much lower carbon footprint, which shows that the future of silicon is likely to involve multiple production routes, as a part of the circular economy.

Electrochemical reduction in molten salt, or directly from molten oxides, has also been proposed [26, 27] but is still in the laboratory scale and at a relatively low technology readiness level at this time.

## Indirect $CO_2$e Emissions from the Electrical Power Source

Many silicon and ferrosilicon plants have historically been located close to hydropower plants, as the availability of a reliable supply of reasonably priced reliable power was the deciding factor. A number of plants were also constructed in France and Spain with access to nuclear energy as a significant part of their energy mix. As the production process and auxiliaries require electric energy in the range of 11.5–13.5 kWh/ton Si, the carbon emissions from the energy source are very consequential for the carbon footprint of the MG-silicon produced.

Mr. Jorn de Linde at CRU has kindly given access to data involving location-dependent energy used for silicon production, as well as specific energy consumption for different regions. This enables estimation of the emissions from the energy used for MG-silicon production in submerged arc furnaces, including auxiliary energy used to drive fans, etc.

In the period from 2000 to 2021 the production of silicon has increased by a factor of 2.6, with most of the increased capacity being built in China. The proportion produced with low emission energy sources such as renewable and nuclear has declined from 76 to 50% in this period. Although the energy mix has had a negative development, it is more favorable than that for the aluminium production, where the portion of renewables in the energy mix has declined from 71 to 33% from 2000 to 2021, while production has, similarly to the silicon production, increased by a factor of 2.6 [28].

A likely reason for a higher portion of new MG-silicon capacity being run on hydropower, is that silicon smelters are small units in comparison to aluminium smelters, for example, and the furnaces can be shut down without permanent damage to the production equipment. This has enabled new silicon production capacity in China to a larger extent being built close to stranded hydropower sources, adapting their power use to possible variations in power supply. In fact, in China, about 5 times more silicon is currently produced using hydro power as compared to 2000, and this is more than 40% of their total silicon production.

The rest of the global energy mix is mostly supplied by coal thermal power, increased from 28 to 46% in the period, and that has great implications for the average carbon footprint for this production, as now around 50% is supplied by fossil fuel.

Figure 1 shows how the energy mix has developed by region and globally since 2000. It can be seen that Europe has slightly increased their silicon metal production and North America has decreased its production, while China has increased its silicon production tremendously. Figure 2 shows how the carbon footprint of MG-silicon production in kg $CO_2e$ /kg Si, including both the emissions from the production process and the indirect emissions from the electrical energy used for silicon production. The figure shows how the carbon footprint for the energy mix per region has changed with time, as well as the global average. The emission factor recommended by IPCC is used and variability of the process carbon footprint by region is neglected. The location-dependent variability in the actual process carbon footprint is likely to be significant due to different shares of biocarbon used in the raw material mix.

From Figs. 1 and 2 it can be seen that the carbon footprint, here described in terms of intensity per kg Si, (kg $CO_2e$/kg Si) from the electrical energy used in silicon production, has decreased significantly in Europe as the energy production has on average shifted mostly to green power. Also the specific emissions have decreased slightly in other regions due to process improvements leading to lower specific energy consumption.

The reduced specific energy consumption is likely to be associated with an improved silicon yield, which decreases emissions per ton produced, but the published information of emission factor from the process does not offer that detail, so such speculations will not be indulged in this paper. The increase in the carbon footprint is due to the shift in the production to China, which currently has a 71% share of the global MG-silicon production, up from 30% in 2000. The indirect carbon footprint was 4.4 kg $CO_2e$/kg Si in 2000 but had risen to 7 kg $CO_2e$/kg Si, or by

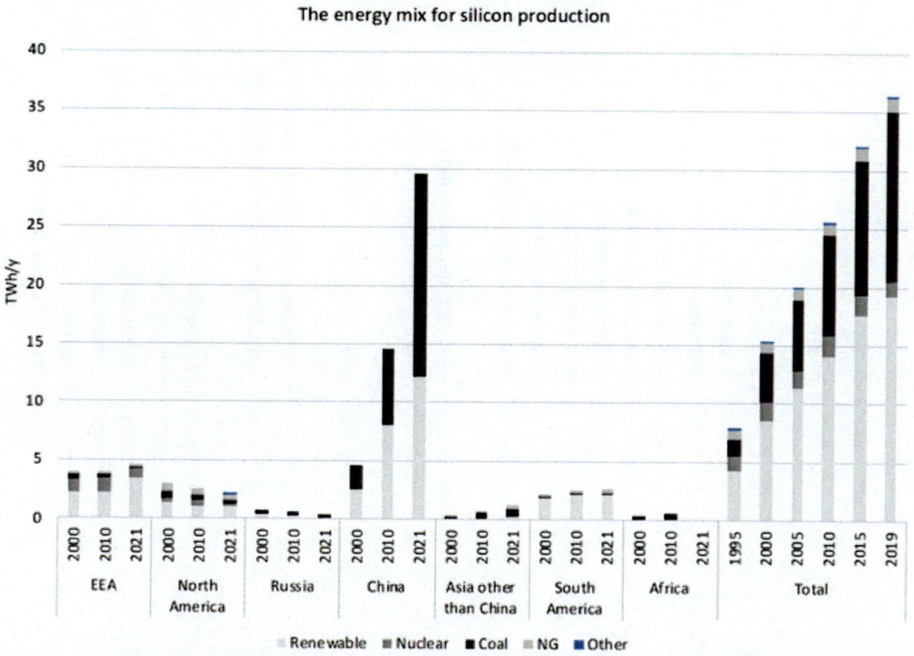


**Fig. 1** Energy mix for MG-silicon production (TWh/y) by region for the years 2000, 2010 and 2021, along with the total energy mix for 1995 to 2021, based on information from CRU

more than 60%. As the process emission factor is assumed constant at 5 kg $CO_2$e/kg Si, this equals an increase in the total carbon footprint from 9.4 to 12 kg $CO_2$e/kg Si, and currently around 58% of the carbon footprint is coming from the electric power production. This can be compared to the development of the carbon footprint for aluminium production, which has increased by 3 kg $CO_2$e/kg Al since 2000, as a result of an increase in production in China, which is based 90% on coal thermal power [29].

## Discussion

With the ongoing initiatives to decarbonize the production process for MG-Silicon, there is a reason for cautious optimism that it will be possible to lower the greenhouse gas emissions from this process considerably in the future. As the current process is carbothermic, and carbon is used, we cannot avoid forming $CO_2$. Possible mitigation methods involve the use of biocarbon, or CCS, and both are technically possible, but expensive.

Alternative production processes, such as the one being developed in the SisAl process, with methane as a reducing agent, or electrochemical methods may give results further down the line.

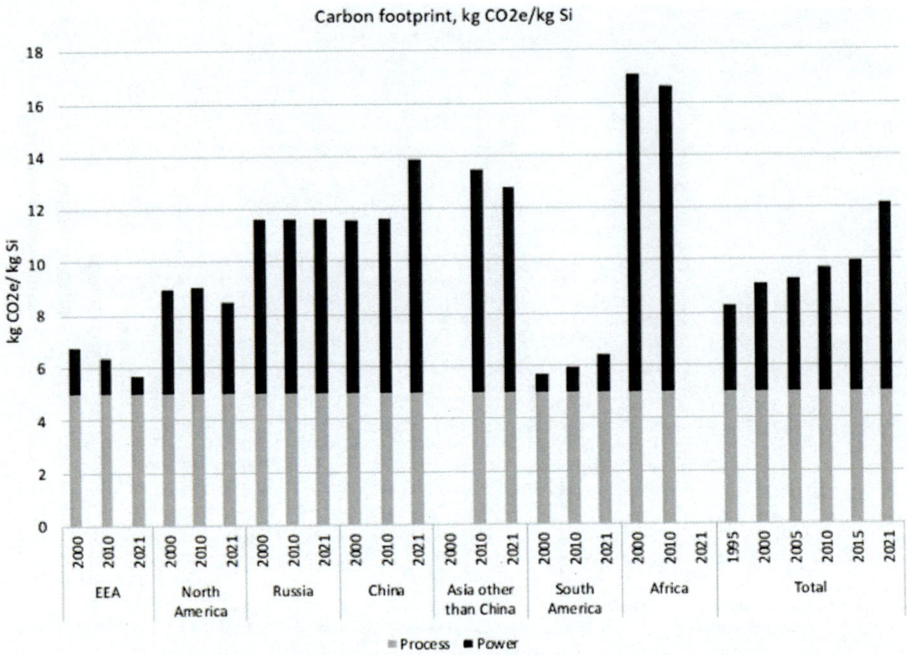


**Fig. 2** The overall carbon footprint of MG-silicon production (kg $CO_2$e/kg Si) based on average emission values attributed to the electrical energy used for silicon production by region in the years 2000, 2010 and 2021 provided by CRU, and the energy emission values published by IPCC [5]. Also the average global emissions for this production are shown for the years 1995 to 2021. Regional variations in the process emissions due to the share of biocarbon in the reductant mix are not accounted for in this work, instead the IPCC emission factor for process emission is used [7]

The main contribution to the carbon footprint is, however, the indirect emissions from the production of the electric power used that can range from practically 0 to 12 kg $CO_2$e/kg Si, lowest if the energy is renewable or nuclear and highest if it is produced using coal. With coal power the energy-related emissions contribute 70% to the global carbon footprint rather than 58% as is currently the case for the global average.

Nuclear energy has historically, and still is, making significant contributions to lower the carbon footprint from the primary production of MG-Silicon. As close to a third of the European production is located in France, benefitting from the low carbon footprint of nuclear energy, this energy source contributes to reducing the average global carbon footprint from the energy used in this process by 0.4 kg $CO_2$e/kg Si or by 5% as compared to if that power was produced by coal power plants, while renewables reduce the global average footprint by additional 4.6 kg $CO_2$e/kg Si.

From Fig. 1 it can be seen that the energy mix for MG-Silicon production in Europe and the Americas is dominated by low-emissions energy sources, mostly hydropower and nuclear. The carbon footprint from the European silicon production energy mix is 0.7 kg $CO_2$e/kg Si, a tenth of the global average, and for Europe and the Americas, the indirect emissions are around 1.5 kg $CO_2$e/kg Si.

When this paper is written, in September 2022, the world, and in particular Europe, is going through an energy crisis, with the price of electric energy having increased by an order of magnitude as compared to the price level before the Covid19 pandemic started. The cause can mainly be attributed to Europe's efforts to transition away from traditional energy sources, including low-emission nuclear. This has caused dependence on natural gas to complement intermittent renewables for electricity production, for heat generation, and for use in industrial processes. At the same time the countries of Europe are not willing to produce the natural gas locally, relying instead heavily on one supplier of natural gas. The future goal is to wean off the use of natural gas as well, but functional alternative technologies to provide the swing power needed for the renewable energy system are still proving elusive.

The current energy situation on the European continent is very difficult for energy-intensive industries without long-term contracts, and even long-term contracts only offer limited protection. If the MG-Silicon industry in Europe was to close their operations, and for this demand-driven commodity that production capacity would be replaced by increased production based on coal power, the global average carbon footprint for electric energy used for silicon production would increase to 8.5 kg $CO_2$e/kg Si. Shifting production outside of Europe while still using the product is called *Carbon Leakage*, as the emissions still occur. In this case the emissions would dramatically increase, the carbon footprint of the imported MG-Silicon would be the new global average of 13.5 kg $CO_2$e/kg Si, instead of the former carbon footprint of 5.7 kg $CO_2$e/kg Si for the domestically produced metal. This would be an extremely counterproductive development, but it is unclear if the mechanisms built into the European Emission Trading System (ETS) to prevent such *Carbon Leakage* are robust enough to prevent such a development given the current circumstance.

## Conclusions

This paper summarizes the development in the carbon footprint from MG-silicon production with focus on the indirect emissions from the electric power used in the process. It is shown that the carbon footprint of MG-silicon can range from 5 kg $CO_2$e/kg Si to 17 kg $CO_2$e/kg Si depending on the source of electric power used in the process, the lowest footprint is obtained if renewable or nuclear energy is used, while the coal thermal power leads to triple carbon footprint. The production of MG-silicon in 2021 has increased by a factor of 2.6 from 2000, mostly in China which currently contributes with 71% of the global production. Although more than 40% of the Chinese production uses hydropower, the increased use of coal power has led to an increase in the total carbon footprint of MG-silicon from 9.4 to 12 kg $CO_2$e/kg Si from 2000 to 2021, and now the indirect emissions from the electric power production contribute around 58% of the overall carbon footprint.

There are many ongoing initiatives seeking ways to decarbonize the MG-silicon production process, but it is clear that it is the source of electric energy used in the process that has the greatest impact on the carbon footprint. Therefore, as mitigation methods and new alternative processes with zero or lower emissions are developed, the energy use must be shifted to low-emission sources such as renewables or nuclear power.

**Acknowledgements** Much of the industrial production data and data on energy consumption used in this paper has been sourced from CRU and were gracefully shared by Mr. Jorn de Linde. His and CRU´s contributions are gratefully acknowledged.

## References

1. Masson-Delmotte V, Zhai P, Pörtner H-O, Roberts D, Skea J, Shukla PR, Pirani A, Moufouma-Okia W, Péan C, Pidcock R, Connors S, Matthews JBR, Chen Y, Zhou X, Gomis MI, Lonnoy E, Maycock T, Tignor M, Waterfield T (eds) (2018) Global warming of 1.5 °C. An IPCC special report on the impacts of global warming of 1.5 °C above pre-industrial levels and related global greenhouse gas emission pathways, in the context of strengthening the global response to the threat of climate change, sustainable development, and efforts to eradicate poverty," © 2019 Intergovernmental Panel on Climate Change., IPCC
2. Monsen B, Lindstad T, Tuset JK (1998) $CO_2$ emissions from the production of ferrosilicon and silicon metal in Norway. In: Electric furnace conference proceedings, vol 56 New Orleans, USA
3. Kero I, Gradal S, Tranell G (2016) Emissions from MG-Si/FeSi production—a summary
4. Schei A, Tuset JK, Tveit H (1998) Production of high silicon alloys. Tapir, Trondheim, Norway
5. Bruckner T, Bashmakov IA, Mulugetta Y, Chum H, de la Vega Navarro A, Edmonds J, Faaij A, Fungtammasan B, Garg A, Hertwich E, Honnery D, Infield D, Kainuma M, Khennas S, Kim S, Nimir HB, Riahi K, Strachan N, Wiser R, Zhang X (2014) Climate Change 2014: mitigation of climate change. Contribution of working group III to the fifth assessment report of the intergovernmental panel on climate change. IPCC
6. World Aluminium (2017) A life-cycle model of Chinese grid power and its application to the life cycle impact assessment of primary aluminium. World Aluminium [Online]. Available: https://www.world-aluminium.org/media/filer_public/2017/06/29/lca_model_of_chinese_grid_power_and_application_to_aluminium_industry.pdf
7. Faerden T, Tranell G, Bubetsky JS, Lindstad T, Olsen SE (2006) 2006 IPCC guidelines for national greenhouse gas inventories. [Online]. Available: https://www.ipcc-nggip.iges.or.jp/public/2006gl/pdf/3_Volume3/V3_4_Ch4_Metal_Industry.pdf
8. Lindstad T, Olsen S, Tranell G, Færden T, Lubetsky J (2007) Greenhouse gas emissions from ferroalloy production 2007. [Online]. Available: https://www.researchgate.net/publication/238745804_greenhouse_gas_emissions_from_ferroalloy_production
9. Riva et al L (2019) LCA analysis of biocarbon pellet production to substitute coke. DEStech Transactions on Environment, Energy and Earth Sciences
10. Si E (2022) Elkem's goal for biocarbon use Elkem. [Online]. Available: https://www.elkem.com/silicon-products/sustainability-in-elkem-silicon-products/biocarbon/
11. Pogge von Strandmann PAE et al (2019) Rapid $CO_2$ mineralisation into calcite at the CarbFix storage site quantified using calcium isotopes. Nat Commun 10(1):1983. https://doi.org/10.1038/s41467-019-10003-8
12. Clark DE et al (2020) CarbFix2: $CO_2$ and $H_2S$ mineralization during 3.5 years of continuous injection into basaltic rocks at more than 250 °C. Geochimica et Cosmochimica Acta 279:45–66. https://doi.org/10.1016/j.gca.2020.03.039

13. Minardi A, Stavropoulou E, Kim T, Ferrari A, Laloui L (2021) Experimental assessment of the hydro-mechanical behaviour of a shale caprock during $CO_2$ injection. Int J Greenhouse Gas Cont 106:103225. https://doi.org/10.1016/j.ijggc.2020.103225
14. Wyns T, Khandeiar G (2022) Metals for a climate neutral Europe A 2050 Blueprint, The Institute for European Studies, Vrije Universiteit Brussel (VUB). The Institute for European Studies, Vrije Universiteit Brussel (VUB), Sep. 2022. [Online]. Available: https://www.ies.be/files/Metals_for_a_Climate_Neutral_Europe.pdf
15. Mathisen A, Normann F, Bierman M, Skagestad R, Haug AT (2019) $CO_2$ capture opportunities in the Norwegian Silicon Industry. Presented at the TCCS–10. $CO_2$ capture, transport and storage, Trondheim, Norway. [Online]. Available: https://sintef.brage.unit.no/sintef-xmlui/bitstream/handle/11250/2637936/SProceedings%20no%204-s49.pdf?sequence=1&isAllowed=y
16. Ember (2022) EU carbon price tracker. Accessed: Sep. 24, 2022. [Online]. Available: https://ember-climate.org/data/data-tools/carbon-price-viewer/
17. CarbFix (2022) CarbFix as a subsidiary. Accessed: Sep. 24, 2022. [Online]. Available: https://www.carbfix.com/carbfix-as-a-subsidiary
18. Snæbjörnsdóttir SÓ, Sigfússon B, Marieni C, Goldberg D, Gislason SR, Oelkers EH (2020) Carbon dioxide storage through mineral carbonation. Nat Rev Earth Environ 1(2):90–102. https://doi.org/10.1038/s43017-019-0011-8
19. Panjwani B, Pettersen T, Wittgens B (2020) Controlling flue gas temperature from Ferro Silicon submerged arc furnaces (SAF) using flue gas recirculation (FGR). Presented at the 14th international conference on CFD in oil & gas, metallurgical and process industries, Trondheim, Norway. [Online]. Available: https://hdl.handle.net/11250/2720852
20. Andersen V, Solheim I, Gaertner H, Sægrov-Sorte B, Einarsrud KE, Tranell G (2022) Pilot scale test of flue gas recirculation for the silicon process. J Sustain Metall, vol In Review
21. Kurrer CM (2020) The potential of hydrogen for decarbonising steel production. EPRS | European Parliamentary Research Service, EU Parliament Briefing PE 641.552, Dec. 2020. [Online]. Available: https://www.europarl.europa.eu/RegData/etudes/BRIE/2020/641552/EPRS_BRI(2020)641552_EN.pdf
22. Han G, Sohn HY (2005) Kinetics of the hydrogen reduction of silica incorporating the effect of gas-volume change upon reaction. J Am Ceram Soc 88(4), 882–888. https://doi.org/10.1111/j.1551-2916.2005.00144.x
23. Monsen B, Kolveinsen L, Prytz S, Tang K (2013) Possible use of natural gas for silicon or ferrosilicon production. In: Proceedings of Infacon 13, Almaty, Kazakhstan, vol 1. [Online]. Available: https://www.pyro.co.za/InfaconXIII/0467-Monsen.pdf
24. Tranell G (2022) SisAL Slag valorization. Accessed: Sep. 24, 2022. [Online]. Available: https://sisalslag.eu/about-project/
25. Philipson H, Wallin M, Einarsrud KE, Tranell G (2021) Kinetics of silicon production by aluminothermic reduction of silica using aluminum and aluminum dross as reductants. Presented at the Infacon XVI, Trondheim, Norway. https://doi.org/10.2139/ssrn.3922132
26. Ge J, Zou X, Almassi S, Ji L, Chaplin BP, Bard AJ (2019) Electrochemical production of Si without generation of $CO_2$ based on the use of a dimensionally stable anode in Molten $CaCl_2$. Angew Chem Int Ed Engl 58(45):16223–16228. https://doi.org/10.1002/anie.201905991.
27. Ustinova Y, Pavlenko O, Gevel T, Zhuk S, Suzdaltsev A, Zaikov Y (2022) Electrodeposition of Silicon from the Low-Melting LiCl-KCl-CsCl-$K_2SiF_6$ Electrolytes. J Electrochem Soc 169(3):032506 https://doi.org/10.1149/1945-7111/ac5a1c
28. Saevarsdottir G, Padamata SK, Velasquez BN, Kvande H (2023) The way towards zero carbon emissions in aluminum electrolysis. To be published 2023.
29. Saevarsdottir G, Kvande H, Welch BJ (2020) Aluminum production in the times of climate change: the global challenge to reduce the carbon footprint and prevent carbon leakage. JOM 72(1):296–308. https://doi.org/10.1007/s11837-019-03918-6

# Alternatives of Copper (I) Oxide Reduction in a Copper Slag Cleaning Furnace

**Goran Vuković, Anton Ishmurzin, Juergen Schmidl, Bojan Zivanović, and Bernhard Handle**

**Abstract** As environmental, energy saving, and legislation considerations are becoming ever more important, there is evidently a demand for further improvement in slag handling and metal recovery practice. The metallurgical industry is initiating its actions into minimizing and processing of slags to achieve sustainable development and circular economy. Reduction of copper oxide is an important step of metal recovery and slag cleaning in copper production. This can be achieved by slag treatment with injected reducing agents. In this paper, it is presented a computational fluid mechanics model of tuyere and purging plug injection of reductants into slag that considers a first-order heterogenic chemical reaction to model copper oxide reduction. Using the model, it is possible to compare and optimize tuyere and purging plug arrangements in terms of time needed to reduce copper oxide content to a given value. Copper oxide reduction accomplished with the injection of different reducing media (oil, natural gas, and hydrogen) for several tuyere/plug arrangements in a slag refining furnace was simulated and the reduction performance using different reduction media was compared. It was shown that purging plugs, which may be installed at the deepest position in the furnace, create an improved gas flow pattern that results in a higher copper oxide reduction rate and decreasing in $CO_2$ emission.

**Keywords** Copper-oxide reduction · Purging plugs · $CO_2$ emission · Decarbonization · Hydrogen · Reduction rate

## Introduction

Metallurgical slags are a huge source of waste if not properly treated [1]. Nowadays the metallurgical industry is targeting to recycle and utilize all their byproducts, to close the sustainable production loop and to reach zero emissions and zero waste

G. Vuković (✉) · J. Schmidl · B. Zivanović · B. Handle
RHI Magnesita GmbH, Kranichberggasse 6, 1120 Vienna, Austria
e-mail: goran.vukovic@rhimagnesita.com

A. Ishmurzin
RHI Magnesita, Technology Center Leoben, Magnesitstrasse 2, 8700 Leoben, Austria



C. Fleuriault et al. (eds.), *Advances in Pyrometallurgy*, The Minerals, Metals & Materials Series, https://doi.org/10.1007/978-3-031-22634-2_13

[2, 3]. However, the non-ferrous metals (NFM) industry is still challenged with achieving efficient utilization of large quantities of generated slags. Generally, due to process complexity as well as lower metal content in concentrates treated relative to those feeds in iron ore processing, a large amount of slag is generated from NFM production. As an illustration, to produce one tonne of copper, approximately 2.0–3.0 tonnes of copper slag are generated as a by-product material [4]. A conservative estimation is that more than 50 million tonnes of NFM slag are produced worldwide each year [5].

Copper slags usually have only very limited technical applications and are typically of interest to a producer only in terms of the final copper content [6]. Forecasts show that copper demand could exceed 35 million tonnes annually by 2030 [7]. Large quantities of slag are being produced and disposed of near the metallurgical plants [6]. With environmental, energy-saving and legislative considerations becoming increasingly important, there is a need for a major improvement in slag handling. The metallurgical industry is directing its efforts into minimizing and processing slags in order to achieve sustainable development and a circular economy [6, 7]. This may be achieved through multi- or cross-disciplinary approaches by changing the process paradigm by designing new, novel, and integrated solutions.

The latest copper smelting technologies make excessive use of oxygen to increase the smelting capacity, thereby enabling a highly efficient and environmentally acceptable smelting process. However, improving the grade of matte has some drawbacks such as increased copper loss in slag. It has been shown that when the copper grade of matte exceeds 70% Cu, the oxygen potential in the slag-matte system increases drastically, leading to copper loss to the slag predominantly in the form of copper (I) oxide [8]. Hence, the recent trends of copper smelting invariably lead to significant copper loss in slag, which needs to be recovered by means of reduction and settling in a separate slag cleaning furnace. The reduction of $Cu_2O$ from slags is carried out by adding carbon/coke or by injection of various reductants (e.g., different types of oil or natural gas) [9]. Nowadays, hydrogen is projected to be not only a source of clean fuel energy but also a reducing agent for metals production in the current industrial decarbonization effort.

Reductants are traditionally introduced into a metallurgical vessel such as a copper slag cleaning furnace by means of a tuyere. There are several drawbacks associated with the use of tuyeres: accretions; formation of "gas shortcuts" (inefficient gas utilization) where the gaseous reductant does not meet slag but bypasses it, thus decreasing the possibility of reduction; as well as refractory wear in the tuyere area associated with its extensive maintenance.

Gas purging in the pyrometallurgical industry has been used for many years [10–12]. Inert or reactive purging gases are used to achieve different effects on the metal bath and it necessitates different plug designs [13]. The operations in slag-cleaning furnaces are improved by nitrogen purging through porous plugs, the main considerations being minimal metal losses to the slag (entrained particles and copper oxide), reduced burner fuel consumption, improved slag skimming, and a reduced accretion layer [14]. On the other hand, reactive gases (e.g., $CH_4$) introduced into the bath through capillary plugs enhance the movement of the liquid melt, and simultaneously act as reactants in metallurgical reactions [9, 13]. This could be an alternative technology to replace tuyeres or minimize their usage. A metallurgical vessel can be upgraded with capillary plugs, allowing an improvement of the mixing and reduction rate, and therefore energy savings. It could also lead to further benefits, including increased refractory lifetime and the minimization of bottom accretions.

In this article, we numerically evaluate the copper (I) oxide reduction performance of the capillary plugs in comparison with tuyeres using computational fluid dynamics (CFD). A parameter study was devised that involves six computational cases illustrating the copper (I) oxide reduction performance of capillary plugs and tuyeres. Using the computational results, we compared the performance of the six different cases in terms of the low-velocity area volumes within the furnace as well as in terms of achieved reduction rates. The differences in the flow patterns created by the tuyeres and capillary plugs and their effects on the reduction rates are discussed. The paper also addresses the potential improvement of the $CO_2$ footprint.

## Specification of Compared Tuyeres and Capillary Plugs Injection Cases

The purpose of this paper is to find differences in the performance of tuyeres and capillary plugs when they are used with high sulphur fuel oil (HSFO), natural gas, and hydrogen for the reduction of copper (I) oxide in a slag cleaning furnace.

We compared six different setups in terms of reductant injection into the liquid slag: two using tuyeres (Fig. 1a) and four using capillary plugs (Fig. 1b, c). The two cases with tuyeres differed in mass flow rate, while the four capillary plug cases differed in the positioning of the plugs and applied reduction medium (Fig. 1b, c, Table 1). A specification overview of considered computational cases is given in Table 1. An example of capillary plug design [13] and its integration into the lining of a slag cleaning furnace is presented in Fig. 2.

In the computational models, the tuyeres were operated using a HFSO-air mixture, whereas capillary plugs were operated using natural gas as well as hydrogen. The total reductant consumption is provided in Table 1.

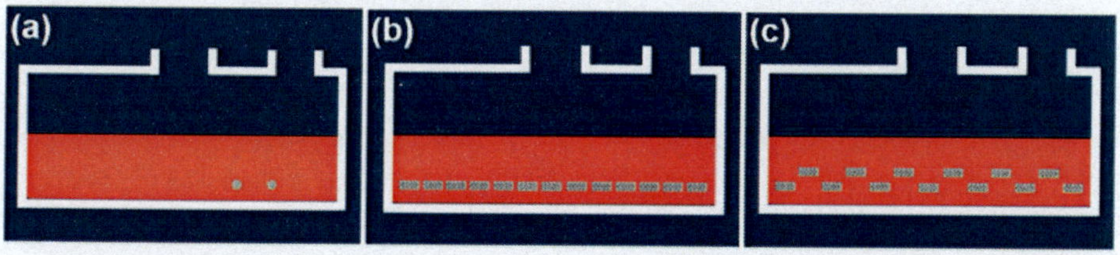


**Fig. 1** Tuyere and purging plug arrangements in a slag cleaning furnace that are considered in the study [9]

**Table 1** Injection setup, reductant injection specifications and reductant consumption. The last four columns are the results of the simulations. The consumption figures given are the total amounts produced either by both tuyeres or by all 13 capillary plugs

| Case | | Injection type | Consumption | Normalized time [s/s] | Reduction time [s] | Consumed amount | Low velocity volume [$m^3$] |
|---|---|---|---|---|---|---|---|
| 1 | Tuyers A HSFO | two tuyeres | 422 kg $h^{-1}$ | 1 | 2000 (estimation) | 466 kg | 8.3 |
| 2 | Tuyeres B HSFO | two tuyeres | 720 kg $h^{-1}$ | 0.54 | 1080 | 432 kg | 0.01 |
| 3 | Plug row natural gas | 13 plugs in one row | 623 $Nm^3$ $h^{-1}$ | 0.71 | 1420 | 3200 $Nm^3$ | 19.0 |
| 4 | Plug zigzag natural gas | 13 plugs in two rows | 623 $Nm^3$ $h^{-1}$ | 0.80 | 1600 | 3605 $Nm^3$ | 21.6 |
| 5 | Plug row hydrogen | 13 plugs in one row | 623 $Nm^3$ $h^{-1}$ | 0.91 | 1820 | 4093 $Nm^3$ | 24.2 |
| 6 | Plug zigzag hydrogen | 13 plugs in two rows | 623 $Nm^3$ $h^{-1}$ | 0.99 | 1980 | 4453 $Nm^3$ | 25.5 |

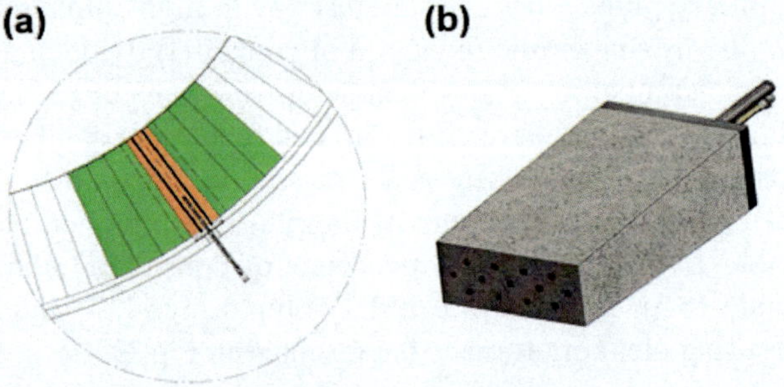


**Fig. 2** **a** Capillary plug integrated into furnace lining. **b** Rectangular capillary plug [9]

Using the developed computational model, we simulated copper (I) oxide reduction from 100% of the initial mass fraction of copper (I) oxide in the slag (e.g., 4 wt.%) down to 25% (e.g., 1 wt.%) in all six cases (Table 1). In each case this took a different amount of time to achieve. The longest reduction time was recorded in the case "tuyeres A". This time in seconds was used to normalize the reduction times in the other five cases, as shown in Table 1.

## Computational Models

The commercial computational fluid dynamics software package Ansys Fluent [15] was used to simulate the two cases of tuyeres and four cases of capillary plug arrangements. The phase interface between slag and furnace atmosphere was tracked using the Volume-of-Fluid (VOF) model with geometric reconstruction volume fraction formulation. The flow of reductants from tuyeres and plugs was modelled using the discrete particle model (DPM) approach [15], for which the average bubble sizes were estimated using the experimental correlations given below in Eqs. 2 and 3. To account for turbulence, the realizable $k$-$\varepsilon$ turbulence model was employed. The computational domain contained approximately 700,000 cells.

The chemical kinetics of copper (I) oxide reduction was modeled as a first-order heterogeneous chemical reaction:

$$\dot{c} = -\boldsymbol{k}\boldsymbol{A}\boldsymbol{c}, \tag{1}$$

where $c(x,y,z,t)$ is the concentration of copper (I) oxide, $k$ is the reaction rate constant, and $A(x,y,z,t)$ is the specific surface of contact between the area of bubbles and the slag (available as a result of DPM-multiphase flow simulation). The kinetic Eq. 1 is implemented with user-defined functions (UDF) in FLUENT [15]. The competing reaction of $Fe_3O_4$ reduction is neglected.

The reaction rate constant $k$ of natural gas was assumed to be higher by 33% than of the HSFO-air mixture to account for the higher reduction efficiency of natural gas, as indicated in FactSage equilibrium. Further, the reaction rate constant $k$ for hydrogen was assumed to be 50% higher than for natural gas.

Since the exact values of the reaction rate constants are not known, the results can be analyzed only relative to each other. For this relative comparison, we report normalized times in Table 1.

On the other hand, due to the exponential nature of the solution of Eq. 1, the change in the value of $k$ will only affect the absolute time of the reduction process, but not the relation between the reduction times in the specified six cases. Thus, despite the absence of exact values of $k$, the relative analysis of reduction times can be made.

## Important Model Parameters

In the case of tuyeres, the average bubble size was estimated using the experimental correlation from [16, 17], which relates the average bubble volume to the outer nozzle diameter $D_{no}$ and the volumetric flow rate $Q$:

$$\boldsymbol{V_b} = 0.083\,\boldsymbol{Q}^{0.867}\boldsymbol{D_{no}}. \tag{2}$$

In the case of a capillary plug, the average bubble size $d$ was estimated using the correlation from [18]:

$$d\left(\frac{g\rho}{\delta\sigma}\right)^{1/3} = 1.65\left(\frac{\text{Fr}}{\text{We}^{0.5}}\right)^{0.160}, \tag{3}$$

Here $g$ is the gravitational acceleration, $\delta$ is the pore diameter, $\rho$ is the density of slag, $\sigma$ is the surface tension of slag, and Fr and We are Froud and Weber dimensionless numbers. Note that the correlation (3) is derived for a porous plug and in case of a capillary plug the pore diameter is significantly larger. The authors are not aware of such a correlation for an array of large pores which would be representative of a capillary plug. However, taking into account the relative character of the present study, the absolute value of the bubble diameter does not seem to be critical.

The surface tension of the slag is estimated based on the experimental data in [19].

In the case of HSFO-gas mixture used in tuyere cases, the mixture density was calculated as [20]:

$$\rho_m = \left(\frac{x_g}{\rho_g} + \frac{x_l}{\rho_l}\right)^{-1}, \text{where} x_l = \frac{\dot{m}_l}{\dot{m}_l + \dot{m}_g}, x_g = \frac{\dot{m}_g}{\dot{m}_l + \dot{m}_g}. \tag{4}$$

Here $\dot{m}_l$ and $\dot{m}_g$ are the mass flow rates of the carrier gas (air) and HFSO.

## Results

The temporal evolutions of the copper (I) oxide concentration during the reduction in the six defined cases are shown in Figs. 3, 4, 5, 6, 7 and 8.

The temporal evolution of the copper oxide shown in Figs. 3, 4, 5, 6, 7 and 8 together with the flow velocity graphs in the middle slice of the slag bath (Fig. 9) provide valuable information about the reactive flow characteristics during the injection of reductants using tuyeres and capillary plugs. Figure 9 shows significantly different velocity distributions caused by injection using tuyeres and capillary plugs.

In the case of tuyeres, strong flow is observed in the vicinity of the tuyeres (see Fig. 9 for tuyeres cases). A reductant gas shortcut is expected in the tuyeres cases. The flow pattern produced by capillary plugs, on the other hand, is characterized by much smaller velocities, the absence of a gas shortcut, and more uniform distribution of velocity across the vessel. The cases with hydrogen exhibit higher velocity magnitudes in the middle slice section in Fig. 9, which on one hand could be attributed to the higher buoyant force that hydrogen bubbles are subjected to due to the larger density difference between slag and hydrogen and on the other hand the hydrogen bubble count is larger than in case of natural gas, since the density of hydrogen is lower and more bubbles of the same diameter are produced during gas introduction.

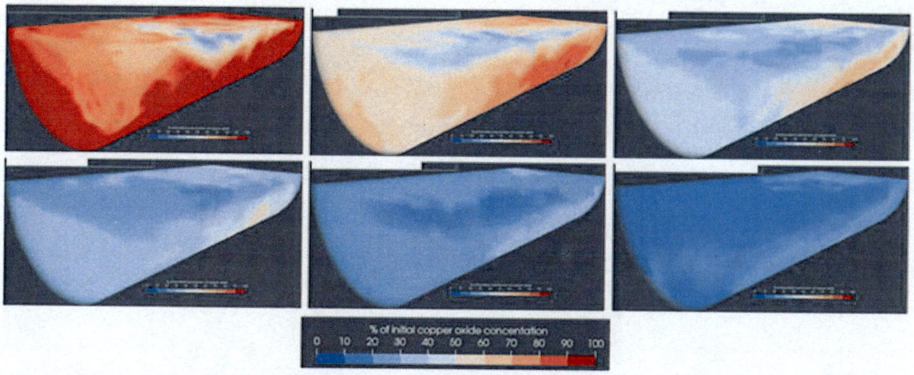


**Fig. 3** Temporal evolution of copper (I) oxide concentration in case "tuyeres A", positioned as shown in Fig. 1a. The first image shows the copper (I) oxide concentration at the beginning of the reduction process. Further 5 images show the copper (I) oxide concentration distribution at times 0.2 T, 0.4 T, 0.6 T, 0.8 T and T, where T is the total time it took to reduce copper (I) oxide from 100 to 25%. T is different in each of the simulation cases (Figs. 3, 4, 5, 6, 7 and 8)

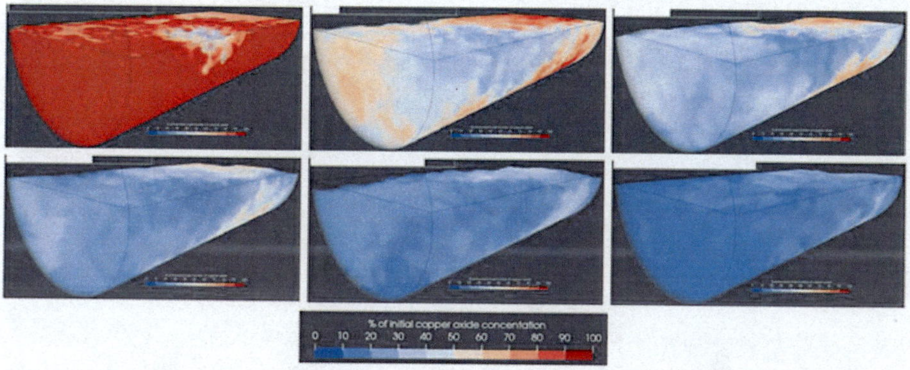


**Fig. 4** Temporal evolution of copper (I) oxide concentration in case "tuyeres B", positioned as shown in Fig. 1a. The first image shows the copper (I) oxide concentration at the beginning of the reduction process. Further 5 images show the copper (I) oxide concentration distribution at times 0.2 T, 0.4 T, 0.6 T, 0.8 T and T, where T is the total time it took to reduce copper (I) oxide from 100 to 25%. T is different in each of the simulation cases (Figs. 3, 4, 5, 6, 7 and 8)

To compare the amount of movement of the slag in the six chosen cases, we can compare the volume of *low velocity zones*. The low velocity zones are defined here as the areas of the slag where the time-averaged root mean square velocity is below 7 cm $s^{-1}$.

The volumes occupied by the low velocity zones (in green) as well as their volumes in $m^3$ are shown in Fig. 10. The low velocity zone volumes are summarized in Table 1.

In Fig. 11 the duration of the relative copper (I) oxide reduction is given as a fraction of the duration of the longest case, "tuyeres A". Assuming a realistic reduction

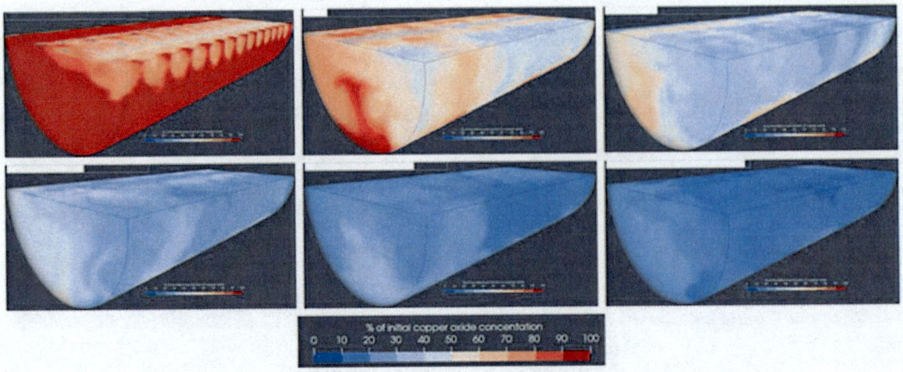


**Fig. 5** Temporal evolution of copper (I) oxide concentration in case "plug row natural gas", positioned as shown in Fig. 1a. The first image shows the copper (I) oxide concentration at the beginning of the reduction process. Further 5 images show the copper (I) oxide concentration distribution at times 0.2 T, 0.4 T, 0.6 T, 0.8 T and T, where T is the total time it took to reduce copper (I) oxide from 100 to 25%. T is different in each of the simulation cases (Figs. 3, 4, 5, 6, 7 and 8)

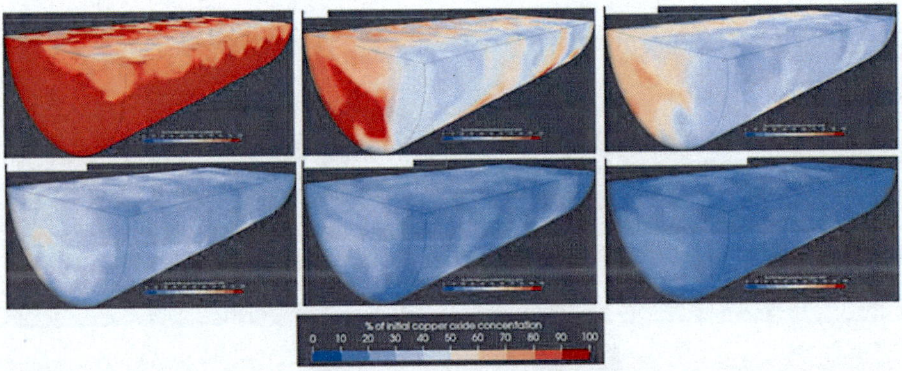


**Fig. 6** Temporal evolution of copper (I) oxide concentration in case "plug zigzag natural gas", positioned as shown in Fig. 1a. The first image shows the copper (I) oxide concentration at the beginning of the reduction process. Further 5 images show the copper (I) oxide concentration distribution at times 0.2 T, 0.4 T, 0.6 T, 0.8 T and T, where T is the total time it took to reduce copper (I) oxide from 100 to 25%. T is different in each of the simulation cases (Figs. 3, 4, 5, 6, 7 and 8)

duration in the case "tuyeres A" of 2000s, we can calculate the total consumption of reductants in the four considered cases. These consumption figures of HSFO, natural gas and hydrogen are presented in Table 1

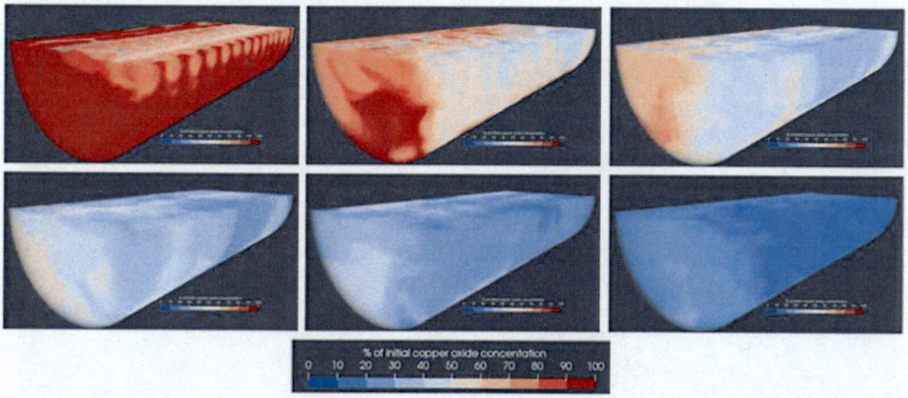


**Fig. 7** Temporal evolution of copper (I) oxide concentration in case "plug row hydrogen", positioned as shown in Fig. 1a. The first image shows the copper (I) oxide concentration at the beginning of the reduction process. Further 5 images show the copper (I) oxide concentration distribution at times 0.2 T, 0.4 T, 0.6 T, 0.8 T and T, where T is the total time it took to reduce copper (I) oxide from 100 to 25%. T is different in each of the simulation cases (Figs. 3, 4, 5, 6, 7 and 8)

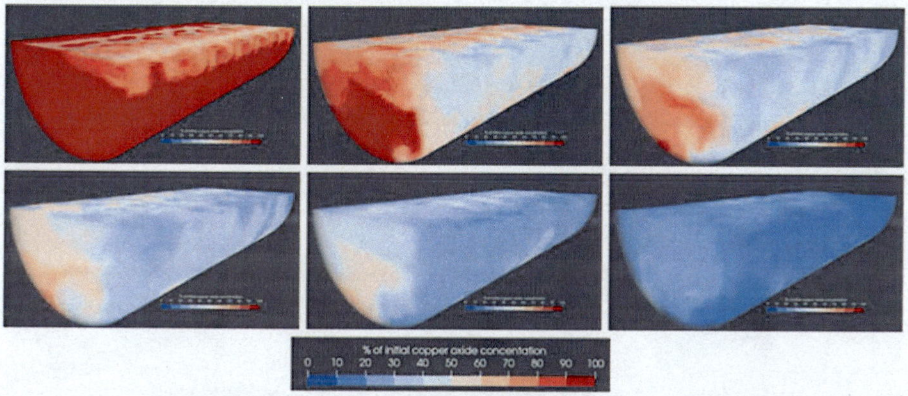


**Fig. 8** Temporal evolution of copper (I) oxide concentration in case "plug zigzag hydrogen", positioned as shown in Fig. 1a. The first image shows the copper (I) oxide concentration at the beginning of the reduction process. Further 5 images show the copper (I) oxide concentration distribution at times 0.2 T, 0.4 T, 0.6 T, 0.8 T and T, where T is the total time it took to reduce copper (I) oxide from 100 to 25%. T is different in each of the simulation cases (Figs. 3, 4, 5, 6, 7 and 8)

## Discussion and Conclusion

The tuyeres cases produced the lowest velocity zone volumes (see Fig. 10, Table 1). Specifically, the low velocity zone volume of the case "tuyeres A" is at least two times smaller than either of the capillary cases. In case "tuyeres A", a much larger part of the furnace is agitated than in either of the cases with capillary plugs.

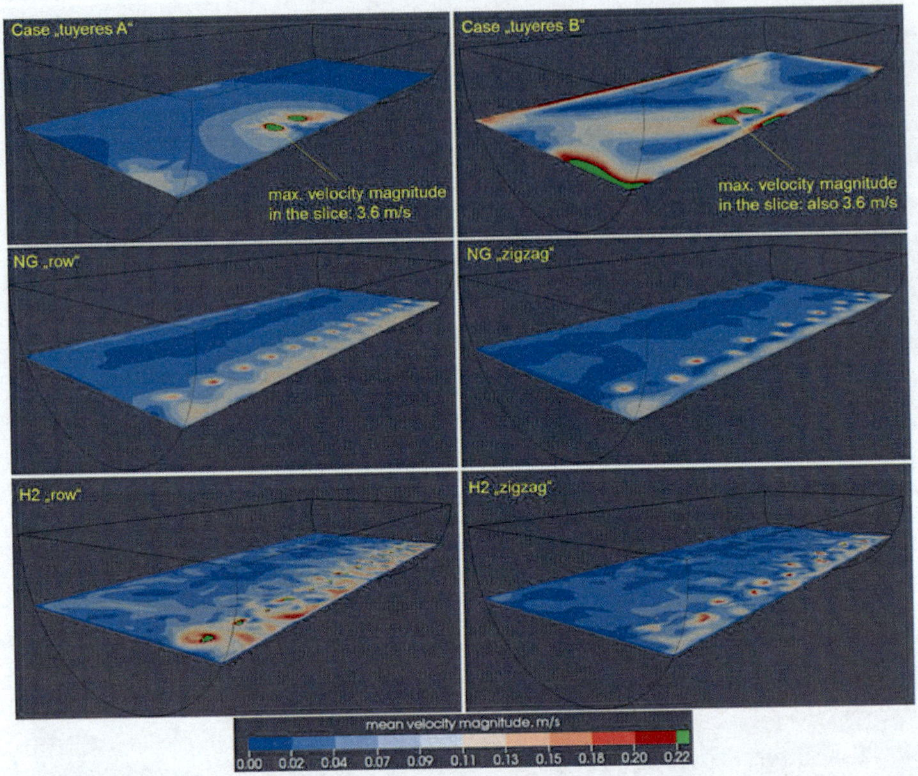


**Fig. 9** Velocity distribution in a horizontal slice for all six cases

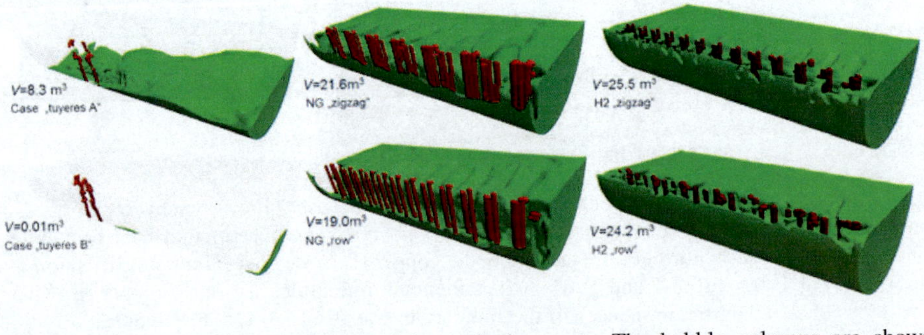


**Fig. 10** Comparison of the volumes of low velocity areas. The bubble columns are shown schematically

On the other hand, a comparison of the reduction curves reveals that the case "tuyeres A" performed worse than either of the capillary plug cases with natural gas and with hydrogen (Fig. 11). This discrepancy should be attributed to the differences in the flow patterns generated by tuyeres and capillary plugs.

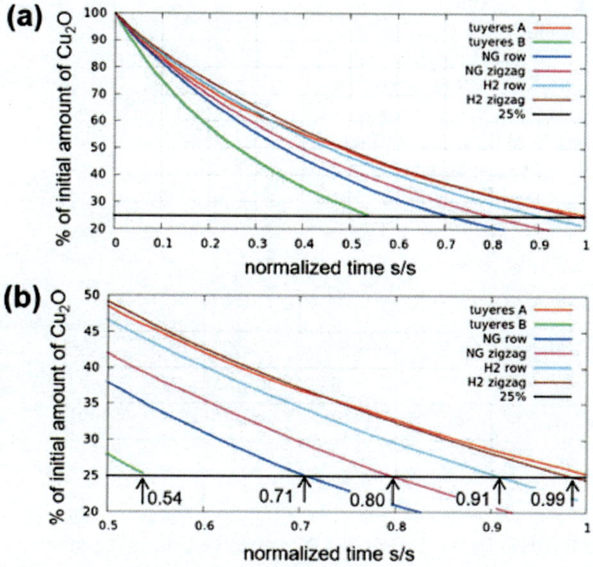


**Fig. 11** Reduction curves of copper (I) oxide for the six selected tuyeres and capillary plug cases. **a** The whole range of copper (I) oxide reduction. **b** Normalized time when copper (I) oxide reaches 25% of its initial amount

The flow patterns produced by the tuyeres and plugs differ significantly. Tuyeres rely on the introduction of high impulse into the bath. Gas shortcuts occur in the vicinity of the tuyeres, reducing the reduction efficiency. The bubble surface available for the reduction reaction is smaller in tuyeres cases due to larger average bubble diameter in this case. Additionally, there are problems with bringing unreacted slag from the corners of the furnace into the reductant-rich zone around tuyeres (bubble columns in Fig. 10).

Capillary plugs produce much smaller velocities, and the velocity distribution is more uniform. The flow pattern resembles solid body rotation, especially when the capillary plugs are arranged in a single row (cases "NG row" and "$H_2$ row"). This flow pattern seems to be effective in bringing the fresh slag to the reductant-rich areas (bubble columns in Fig. 10). In case of zigzag capillary plug arrangement, both when using natural gas and hydrogen, this solid body resembling rotation is less pronounced, probably due to a more complex interaction of bubble plumes.

The flow pattern produced by capillary plugs with hydrogen resembles the flow pattern produced in cases with natural gas. The significant differences are higher velocities in the middle slice in Fig. 9 and higher volume of low velocity zones in Fig. 10. The higher volume of low velocity zone is probably due to the smaller mass of a single $H_2$ bubble which produces less drag and thus agitates less liquid than an NG-bubble.

The higher reduction rate in the case of capillary plugs with natural gas and hydrogen qualitatively agrees with the experimental results and conclusions given in [21]. Although [21] compared tuyeres and porous plugs, we can assume that capillary plugs should lie between tuyeres and porous plugs in terms of gas utilization (Fig. 12).

Since hydrogen and natural gas are injected at the same norm volumetric rate, the number of $H_2$-bubbles is higher than the number of NG-bubbles due to lower density

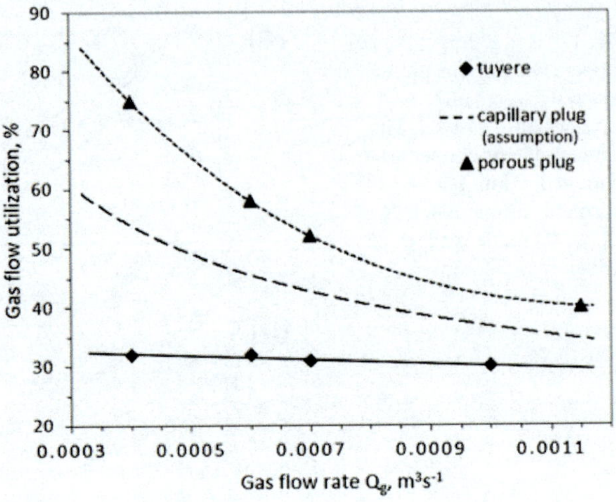


**Fig. 12** Gas utilization versus flow rate comparing the $CO_2$ gas utilization in a NaOH-water solution using a tuyere and a porous plug, taken from [21]. The dashed line is the assumed gas utilization for a capillary plug [9]

of hydrogen, which provides more bubble-slag surface where the reduction reaction takes place. It can explain the higher reduction yield per unit mass of hydrogen compared to natural gas.

Additionally, according to an equilibrium thermodynamic calculation, the use of natural gas reduces the emission of $CO_2$ by 42% in comparison with HSFO, as can be seen from Fig. 13. The reduction with hydrogen represents carbon free process without evidence of emission of $CO_2$ (Fig. 13). The equilibrium calculations were performed using FactSage 8.1 [22] with the UQPY copper-consortia database [23] for a typical slag composition. It was assumed that all copper is present in the form of copper (I) oxide (4 wt.%) and it is accompanied by 18 wt.% of $Fe_3O_4$.

The capillary purging plugs, which may be installed at the deepest position in the furnace, create an improved gas flow pattern that results in a higher copper oxide reduction rate and in significant decreasing or complete elimination of $CO_2$ emission. This property of plugs in combination with the possible usage of alternative reductants establishes capillary plugs as a complementary solution for slag cleaning furnaces instead or even in combination with tuyere technology.

**Fig. 13** Mass of carbon dioxide in kilograms produced per metric tonne of slag when using HSFO versus natural gas and hydrogen

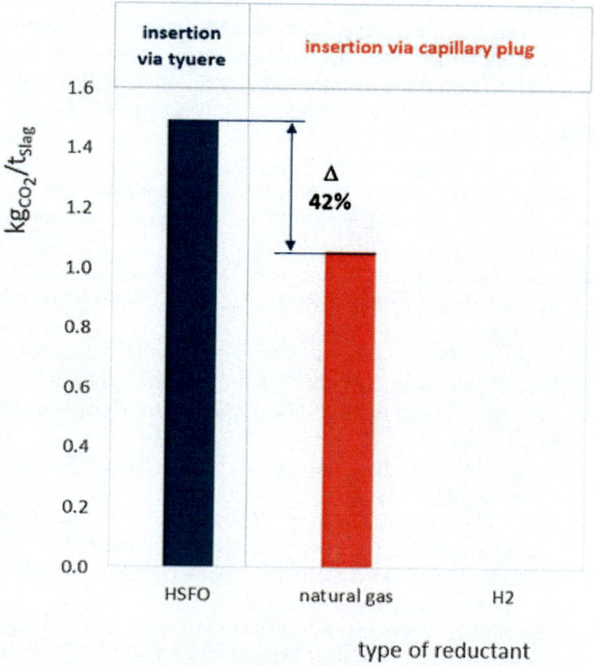


## References


1. Reuter M, Xiao Y, Boin U (2004) Recycling and environmental issues of metallurgical slags and salt fluxes. Paper presented at the II International Conference on Molten Slags Fluxes and Salts. The South African Institute of Mining and Metallurgy
2. Sykes C, Brinson A, Tanudisastro G, Jimenez M, Djohari J (2020) Report: zero emission copper mine of the future. The Warren Centre for Advanced Engineering. Zero Emission Copper Mine of the Future (usyd.edu.au).pdf
3. World Steel Association (2020) Public policy paper: steel industry co-products. World Steel Association https://www.worldsteel.org/publications/position-papers/co-product-position-paper.html
4. Sim S, Jeon D, Kim DH, Yum WS, Yoon S, Oh JE (2021) Incorporation of copper slag in cement brick production as a radiation shielding material. Appl Radiat Isot 176:109851. https://doi.org/10.1016/j.apradiso.2021.109851
5. Wang, GC (2016) The utilization of slag in civil infrastructure construction. Woodhead Publishing
6. Sánchez M, Sudbury M (2013) Physicochemical characterization of copper slag and alternatives of friendly environmental management. J Min Metall Sect B-Metall 49(2):161–168
7. Kulczycka J, Lelek Ł, Lewandowska A, Wirth H, Bergesen JD (2017) Environmental impacts of energy-efficient pyrometallurgical copper smelting technologies. J Ind Ecol 20(2):304–316
8. Reddy RG, Prabhu VL, Mantha D (2003) Kinetics of reduction of copper oxide from liquid slag using carbon. High-Temp Mater Proc 22(1):25–33
9. Ishmurzin A, Vukovic, G, Schmidl J, Zivanovic B, Handle B (2022) Modelling of copper (I) oxide reduction in a copper slag cleaning furnace. Paper presented at the Copper-Cobre 2022 International Conference, Santiago, Chile, 13–17 November 2022 (paper accepted)

10. Rigby AJ, Lanyi MD (1999) Porous plugs in molten copper production and refining. Paper presented at the copper 99-Cobre 99 international conference—the minerals, metals and materials society, Phoenix, Arizona, 10–13 October 1999
11. Lee S-S, Kim BS, Choi SR (2003) Application of the porous plug system in the anode furnace at onsan smelter. Paper presented at the 132nd TMS annual meeting, San Diego, California, 2–6 March 2003
12. Acuna M, Sherrington M (2003) Efficiency of porous plugs in fire refining of crude copper. Paper presented at the 132nd TMS annual meeting, San Diego, California, 2–6 March 2003
13. Vukovic G, Gamweger K (2017) A gas purging system for copper and aluminum furnaces. JOM 69(6):1007–1012
14. Handle B, Luidold S, Zivanovic B, Mnich M, Marquez T (2008) COP KIN bottom gas purging: a potential application for slag cleaning processes in the primary copper industry. RHI Bullet 2:6–14
15. Ansys Inc. (2009) Ansys FLUENT 12 Manual. Ansys Inc.
16. Irons GA, Guthrie RI (1978) Bubble formation at nozzles in pig iron. Metall Trans B 9:101–110
17. Sano M, Mori K (1976) Bubble formation from single nozzles in liquid metals. Trans Jap Inst Met 17:344–352
18. Koide KO, Shinji KA, Tanaka YU, Kubota HI (1968) Bubbles generated from porous plate. J Chem Eng Jap 1:51–56
19. Zhang H, Fu L, Qi J, Xuan W (2019) Physicochemical properties of the molten iron-rich slags related to the copper recovery. Metall Mater Trans 50:1852–1861
20. Awad M, Muzychka YS (2008) Effective property models for homogeneous two-phase flows. Exp Therm Fluid Sci 33:106–113
21. Stapurewicz T, Themelis NJ (1987) Mixing and mass transfer phenomena in bottom-injected gas–liquid reactors. Canad Metall Q 26:123–128
22. Bale CW, Bélisle E, Chartrand P, Decterov SA, Eriksson G, Gheribi AE, Hack K, Jung IH, Kang YB, Melançon J, Pelton AD, Petersen S, Robelin C, Sangster J, Spencer P, Van Ende MA (2016) FactSage thermochemical software and databases 2010–2016. Calphad, vol 54, pp 35–53
23. Jak E et al (2021) apha-UQPY_5299 database for FactSage. Retrieved from https://chemeng.uq.edu.au/research/metallurgy-and-minerals-processing/pyrometallurgy-innovation-centre/pyrosearch-facilities/factsage

# Towards Bio-carbon Substitutes in the Manufacture of Electrodes and Refractories for the Metallurgical Industries: A Science and Technology Review

**Jesse Franklin White, Natalia Skorodumova, and Björn Glaser**

**Abstract** The unique structural versatility and chemical and thermophysical properties of carbon make it essentially irreplaceable for non-reductant uses in many high-temperature metallurgical processes. At present, bio-carbon substitutes are not technically feasible for large-scale use in electrode and refractory materials that are vital consumables in the steel, aluminum, and non-ferrous metal industries. Carbon electrodes of all types (including Soderberg, prebaked, and anodes/cathodes for Al) as well as carbon lining pastes are all similar in that they are comprised of a granular carbon aggregate and a carbon-based binder. Similarly, refractories such as MgO-C utilize both natural (mined) graphite and carbon-based binders. Replacements of fossil materials with equivalent bio-carbon substitutes have the potential to dramatically reduce the carbon footprints of these products. However, there are still considerable materials engineering challenges that must be surmounted. The properties of bio-carbon materials and technological obstacles are explored, including catalytic graphitization and development of bio-pitch materials.

**Keyword** Pyrometallurgy · Sustainability · Carbon materials · Bio-carbon · Electrodes · Refractories

## Extended Abstract

Carbon is employed as both a refractory material (resistant to both heat and chemical corrosion), in refractory furnace linings, and as an electrical conductor in electrodes. Replacement of fossil carbon materials with equivalent bio-carbon substitutes in electrodes and refractories has the potential to dramatically reduce the carbon footprints of these products. However, it is an enormous material engineering challenge to find viable bio-based carbon replacements in these applications.

Carbon materials are heavily relied upon in both ironmaking and secondary steelmaking in refractories and electrodes. Graphite blocks comprise blast furnace

J. F. White (✉) · N. Skorodumova · B. Glaser
KTH Royal Institute of Technology, Stockholm, Sweden
e-mail: jfwhite@kth.se



C. Fleuriault et al. (eds.), *Advances in Pyrometallurgy*, The Minerals, Metals & Materials Series, https://doi.org/10.1007/978-3-031-22634-2_14

hearths. Carbon-containing refractories such as magnesia-carbon are in prevalent use in converters and ladle linings. In secondary, scrap-based steelmaking, large electric arc furnaces consume enormous quantities of graphite electrodes.

In electrolytic aluminium production, carbon anodes and cathodes in the cells generate Joule heating as well as electrical current and carbon for reduction. While the cathodes last the lifetime of the cells, very large quantities of anodes are continuously consumed in the process as oxygen released by electrolysis reacts with anodes to generate $CO_2$ gas.

In silicon and ferroalloy production, large electric submerged-arc furnaces consume Söderberg, prebaked, and composite electrodes. The use of continuous Söderberg electrodes is still prevalent today, with electrode diameters exceeding 2 m in modern furnaces. Furnace linings in silicon and ferroalloy furnaces are partially constructed of either carbon blocks or carbon lining paste.

Carbon electrode materials of all types (this includes Söderberg, prebaked, and anodes and cathodes for Al), and carbon ramming and lining pastes are similar in that they are composite materials comprised of a granular carbon aggregate along with a carbon-based binder. Traditional Söderberg electrode paste is composed of 70–80% carbon aggregate, and 20–30% coal tar pitch binder. A typical prebaked anode in Al production contains about 65% calcined petroleum coke, 20% recycled anode butts, and 15% coal tar pitch binder.

The vast majority of bio-derived carbon materials are hard carbons that do not graphitize with simple heat treatment. Bio-char, a solid product of pyrolysis, is characterized by low density (high porosity), and low electrical conductivity. As a result, unmodified bio-char as a wholesale substitute for the fossil carbon aggregate in carbon electrodes is not technically feasible. Catalytic graphitization of charcoal with iron for example could be one approach to transform charcoal into a material suitable for electrode and refractory applications, though there are considerable technical challenges such as efficient catalyst separation, reactivity with oxygen and carbon dioxide, and densification.

The replacement of fossil-derived binders by bio-binders alone in carbon electrode and refractory materials would dramatically reduce the carbon footprints of carbon electrode and refractory materials. The aluminium industry is leading the way in bio-pitch development, though bio-pitches derived from pyrolysis oils could also potentially be used in Söderberg electrode paste, lining, and ramming pastes. Key difficulties that remain are shrinkage, high oxygen content, low coking value, reactivity, and low graphitizibility. Considerable additional development is still necessary from the laboratory scale and upwards to enable the large-scale use of bio-carbon materials in these demanding applications.

# A Pilot Trial Investigation of Using Hydrochar Derived from Biomass Residues for EAF Process

Chuan Wang, Yu- Chiao Lu, Liviu Brabie, and Guangwei Wang

**Abstract** Biocarbon will play an important role to achieve a carbon neutral and sustainable steel industry. In this study, three hydrochars (one type of biocoal produced via the hydrothermal carbonization process) derived from orange peel, green waste, and rice husk were tested in a 10-ton test-bed EAF (electric arc furnace). These hydrochars were added to EAF via injection and top-charge as carburizer to substitute anthracite. The obtained liquid slag composition after scrap meltdown is favorable for the desulphrization process. Moreover, a higher carburization yield was achieved by top charging of hydrochar into EAF at the beginning of the heat. The final *P* and *S* of liquid steel with addition of hydrochars were controlled to acceptable levels. Some perspectives of using hydrochar for EAF steelmaking are also presented.

**Keywords** EAF · Hydrochar · Carburizer

## Introduction

The EAF steel production route from scrap or from hydrogen-reduced direct reduced iron (DRI) will in future play at least an important if not a fundamental role to contribute to the GHG emission reduction in the iron and steel industry. To produce a fully green and carbon-neutral steel, it will be necessary to use alternative carbon sources in the EAF that are renewable such as biomass. In modern EAFs, the share of energy input from fossil fuels, for instance, natural gas and coal, is over 40% of the total energy input [1]. In addition to their energetic use as a substitute for

C. Wang (✉) · L. Brabie
SWERIM AB, Process Metallurgy, 971 25 Luleå, Sweden
e-mail: Chuan.Wang@swerim.se

C. Wang · Y.-. C. Lu
KTH Royal Institute of Technology, Material Science and Engineering, 100 44 Stockholm, Sweden

G. Wang
University of Science and Technology Beijing, Beijing 100083, China



C. Fleuriault et al. (eds.), *Advances in Pyrometallurgy*, The Minerals, Metals & Materials Series, https://doi.org/10.1007/978-3-031-22634-2_15

electrical energy, carbon sources are used in particular as slag foaming agents [2]. Solid carbon sources, like coal, petrol coke, etc., are used in the EAF in two ways. Firstly, the bulk carbon is charged together with the scrap or other iron sources and additives at the beginning of the heat. This carbon serves to carburize the melt and contributes thereby to the slag foaming and by direct oxidation during meltdown realizes a chemical energy input. Secondly, fine carbon powder is injected into the EAF via lances or injectors together with oxygen to generate CO bubbles within the slag and thereby to foam the slag. The average fossil carbon addition is about 12 kg per ton steel, which is corresponding to 40–70% of the direct $CO_2$ emissions of the EAF steelmaking process without consideration of the use of natural gas and graphite electrodes [3]. When using alternative carbon sources, it must generally be considered that the properties of these materials can differ significantly from those of common fossil coals and natural gas. These differences can have an impact on the steelmaking EAF process. Most research concerning the use of biocarbon in the steelmaking processes focused on biochar produced from pyrolysis or torrefaction. These conversion processes have disadvantages such as high energy consumption and low char yield, and are limited to processing low-moisture feedstocks, such as forest wood. In addition, there are some other concerns such as high price compared to coal and natural gas; low or no availability of biomasses (e.g. dry wooden biomass) to produce biochar; and high alkali and phosphorus contents in biochar.

Recently, hydrochar has been proposed as a new alternative biocarbon source that can be used in iron- and steelmaking [4]. Hydrochar, a type of biocarbon, is produced by the hydrothermal carbonization (HTC) process, which is a thermo-chemical biomass conversion process in sub-critical water. As HTC is a water-based process, this technology is suitable for upgrading wet biomass residues and wastes which are abundant and locally available compared to forest wood, leading to a low production cost due to reduced transportation cost and low raw material cost. Compared to pyrolysis and torrefaction, no drying of feedstock is required for HTC process, and this leads to higher energy efficiency of the process. Furthermore, HTC facilitates the convenient removal of unwanted elements such as alkali metals, as these elements are dissolved in the solution and, therewith, can be removed from the carbonaceous solid. The physical ash reduction process (based on gravity/density) can further reduce the inorganic part of the solid product.

In this study, several different types of hydrochar were tested in a pilot scale EAF to investigate their performance as a carburizing agent. The characterization of studied hydrochars was presented, followed by an introduction of the EAF test-bed used in the pilot trials. Some testing results were presented and discussed in Sect. 3, and finally some concluding remarks were given in Sect. 4.

## Material Preparation and EAF Trials

### *Hydrochar Production*

Hydrochars were produced at the HTC plant owned by Ingelia, an HTC technology developer from Spain, as shown in Fig. 1. The Ingelia HTC technology includes separation equipment for impurities that are present in the waste such as sands, stones, pieces of metals, or glass. However, there are some inorganic components in the carbon structure, such as *Ca*, *K*, or *P*, that can be reduced by specific washing and chemical post-treatment steps. As a result of the HTC process, most of the carbon content of different wet organic waste streams is concentrated and retained within the obtained hydrochar. The HTC process acts as an acceleration of the natural coal formation process, working at moderate pressure and temperature (20 bar and 210 °C), allowing the dehydration of the organic matter and increasing the C-content up to 60 wt%.

**Fig. 1** The layout of two reactors at Ingelia's HTC plant

### *Hydrochar Products*

Three different feedstocks were used for the hydrochar production, namely, orange peel, green waste (e.g. leaves and branches) and rice husk. The chemical analysis and heating value for the various hydrochar are listed in Table 1. After the HTC upgrading, hydrochars are more like lignite as shown in the Van Krevelen diagram, Fig. 2.

**Table 1** Chemical analysis and heating value of hydrochars

| | | OPHF | OPHP | GWHF | GWHP | RSHP | Anthracite |
|---|---|---|---|---|---|---|---|
| C | wt% | 63.2 | 62.7 | 52.9 | 52.4 | 51.3 | 90.5 |
| H | wt% | 5.5 | 5.4 | 4 | 4.9 | 4.7 | 1.3 |
| O | wt% | 26.4 | 26.5 | 20.7 | 22.6 | 21.7 | 2.7 |
| N | wt% | 1.63 | 1.65 | 1.37 | 1.35 | 1.39 | 0.68 |
| S | wt% | 0.10 | 0.11 | 0.14 | 0.13 | 0.15 | 0.18 |
| P | wt% | 0.19 | 0.25 | 0.22 | 0.23 | 0.14 | 0.01 |
| Heating value | MJ/Kg | 25.6 | 25.5 | 21.9 | 21.7 | 21.2 | 32.96 |
| Volatile content | wt% | 64.6 | 63.1 | 58.8 | 63.5 | 56.2 | 1.8 |
| Ash content | wt% | 3.9 | 4.5 | 20.9 | 18.4 | 20.6 | 4.6 |
| Fixed carbon | wt% | 31.5 | 32.4 | 20.3 | 18.1 | 23.2 | 93.6 |

*Abbreviations* OPHF—orange peel hydrochar fines; OPHP—orange peel hydrochar pellets; GWHF—green waste hydrochar fines; GWHP—green waste hydrochar pellets; RSHP—rice husk hydrochar pellets

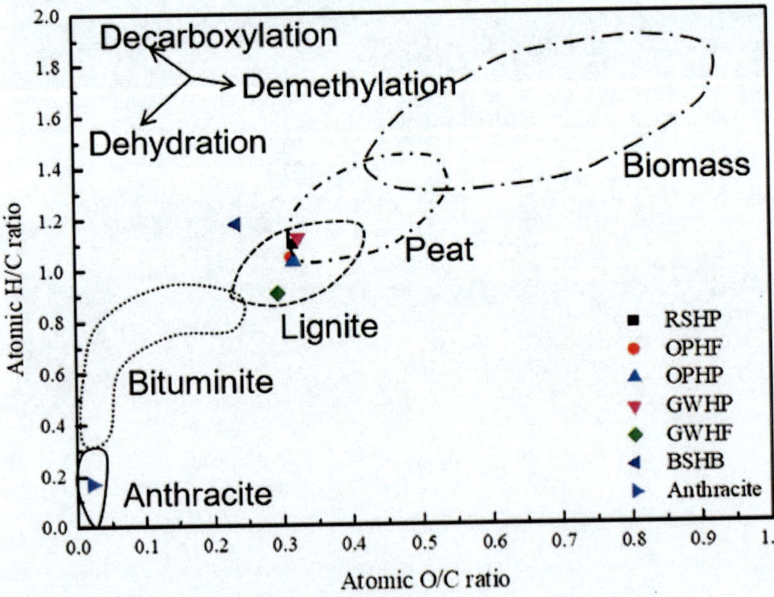


**Fig. 2** The Van Krevelen diagram of hydrochars

## *EAF Test-Bed*

The pilot trials were performed in the EAF test-bed at Swerim. Figure 3 illustrated the EAF furnace and the layout of the control system. The furnace heat size is 10 tons with an inner diameter of 2.1 m (after lining) and the electrode diameter of 250 mm.

The furnace gases are delivered to the gas cleaning plant in water- and/or air-cooled ducts. Continuous analysis of hot furnace off-gases (post combusted) is a normal procedure enabling constant dynamic heat and mass balances during melting and reduction/oxidation periods. The auxiliary equipment includes a bag-house filter (7000, $Nm^3/h$) for off-gas, including a weighing bridge for produced dust (collected in 180 l barrels). A probe for off-gas analysis of CO, $O_2$, $SO_2$, $CO_2$, and NO are positioned before the gas cooler. Off-gas flow and temperature are also measured. Furnace gas temperature and pressure measurements are also included.

Metal was sampled through the inspection-hole with an argon probe during the power-off time. Slag samples for regular analysis were taken by spoon. Carbon and sulphur in the metal were analysed in-house using the LECO instrument, the response time is normally 5 min. Full chemical analysis of metal and slag were analysed by the nearby laboratory at SSAB (the Swedish Steel plant), and samples were conveyed by a pneumatic tube system with a response time of around 20 min. The production data (e.g. power and anthracite consumption, slag amount and analysis, liquid steel amount and analysis, dust amount and analysis, fuel gas volume, temperature, and analysis, etc.) were collected for further analysis. The layout of the EAF test-bed and the control system is shown in Fig. 3.

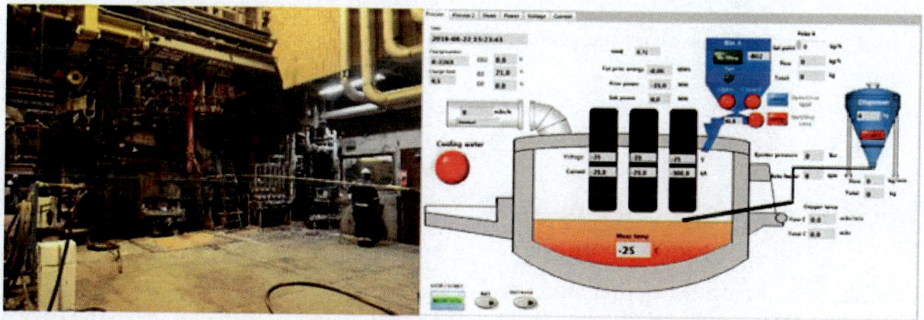

**Fig. 3** Swerim's pilot EAF furnace, the layout (left) and the control system (right)

# Results and Discussion

## Combustion Performance of Hydrochars

The combustion performance was evaluated by TGA, as shown in Fig. 4. In general, all hydrochars have low initial combustion temperature, indicating that hydrochar have high reactivity due to their high volatile content. The TGA curve shows that the combustion process of hydrochar samples is more complex, which consists of two or three weight loss peaks, meanwhile there is only one peak for the anthracite. For the OPHP sample, two peaks are located in the low temperature region, and one peak appeared in the high temperature region; while for GWHP and RHHP, there is one peak in the low temperature region and one peak in the high temperature region.

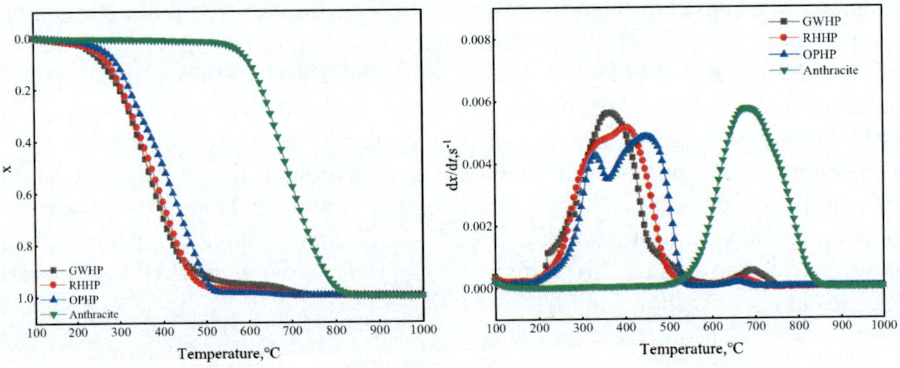


**Fig. 4** TGA curve of hydrochars and coal samples

## Comparisons of the Carbon Yield

Carbonaceous materials were added to EAF both via injection in the form of powder and top-charging (at the beginning of a heat) as pellets in order to carburize the melt. To understand the evolution of the slag/melt composition, sampling was performed at three different process stages. In some heats, carbonaceous materials were only top-charged whereas in other heats both top-charging and injection of hydrochar were performed. The first slag and steel sample were taken after scrap-melt down and dissolution of top-charged carbon (before the injection of carbon); the second slag and steel sample were taken when the injection of the carbonaceous material was halfway. The third sample was taken just before tapping. Slag conditions were observed through the inspection/slag-tapping door during each heat. The aim of the pilot trials is to evaluate and compare the carburization yield of three types of hydrochar and anthracite. It has to be pointed out that oxygen lance injection was not used for all trials.

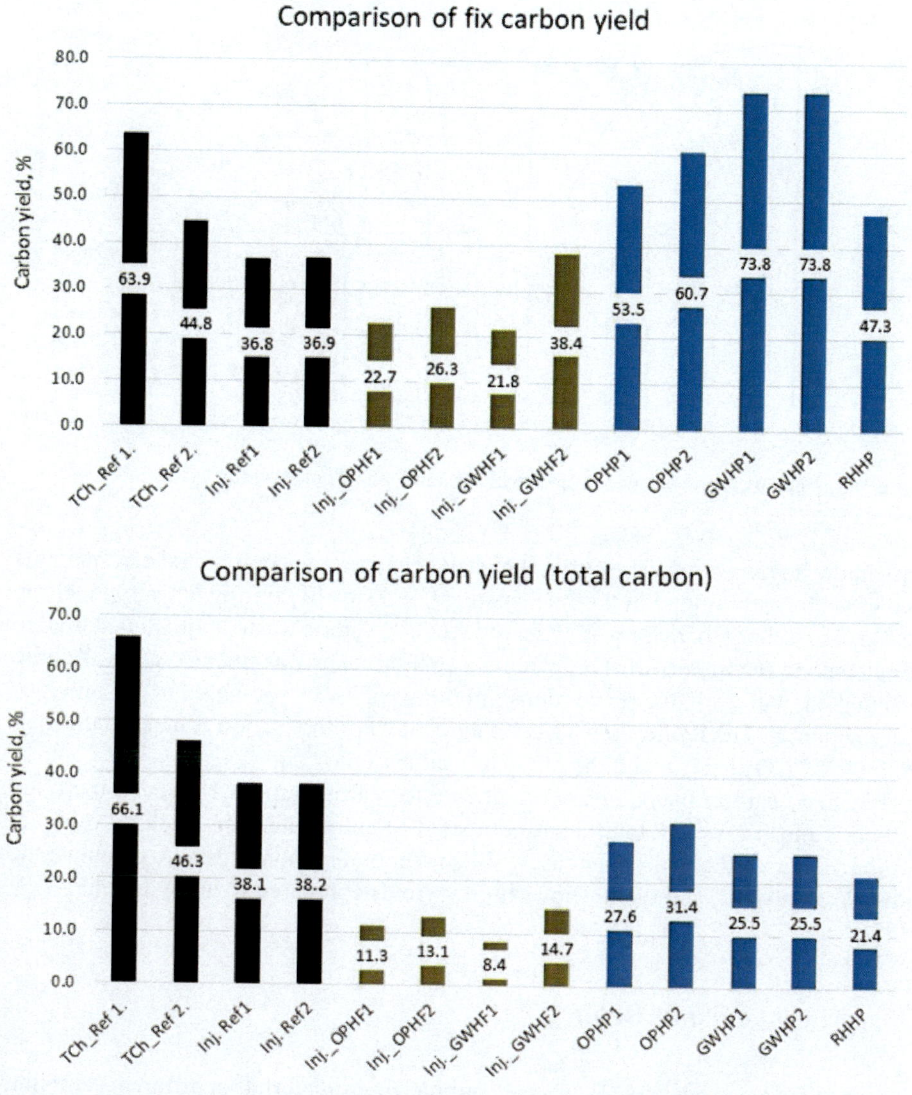


**Fig. 5** Comparison of carbon yield: fixed carbon (upper); total carbon (lower)

Figure 5 shows the calculated carburization yield in the view of both fixed carbon and total carbon. It can be seen that a higher carburization yield can be achieved when carbon materials is top charged into the EAF compared to when injected. The same observation can be noticed for the reference cases when anthracite was used. This can be explained in Fig. 6 where the green wastes were added to EAF via both

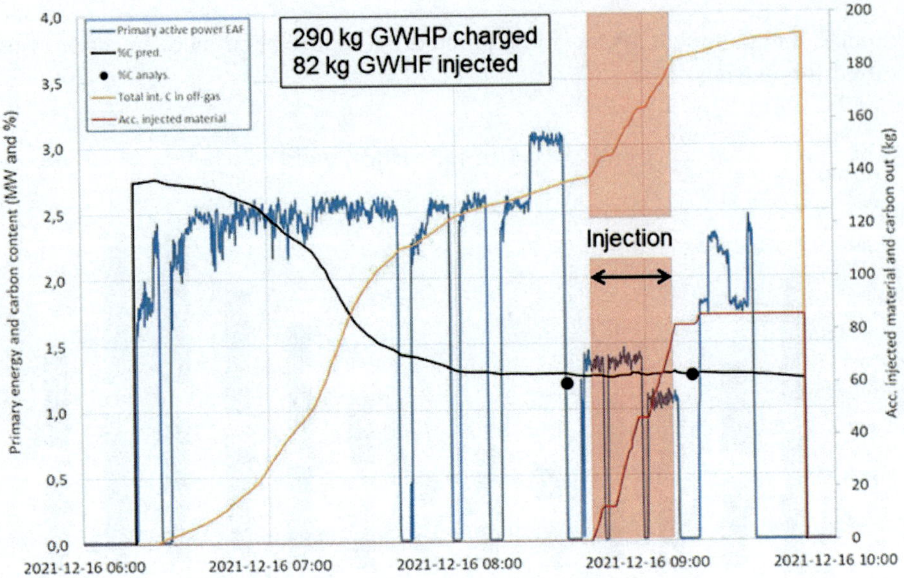


**Fig. 6** Example of carbon balance in EAF with the addition of green waste

top-charge and injection. The curve shows the changes of accumulated carbon in the flue gas. The carbon load in the flue gas increased steadily during the scrap melting period, and a sharp increase was observed when the green waste hydrochar started to be injected to the melt bath. The green waste hydrochar was injected after all scrap was melted, therefore, the power consumption was lower compared to the previous melting period. The figure shows a good agreement of the carbon content in the melt from the LECO analysis and the predicted value in the control system.

The fixed carbon plays an important role for carburization. However, the fixed carbon is much lower in hydrochar compared to anthracite, thus leading to higher carbon yields for the hydrochar cases. When the total carbon is used to calculate the carburization yield, anthracite shows higher yields compared to all hydrochars.

## *Observation of Slag Behavior*

The calculated slag basicity (B2) for all carbonaceous materials at different sampling points are summarized in Fig. 7. In general, the slag basicity is lower at the end as more *Si* will be oxidised from the metal into the slag in an oxidation atmosphere. One exception was noticed for GWHP2 for the second sampling point, in which B2 is increased. The reason was that 84 kg anthracite was injected to EAF, hence some amounts of $SiO_2$ in the slag were reduced back to the metal. This was confirmed by an extra sampling point of GWHP2 (3), showing a high B2 as well.

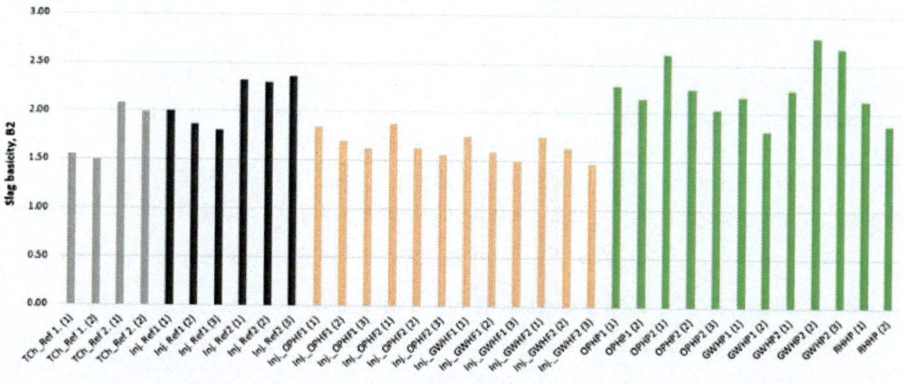


**Fig. 7** Basicity B2 ($CaO/SiO_2$) for trials with all carbonaceous materials

The basicity of final slags with hydrochar injection was lower than 2.0, and for most top charging cases, the slag basicity was above 2.0. This might indicate that a more oxidation condition occurred when hydrochars were injected into EAF compared to the top charging. Figure 8 shows the observed slag from the tapping/inspection door as one example. The slag in most pilot trials were in normal condition.

**Fig. 8** Slag observation from the tapping door

## *P* and *S* Content in the Metal

The *P* and *S* content in the melt are shown in Fig. 9 for both hydrochar additions via top-charge and injection. As it can be seen, the *S* level for the heats with hydrochar addition is at a similar level as the reference, or a bit lower for some cases, meaning a good sulphur removal. The conditions which are favorable for sulphur removal are for instance, high slag basicity, high temperature, good slag fluidity, and low

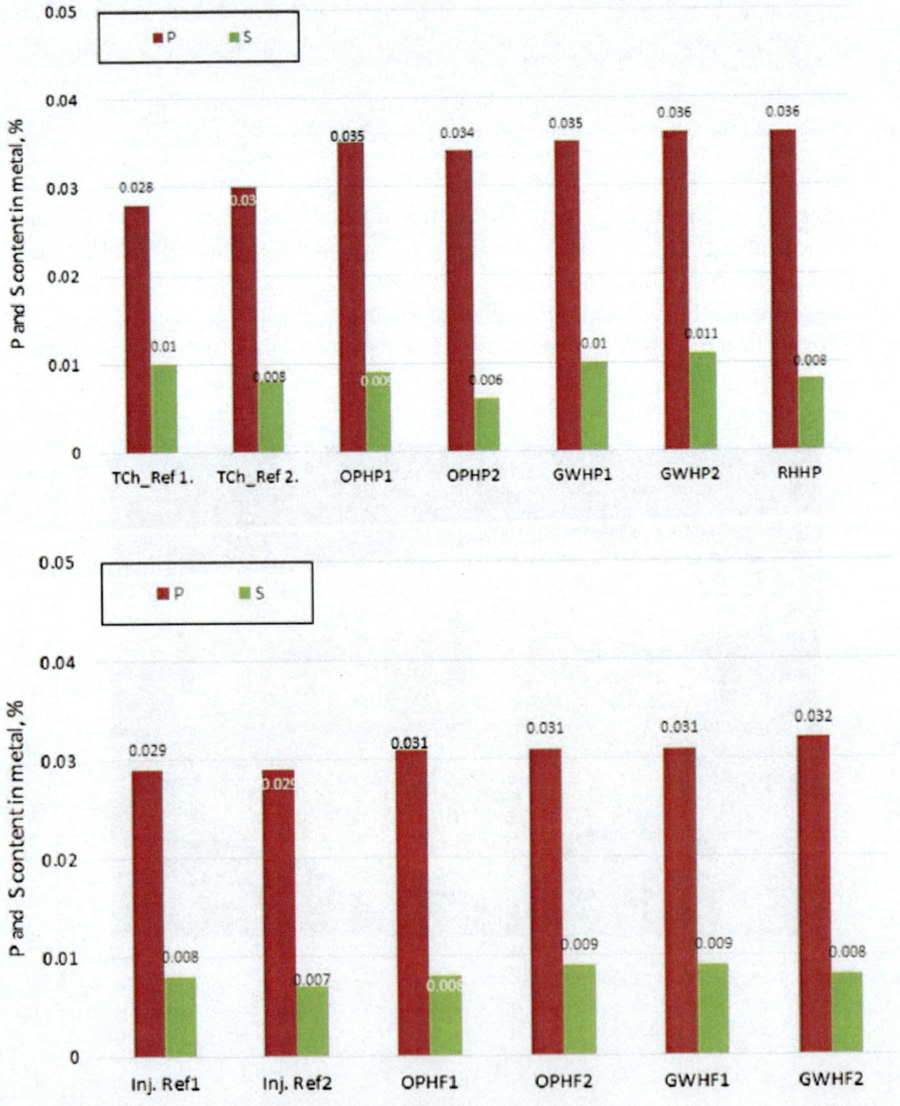


**Fig. 9** Phosphorus and sulphur levels in the metal. Top-charge (upper) and injection (lower)

dissolved oxygen in the melt bath. Main conditions that are favorable for *P* removal are opposite to that of *S* removal, for example, it prefers a high oxygen content in the melt, ferrous oxide of 15–20% in the slag and a lower temperature. Given the conditions obtained during the pilot trials, it was more favorable for *S* removal than *P* removal. The trials with addition of hydrochar shows a slightly higher *P* level in the melt compared to reference cases using anthracite, although the *P* level is still within the acceptable range. The EAF operating conditions should be further optimized for a better *P* and *S* removal when hydrochar is used.

## Concluding Remarks

A series of EAF pilot trials were carried out to investigate the carburization yield for three types of hydrochar. Following remarks are highlighted from this study,

- Hydrochar can be added into EAF via both top charging and injection as carburizing agents.
- Higher carburization yields were achieved by top-charging hydrochar (added at the beginning of a heat) compared to the injection into the melt.
- *P* and *S* levels in the melt could be controlled within acceptable ranges for all three types of hydrochar.
- The slag conditions obtained during the trials were good for desulphurization. However, the phosphorus removal was not favorable in terms of melt bath temperature and slag compositions, requiring deep investigations in future research work.
- The strategy of adding hydrochars into EAF should take into account the additional amount and the additional approach.

**Acknowledgements** This research was funded by VINNOVA (Swedish Governmental Agency for Innovation Systems) and some of the participating companies in OSMET 3.0 project (dnr: 2020-04140).

## References

1. Bianco L et al (2013) BHM Berg-und Hüttenmännische Monatshefte 158:17–23
2. Jellinghaus M (1994) Stahlerzeugung im Lichtbogenofen, 3rd ed. Verlag Stahleisen GmbH, Düsseldorf, Germany
3. Echterhof T (2021) Metals 11(2):222
4. Lu Y-C et al (2022) Sustainability 14:5383

# Biocarbon Materials in Metallurgical Processes—Investigation of Critical Properties

**Nicholas Smith-Hanssen, Gøril Jahrsengene, and Eli Ringdalen**

**Abstract** The silicon, ferroalloy, and aluminum industries have mostly been dependent on fossil carbons for their respective process. However, efforts to reduce their fossil $CO_2$ emissions, the switch to biocarbon has already begun and targets of 25–40% biocarbon by 2030 have been set by various producers in Norway. To achieve this transformation a better understanding of the effects of physical properties of the carbon on the process must be obtained so that the transformation can occur with minimal process interruptions. For the silicon, ferrosilicon, and ferromanganese industries the effects of biocarbon reductants are the primary interest whereas for the aluminum industry, use of biocarbon to replace packing coke used in anode baking is desired. In this work, an overview over relevant carbon properties and methods to characterize these are presented together with an evaluation of how these properties may affect the different processes when introducing biocarbon.

**Keywords** Biocarbon · Stage-gate matrix · Silicon · Ferroalloy · Aluminum

## Introduction

The silicon, ferroalloy, and aluminum industries all use carbon materials in their processes. Historically, this carbon has come from fossil-based sources (coal, metallurgical coke, petroleum coke), with the exception of some biocarbon use in ferrosilicon and silicon production. Replacing the traditional fossil carbons with biocarbons that have a significantly smaller fossil $CO_2$ footprint is desirable from an environmental viewpoint. Targets of 25–40% biocarbon by 2030 have been set by various Si and ferroalloy producers within Norway [1, 2].

As the industries have historically relied upon fossil-based carbon sources their processes are tailored to utilize a carbon material within this range of properties. Utilization of carbon materials that do not meet these requirements, while theoretically possible, could require a significant change in the operation of the furnaces.

N. Smith-Hanssen · G. Jahrsengene (✉) · E. Ringdalen
SINTEF, Trondheim, Norway
e-mail: Goril.Jahrsengene@sintef.no

 
C. Fleuriault et al. (eds.), *Advances in Pyrometallurgy*, The Minerals, Metals & Materials Series, https://doi.org/10.1007/978-3-031-22634-2_16

The specific requirements will depend highly on the quality of the produced alloy as well as the other raw materials that is used, furnace specifications, situation at the actual production site, and operational strategies for the different producers. It is thus not possible to set a single set of requirements, but rather discuss the effects of the different carbon properties on the processes.

The search for new non-fossil-based carbon materials is focusing both on development of raw materials that are tailor made for a particular process, and how to adopt existing processes to new raw material properties. In this work, the carbon properties and their effects on different processes are described. Some evaluations of biocarbon are also included with this in mind. Typical characterization methods are also described and evaluated. A suggested sequence of these tests to evaluate new materials is also included in the form of a stage-gate matrix. The goal of this matrix is to outline a proposed investigation route such that the most critical properties are studied first. If the material does not meet the critical requirements, it can be discarded for use in the investigation process. Since critical properties also depend on the context, the stage-gate matrix can vary from case to case. However, it is regarded as important to have a general test matrix as a starting point. For Si and ferroalloys, the properties are discussed based on use in submerged arc furnaces (SAF), but it might also be relevant for other production methods. The discussion and evaluations are based on the experience, raw materials, furnaces, and furnace operation in Norway.

## Alloy Specific Requirements

In general, the different metallurgical industries have specific physical and chemical property requirements that are particular to the specific alloy being produced and to a particular furnace, and also to some extent a part decided by the company's operation strategy. However, there are some requirements that are common to most producers. One of these is the requirement for low amounts of fines in the raw materials, both in cold materials before their processing and in the alloy producing process. A high mechanical strength is thus desired, regardless if the end user is a Si, ferroalloy, or aluminum producer. Mechanical strength refers to two important factors. First, it is the ability of the material to resist fines formation that can cause issues furnace's stability and operation. Friable materials such as charcoal are generally regarded to be more prone to fine formation compared to coal and metallurgical coke. Second, the ability of the material to resist dust formation is critically important. If the material produces a large amount of airborne dust during transport and charging this can have a negative impact on the air quality on the plant as well as pose certain safety risks.

Beyond the mechanical strength, a high carbon content, fixed C, and a high density are in many cases desirable for production in submerged arc furnaces, as in the silicon and ferroalloy industries as it minimizes the volume of reductant that is required to produce 1 ton of metal. The fixed carbon density (which gives kg C/m$^3$) can be calculated from the bulk density and fixed C. From Table 1 it can be seen that coal

**Table 1** Typical ranges for different carbon reductants available today [3–7]

| | Coal | Metallurgical coke | Charcoal | Woodchips |
|---|---|---|---|---|
| Fixed carbon (%) | 60 | 87 | 65–87 | 12 |
| Density (kg/m$^3$) | 800 | 500 | 180–300 | 250 |
| Volatiles (%) | 40 | 2 | 10–30 | 35 |
| Fixed carbon density (kg/m$^3$) | 480 | 435 | 117–261 | 30 |

and metallurgical coke have notably higher fixed C densities than typical charcoals. A range is given for the charcoals as different wood types and pyrolysis parameters create a large range of charcoal properties and it is of interest to see the range for charcoal. In the best case, the volume of carbon material required to be charged into the furnace to have the required amount of fixed C must double if charcoal is used instead of coal or coke.

A low fixed C density is not necessarily negative for all carbon materials since some Si producers use a small percentage of woodchips in the charge mixture to improve the behavior in the furnace. Since it is not added primarily as a carbon source, and the physical and chemical properties vary significantly from the other reductants, it cannot be compared with other carbon sources and is not included in further discussion [7].

## *Si and FeSi Production*

Generally speaking, today the Si and FeSi industry in Norway utilize carbon reductants that have the following properties:

- High mechanical strength
- Medium to high volatiles
- High SiO-reactivity
- High density.

As the coals used today can have up to 40% volatiles, the furnaces and off-gas systems can handle biocarbons which normally have a lower volatile content than this. Reductants with lower volatile contents such as coke are also used by some smelters thus there are no specific requirements on volatiles can be established.

A cross-section of the Si furnace is shown schematically in Fig. 1. In the production of silicon metal, the carbon is not used to directly reduce the $SiO_2$, rather the carbon is reacted with SiO gas in the upper part of the furnace to form SiC per Eq. 1. This SiC travels down to the lower part of the furnace where it reacts to produce Si and SiO gas per Eq. 2. This process is shown schematically in Fig. 1. Any SiO gas that is not captured by the carbon exits the top of the furnace and results in a loss of Si giving a lower Si yield. The ability of a carbon to capture the SiO gas is critically important and is known as the SiO reactivity [8]. The temperatures at the top of

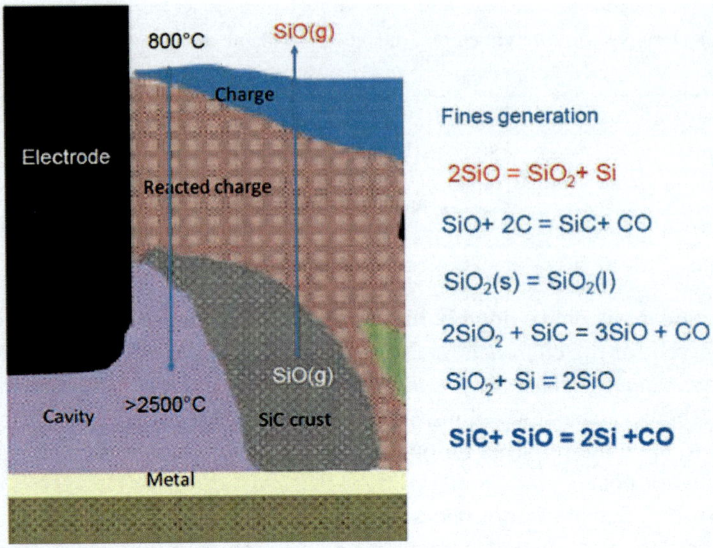


**Fig. 1** Schematic of the industrial Si furnace [9]

the charge have been found to be at least 800 °C on the charge top. This results in significant thermal stresses in the materials when they are charged into the furnace thus the thermal shock strength is of interest [9].

$$SiO + 2C = SiC + CO \tag{1}$$

$$SiC + SiO = 2Si + CO \tag{2}$$

## *FeMn and SiMn Alloy Production*

In general, the manganese alloy industry in Norway looks for a carbon reductant that has the following properties:

- High mechanical strength
- Low volatiles
- Low $CO_2$ reactivity
- High density.

The FeMn furnace can be divided into 2 zones: a prereduction zone and a coke bed zone as shown schematically in Fig. 2. Mn ore is far from homogeneous and contains a range of Mn oxides including $MnO_2$, $Mn_2O_3$ and $Mn_3O_4$. These oxides

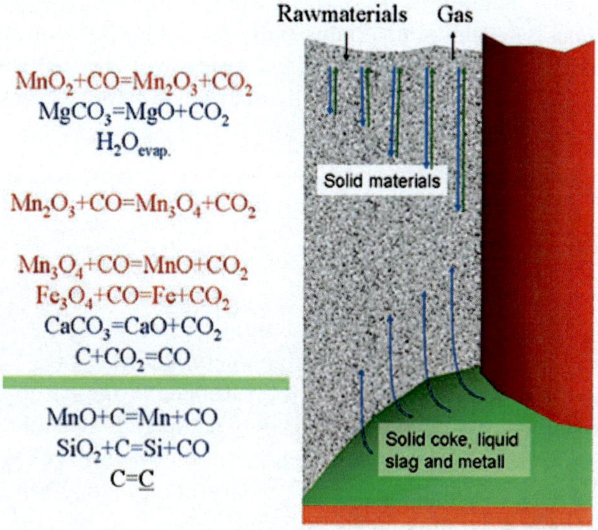


**Fig. 2** Schematic over the FeMn furnace [6]

are ideally reduced by CO gas to MnO via Eqs. 3–5 in the prereduction zone of the furnace, producing $CO_2$ gas in the process. In the coke bed zone, the final reduction of MnO and any remaining $Mn_3O_4$ to Mn takes place via Eqs. 6 and 7 [6].

In addition to the reactions mentioned above, the Boudouard reaction given in Eq. 8 can also occur at higher temperatures. While the exact onset temperature for the reaction is dependent on the carbon reductant, it is typically stated to start at 800 °C for metallurgical coke [6]. Under ideal conditions, the ore will be entirely reduced to MnO before the temperature reaches the onset temperature of the Boudouard reaction as it will reduce the amount of $CO_2$ available for the Boudouard reaction. As the Boudouard reaction is both carbon consuming and endothermic, the FeMn industry aims to minimize the extent of the reaction. Thus, selecting a carbon reductant that has a high onset temperature and the slow reaction rate is seen as important [6].

$$MnO_2 + \tfrac{1}{2}CO = \tfrac{1}{2}Mn_2O_3 + \tfrac{1}{2}CO_2 \quad (3)$$

$$\tfrac{1}{2}Mn_2O_3 + \tfrac{1}{6}CO = \tfrac{1}{3}Mn_3O_4 + \tfrac{1}{6}CO_2 \quad (4)$$

$$\tfrac{1}{3}Mn_3O_4 + \tfrac{1}{3}CO = \tfrac{1}{3}MnO + \tfrac{1}{3}CO_2 \quad (5)$$

$$\tfrac{1}{3}Mn_3O_4 + \tfrac{1}{3}C = Mn + \tfrac{1}{3}CO \quad (6)$$

$$MnO + C = Mn + CO \quad (7)$$

$$\tfrac{1}{2}C + \tfrac{1}{2}CO_2 = CO \quad (8)$$

The metallurgical cokes that are in use today typically have a volatile content under 2%, while most industrially available charcoals have a volatile content of 10–20%, thus the increase in volatiles may overwhelm the off-gas system. In opposition to this there are indications that certain volatile species, such as $H_2$, may aid in the prereduction and could therefore have a positive effect on the furnace operation [10].

### *Aluminum Production*

Today carbon is used by the aluminum industry in the form of anodes. These anodes are typically made from calcined petroleum coke (CPC), coal tar pitch and recycled anode butts in an anode factory. Replacing the carbon in the anode with biobased materials is still a ways from being technically feasible. However, it may be possible to use biocarbon as a replacement for the packing coke used during the anode baking process. Green anodes must be baked before they can be used to produce Al metal. To help hold the anode shape and reduce air burnoff the anodes, they are packed with packing coke during baking. Two furnace designs are used: open and closed top, giving some different options of packing materials [11]. It may be possible to use biocarbon as the packing coke, or partially replace the packing coke, especially in closed top furnaces where the burn-off of the CPC packing material is significant.

Today the packing material is traditionally the same as the CPC used in the anodes and is often recycled back into the baking furnace or into the anodes themselves. Thus, the requirements for packing coke follow the anode requirements. When considering other materials than CPC for packing the requirements may not be as strict. The major requirements are:

- Relatively small particle size
- Non-spherical shape
- Low rate of air burn
- Good mechanical strength.

## Possible Methods to Determine Critical Parameters

Based on the above descriptions of the effect of various carbon properties in the different processes, different methods for investigating carbon properties are evaluated below. Specific requirements for each carbon parameter have not been given as the requirements of carbon properties vary between company, smelter, and alloy. Carbon materials must thus be evaluated with a specific site in mind to determine if they are suitable. However, some openly published results are given as a basis for comparison. A variety of methods exist to test each property, these methods may be governed by ISO/ASTM standards or custom methods that are developed to better understand a specific property in relation to a specific process. Many of the standard ISO/ASTM methods such as ISO 18894, "Determination the coke reactivity index

(CRI) and coke strength after reaction (CSR)", have been specifically developed for the iron and steel industry and especially the blast furnaces to test metallurgical coke and other carbon materials for this use. They are as such in many cases neither suitable for characterization of carbon materials for other industries as those discussed here, or suited to investigate properties of biocarbons that have a higher reactivity. The methods must therefore be adopted to allow the critical properties to be evaluated in a manner that gives relevant results. A brief overview of each of the characterization methods is given below and where necessary references to more detailed descriptions and previous results.

### *Particle Shape*

The particle shape is a simple visual check to see if the material is suitable to be used as Al packing coke. Spherical particles are not ideal as they are difficult to pack effectively and must be avoided; thus, any new packing coke must not be spherical in shape, and if it, is it must be discarded as a suitable material for packing coke. In the other processes, as new biocarbons may have significantly different shapes compared to the traditional lumpy coke and coal, this may influence the furnace behavior. For example, some charcoals have been seen to have a large aspect ratio where the cross-section is small compared to the length. This can cause issues in the final sieving as particles with a high aspect ratio behave differently in the sieve than particles with a ratio close to one. Thus, it is good to note any deviations in the particle shape from what is typically used as if can be useful in the final conclusion to use or discard a material.

### *Cold Strength*

The cold strength is critical for all industries and is currently a significant obstacle to increasing the percentage of biocarbon used. It has been shown that up to 25% of charcoal can be lost with transport from the pyrolysis plant to the furnace [12]. Given the high cost of charcoal, a loss of 25% can have a significant cost impact that must be addressed. Further, the lost material can have a negative impact on the surrounding environment. The mechanical strength can be investigated with multiple methods such as tumble/abrasion, drop and compression tests. Tumble strength is the main method used to estimate the mechanical strength as it attempts to replicate the mechanical degradation that occurs when the material is transported and charged into the furnace. For this method, the sample with a known particle size is placed in a steel drum and tumbled for a set time and speed. After tumbling the particle size is measured and the number of fines formed is used to estimate the mechanical strength. Regarding biocarbons, a significant increase in strength may be achieved by pelletization, which an increasing amount of producers and consumers are looking into.

## *Composition*

The composition of the reductant is critical to the process. This includes both the proximate analysis (Fixed C, volatiles, and ash) and the ash composition. The proximate analysis is typically done to the relevant ISO standard (ISO 17246), where the ash and volatiles are measured from the mass loss after heating and the fixed C is the remainder of the sample that is not ash or volatiles. The ash composition can be determined by ICP-MS or XRF. The acceptable composition must be evaluated based on the smelter where the reductant will be used as different smelters have different limits on the amount of ash and volatiles the furnaces can handle. Select trace elements such as P, B, Zn, and K must be limited to ensure safe operation, low emissions, and suitable metal quality.

## *Bulk Density/Porosity*

A high porosity has been proposed as an indicator of high gas reactivity of the carbon material. Although this does not seem to be a parameter that can be used to distinguish between carbon materials and select best suited materials, it will give valuable information about the main level of gas reactivity and may, when more knowledge has been generated, be used to eliminate unsuited materials [13]. It is thus important to know. The bulk density is related to the porosity and is typically given in units of mass per volume such as kg per $m^3$. The bulk density is used to estimate the fixed C density and the volume of raw materials that need to be fed into the furnace for a given production. Typical values for the bulk density and fixed C density are given in Table 1. A variety of standard methods and equipment such as helium pycnometry exist to measure the porosity. However, each of these methods have different range of measurable pore sizes as such the same method must be used when comparing different samples.

## *Methods Specific for Materials Used in Si and FeSi Production*

### Shock Heating Test

As the top of the industrial Si furnace is typically 800–1000 °C, the raw materials are rapidly heated when they are charged into the furnace. This rapid heating can cause significant thermal stresses and result in the fracturing of the material [3]. The ability of the reductant to withstand this rapid heating can be estimated by dropping a sample with a known particle size into a preheated crucible at 1000 °C and holding it until the volatiles have been removed and it reaches a stable temperature. The change in particle size distribution can be used to evaluate the thermal shock strength.

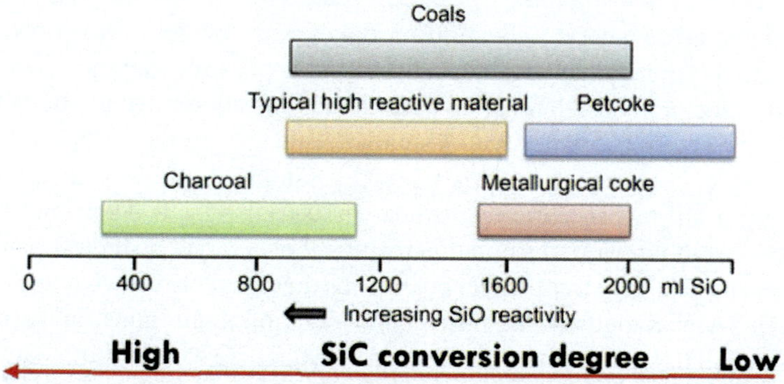


**Fig. 3** Typical SiO reactivities depending on material type [14]

### SiO Reactivity

The SiO reactivity is a critical parameter to have a high Si yield. To compare the reactivity a standardized test method has been developed, the SiO-reactivity test. In this test, the ability of the carbon reductant to capture SiO gas is measured by exposing the carbon material to SiO gas at a temperature of 1650 °C. The SiO will react with the carbon per Eq. 1. The amount of SiO gas passing through the carbon material unreacted is measured in ml based on online off-gas analysis. As the goal is to capture as much of the SiO gas as possible a low volume of unreacted SiO gas is desired and the lower the SiO reactivity number the more reactive a sample is. Typical ranges for different reductant types are given in Fig. 3. More details on the SiO reactivity can be found in a recent paper by Jayakumari [13].

### Additional Properties of Importance for Si and FeSi

The formed SiC is only an intermediate product, thus the reactivity of the SiC in the lower portion of the furnace can be important and can be investigated as was done in the doctoral work of Jayakumari [14].

## *Methods Specific for FeMn Production*

### Measurement of $CO_2$ Reactivity of Materials for Mn-Alloy Production

The CRI, which is the traditional method for measuring $CO_2$ reactivity for carbon materials used for iron production, is not optimal to distinguish between and select carbon materials for Mn-alloy production in SAFs. As the gases in Mn-alloy furnaces will have a high CO-content that will affect the rate of the Boudouard reaction, CO must be included in investigations of carbon materials for this purpose. Additionally, the CRI test gives a good indication for cokes with low reactivity, however, for biocarbon which typically has a higher reactivity it is possible that the entire sample is consumed in the defined 2-h reaction period and thus alternative methods for estimating the $CO_2$ reactivity are needed. A smaller scale test similar to the CRI is using a milligram scale TGA which utilizes a powder sample in 100% $CO_2$ atmosphere [15]. It is important to use particles in near industrial size to determine the $CO_2$ reactivity to have the most representative results, thus a larger scale test is needed.

For measuring $CO_2$ reactivity under conditions that are more relevant for Mn-alloy production in SAFs, a method that mimics the situation in an industrial furnace has been developed. The measurement is done in a macro TGA (30–100-g sample size depending on bulk density). Lumpy material with a known particle size (5–10 mm) can be placed in the crucible of the TGA. The sample is heated under Ar and when the desired reaction temperature is reached the Ar is switched to a 50:50 $CO_2$:CO mixture. When 20% of the fixed carbon mass has been lost the CO-$CO_2$ gas flow is switched back to Ar. A mixture of CO-$CO_2$ is used for evaluating materials for the FeMn process as it better represents the atmosphere inside the industrial furnace. Furthermore, stopping at 20% fixed C reacted for all materials means the mechanical properties afterwards can be compared at a similar degree of reaction [4, 7]. Typical values based on the method above are 0.02–0.032% FixC/s for charcoal and 0.002–0.005% Fixed C/s for metallurgical coke samples. Here it can be seen that charcoals have approximately 10 times higher reactivity than metallurgical coke [4, 16]. The effect of impurity elements such as potassium can and should also be investigated as they have a known catalytic effect on the Boudouard reaction [17], which is important as potassium is known to accumulate and build up to high levels of 4–5% in FeMn furnaces.

#### Additional Test for FeMn

Additionally, it can be of interest to measure the bulk electrical resistivity of carbon materials with near industrial particle sizes. A method for measuring electrical resistivity under relevant conditions has previously been suggested, based on the 4-point method [18]. Additionally, the measurement of the ability of the solid carbon to reduce MnO slag to Mn can be of interest as discussed in greater detail in the book by Suarez and Ruiz [7].

### *Methods Specific for Al Production*

For packing materials, the suitability should be possible to evaluate based on the above-mentioned general tests, with possibly an evaluation of $CO_2$ reactivity (similarly as for FeMn, but with a different atmosphere), and air reactivity. The calcined carbon materials currently used have already been heated to higher temperatures than what is reached during the baking process, and as such, the calcination behavior and warm strength are also important to understand. During the baking process, biocarbons may form an ash-layer that works as a protective layer on top of the baking furnace, thus giving an important reduction in the burn-off of the material beneath. However, the higher amount of ash components in bio-carbons compared to traditional CPC materials may have a negative effect on the lining materials in the anode baking furnace and is something that needs to be studied before setting specifications on what composition is acceptable in the materials.

## Proposed Sequence of Investigations

Based on the similar and specific critical parameters for the three processes (Si, ferroalloy, and aluminum), a stage-gate matrix has been developed to aid in investigations of new carbon materials. A proposed test matrix is given in Fig. 4. This is a part of an ongoing work that is currently improved with new input from ongoing projects. The goal of this matrix is to serve as a guideline for the selection of new carbon materials and for feedback to producers of novel biocarbons about the suitability of the material. It will reduce the need to carry out excessive testing on unsuitable materials. While charcoal (and woodchips) is the only known commercially available biobased reductants on the market today, there is an increasing focus on developing biocarbons that are tailormade to meet the specific requirements an individual company or smelter through methods such as pelletization [19], hybrid coal-biomass blends [20], and densification through carbon deposition from methane [16]. These materials will need to be characterized to evaluate their suitability for use by their respective industries. Often the amount of these materials is limited so carrying out a full range of tests may not be possible. With the stage-gate matrix, the most critical property is tested first and any carbon material that does not meet the requirements can be discarded, if the material passes the first test, it can move to the next gate and so on.

As mentioned before, specific criteria for each test have not been given as the criteria vary between company, smelter, and alloy, thus the results must be evaluated with a specific site in mind to determine if they are suitable. Additionally, these criteria will also be expected to change with time, as process improvements may allow more variation in the new materials.

This matrix is designed to help investigate specific properties of a carbon material. However, it is impossible to determine exactly how a new carbon material will behave in the industrial environment. This can be indicated by comparing materials in pilot scale, but will first be seen in an industrial test campaign.

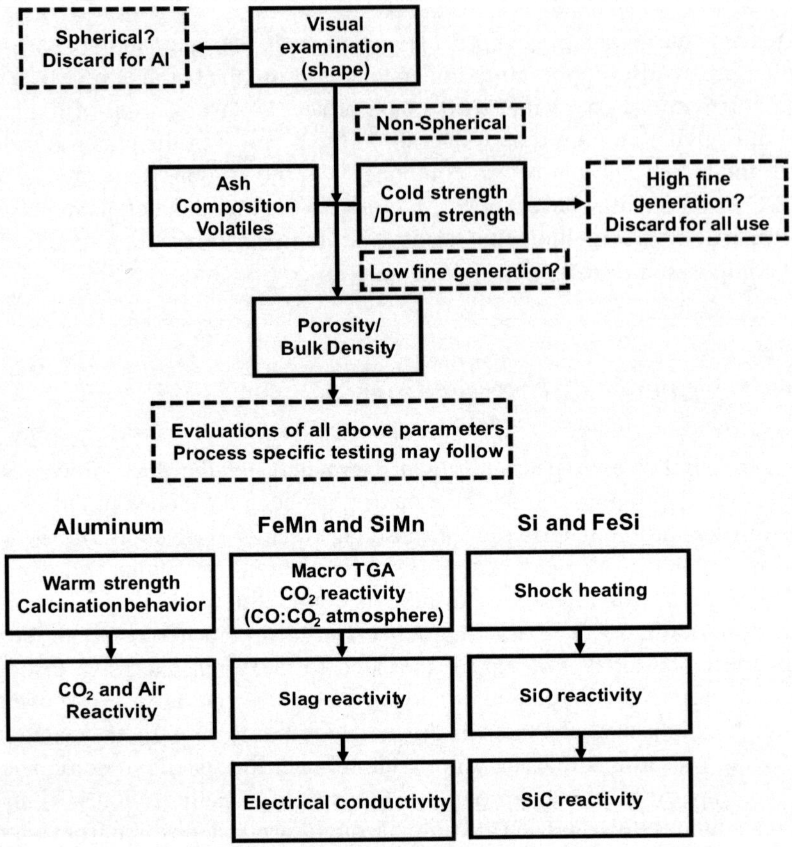


**Fig. 4** Stage-gate matrix for testing of new carbon materials for the silicon, ferroalloy, and Al industries

# Conclusions

Understanding how the carbon properties affect the different metallurgical processes is crucial to effectively introduce new, $CO_2$-neutral biocarbon when producing silicon, ferroalloys, and aluminum. This work presented an overview of the critical parameters needed for evaluation and an evaluation of how biocarbon may be evaluated compared to currently used materials. Based on the importance of the parameters, a general stage-gate matrix has been suggested to provide a guide for testing the suitability of new biocarbon materials for use as a replacement of traditional fossil carbons. The matrix shows a route to investigate the most critical properties, composition, and mechanical strength, first, and if the material fails to meet the criteria it can be discarded as it is unsuitable for use in the furnace today. The specific criteria will need to be individually determined and are likely to change over time.

**Acknowledgements** This research was funded by Research Council of Norway grant numbers 294679 (KPN BioCarbUp) and 280968 (KPN Reduced $CO_2$).

## References

1. Elkem (2020) ELKEM ESG report 2020, Elkem
2. Eramet (2021) Eramet Bærekraftrapport 2020, Eramet
3. Schei A, Tuset JK, Tveit H (1998) Production of high silicon alloys. Tapir Academic Press, Trondehim
4. Monsen B et al (2007) Charcoal for manganese alloy production. In: Paper presented at INFACON XI, New Delhi, India, 18–21 Feb 2007
5. Nordhaugen E, Gjølsjø S (2013) Flis og Flisegenskaper, NIBIO
6. Olsen S, Tangstad M, Lindstad T (2007) Production of manganese ferroalloys. Tapir Academic Press, Trondheim
7. Tangstad M, Beukes JP, Steenkamp J, Ringdalen E (2018) Coal-based reucing agents in ferroalloys and silicon production. In: Suarez-Ruiz I, Rubiera F, Diez MA (ed) New trends in coal conversion combustion, gasification, emissions, and coking. Woodhead Publishing, Duxford
8. Lindstad T, Gaal S, Hansen S, Prytz S (2007) Improved SINTEF SiO reactivity test. In: Paper presented at INFACON XI, New Delhi, India, 18–21 Feb 2007
9. Ringdalen E, Tveit H, Bao S, Nordnes E (2019) Melting properties of Quartz and their effect on industrial Si and FeSi production. In: Paper presented at international congress & exhibition "Non-ferrous Metals and Minerals", Krasnoyarsk, Russia, 16–19 Sept 2019
10. Schanche T, Tangstad M (2021) Prereduction of Nchwaning ore in $CO/CO_2/H_2$ gas mixtures. Minerals 11(10):1097–1114
11. Menard Y (2021) Packing material in carbon baking furnaces. TRAVAUX 50, proceedings of the 39th international ICSOBA conference, presented online 22–24 Nov 2021, pp 537–549
12. May PH (1999) Natural resource valuation and policy in Brazil: methods and cases. Columbia University Press, New York
13. Jayakumari S, Ringdalen E (2022) Effect of varing SiO contents on Si and FeSi production. In: Paper presented at Silicon for the chemical and solar industy XVI, Trondeheim, Norway, 14–16 June 2022
14. Jayakumari S (2020), Formation and characterization of β-and α-Silicon Carbide produced during Silicon/Ferrosilicon process. Ph.D. thesis, NTNU Norwegian University of Science and Technology
15. Wang L et al (2016) $CO_2$ Reactivity assessment of woody biomass biocarbons for metallurgical purposes. Chem Eng Trans 50:55–60
16. Kaffash H, Surup G, Tangstad M (2021) Densification of biocarbon and its effect on $CO_2$ reactivity. Processes 9(2):193
17. Linstad T et al (2004) The influence of alkalis on the Boudouard reaction. In: Paper presented at Infacon X, Cape Town, South Arica 1–4 Feb 2004
18. Surup G et al (2020) Electrical resistivity of carbonaceous bed material at high temperature. Processes 8(8):933
19. Riva L (2020) Production and application of sustainable metallurgical biochar pellets. Ph.D. thesis, University of Agder
20. Książek M et al (2022) Bio-coke for manganese alloy production. In: Proceedings of sustainable industrial processing summit, presented at 8th International Symposium on Advanced Sustainable Iron & Steel Making, Phuket, Thailand, 27–30 Nov 2022

# Characterizing Bio-carbon for Metallurgical Processes Using Micro X-ray Computed Tomography with High Temperature Experiments

**Stein Rørvik, Nicholas Smith-Hanssen, Sethulakshmy Jayakumari, and Liang Wang**

**Abstract** An important path to the goal of reducing the metal producing industries' $CO_2$ footprint is to replace fossil carbon sources with bio-based carbon sources for the electrodes and reductant agents. Since the structure of bio-carbon is substantially different from fossil carbon, characterizing the bio-carbon structure and examining its behaviour during the relevant processes are important. Focusing on the silicon and ferroalloy industries in Norway, micro X-ray computed tomography (μCT) has been used to analyse and compare single grains of bio-carbon before and after various experimental procedures. These procedures consist of high temperature treatment under different conditions for $CO/CO_2$ and SiO gas reactivity test, K-impregnation and $CH_4$-based carbon deposition. This paper shows examples on results from μCT measurements before and after the experiments, and describes briefly the data processing methods applied. The relevance to the experiments and industrial applications is also discussed.

**Keywords** $CO_2$-neutral metal production · Charcoal · Bio-carbon · SiO-reactivity · μCT

## Background

A majority of the industrial metal producing processes today (Al, Si, ferroalloys) are based on fossil carbon as a raw material, both as a reductant and a conductor (electrode). An alternative to using fossil carbon is using carbon from biological sources (charcoal). Unfortunately, bio-carbon has a very different structure from fossil carbon, so it cannot replace fossil-carbon as is. Because of this, considerable effort has been started on this topic in recent years. As part of this research, micro

S. Rørvik (✉) · N. Smith-Hanssen · S. Jayakumari
SINTEF Industry, Trondheim, Norway
e-mail: stein.rorvik@sintef.no

L. Wang
SINTEF Energy Research, Trondheim, Norway



C. Fleuriault et al. (eds.), *Advances in Pyrometallurgy*, The Minerals, Metals & Materials Series, https://doi.org/10.1007/978-3-031-22634-2_17

X-ray computed tomography (μCT) has been applied to characterize the bio-carbon structure.

A major advantage of the μCT imaging method is that it is non-destructive, thus allowing imaging of the very same sample before and after some type of treatment. The treatment can be part of an industrial process, or a laboratory experiment involving heat treatment combined with some type of controlled gas exposure.

## Imaging Method

The μCT data presented in this paper was acquired by a Nikon XT H 225 ST instrument (cone beam volume CT) at the Department of Physics, NTNU, Trondheim, Norway. A tungsten reflection target was used, with an acceleration voltage of 150 kV and a current of 150 μA. The X-rays were not filtered. The imaging was done with an integration time of 708 ms, signal amplification of 18 dB, with 6283 projections per 360° rotation. The detector panel in the instrument is a PerkinElmer 1620 AN CS model with 2000 × 2000 pixels sized 200 μm × 200 μm; total panel size 40 × 40 $cm^2$.

The distance from the source to the detector was fixed at 1124.22 mm. The source to sample distance was adjusted according to the object size, letting the projection of the sample fill as much as possible of the detector width for maximum resolution, without exceeding it. For the scans presented in this paper, the distance from the source to the sample varied from 32.08 mm to 126.08 mm, resulting in a voxel size varying from 5.7 μm to 22.4 μm. The images were exported as 16-bit TIFF and processed in the public domain software ImageJ [1, 2] using scripts developed at SINTEF. A general description of the method was published in TMS 2017 [3].

The relationship between X-ray intensity and material attenuation (Beer-Lambert's law) is given by Eq. (1) where $I_0$ is the initial X-ray intensity; $I$ is the transmitted X-ray intensity, $\mu$ is the linear attenuation coefficient and $x$ is the distance travelled through the material. The linear attenuation coefficient depends on both the density and the atomic weight of the material; where the atomic weight has a higher influence than the density. The linear attenuation coefficient is directly proportional to density if the atomic weight is constant, which is often the case with carbon. In our workflow, the μCT reconstruction software supplied by the instrument vendor has been set up to calculate $\mu$ for any point in the volume, and export these values as TIFF images where the greyscale values are proportional to the linear attenuation coefficient $\mu$.

$$I = I_0 e^{-\mu x} \tag{1}$$

After exporting the data, a custom written macro in ImageJ reads the metadata files associated with the images, and from this calibrates the voxel values such that they represent the linear attenuation coefficient $\mu$ with unit $cm^{-1}$. The measured values are therefore in principle identical to those reported in the NIST database tables

[4, 5], but in practice the relationship is more complicated as the X-ray radiation is polychromatic with a certain spectrum, and the detected intensity varies with X-ray energy. The spectral sensitivity of the detector has not been disclosed by the panel vendor, so a quantitative relationship between the measured $\mu$ values and material properties (chemical formula) would require empirical data which is outside the scope of this paper.

In simple terms, the greyscale level in the images is roughly proportional to the average atomic weight of the material at each point of the volume. Air is thus black, carbon is medium grey, and the inorganic impurities (mineral inclusions or reaction products) are white. Since charcoal has a lot of micro-porosity, it will appear substantially darker than e.g. metallurgical coke or graphite.

Figure 1 shows an example μCT image of a calcined charcoal grain. The diagonal size of the grain is approx. 17 mm, and it is scanned inside a plastic container of approx. 22 mm inner diameter. The brightest parts are various inorganic deposits related to the nutrient transportation mechanisms of the plant. These are often observed along bands of higher porosity channels in the wood structure. A detailed assessment of the relationship between the charcoal structure and plant biology has not been done, as it is outside the scope of this work. The charcoals presented in this paper are in most cases commercial products obtained from the metallurgic industry.

Figure 2 shows a charcoal grain where the differentiated growth layers from the outer part of the wood are clearly visible. The top part corresponds to the phloem ("inner bark"), the lower part corresponds to the xylem ("sapwood") and in between these lies the thin cambium ("stem cell") layer. Xylem transports water and water-soluble nutrients upwards in the tree, while phloem transports sugars, proteins, and other organic molecules downwards. The large difference in the type of components these layers carry causes the phloem to have a very different structure from the xylem after pyrolysis. In future work, it could be of interest to characterize the elemental content of these layers separately, to assess if a more thorough pre-treatment of the

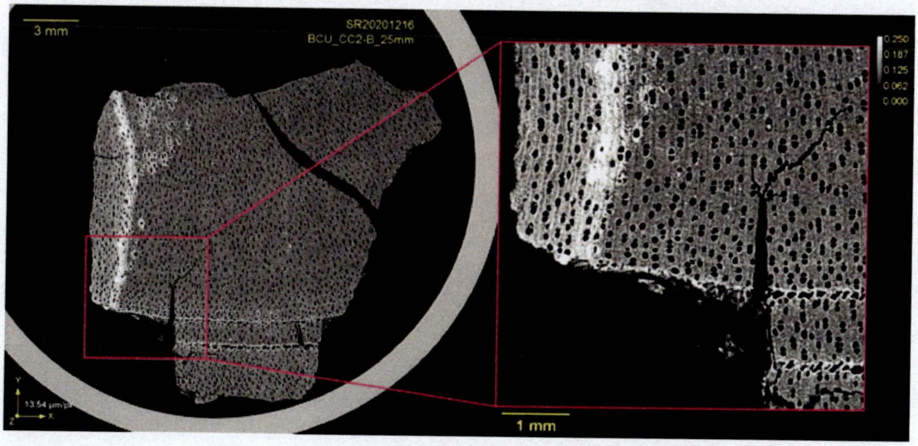


**Fig. 1** Example μCT image of a charcoal grain

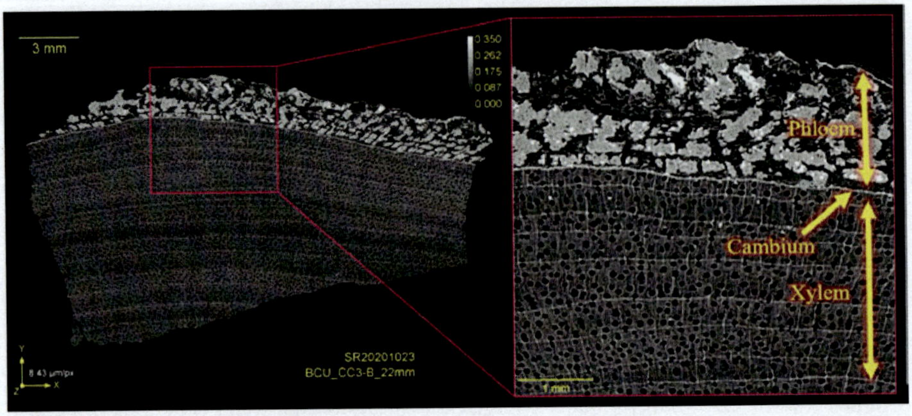


**Fig. 2** Example μCT image of a charcoal grain, with inner bark

wood (removal of bark and inner bark) can enable the use of bio-carbon in metallurgic processes sensitive to certain elemental impurities.

## Data Processing

The main goal of this work has been to observe and measure the changes to the carbon structure from before and after some kind of experimental treatment in the laboratory. This is possible because μCT imaging is completely non-destructive, so the very same grain can be analysed before and after the experiment. Automatic alignment of the pre- and post-experiment datasets turned out to not be straightforward as the commonly used registration methods available are based on landmarks (features that can easily be marked or automatically recognized). Since the charcoal structure usually consists of predominantly carbon, landmarks based on bright spots can often not be found. Landmarks based on pores can also not be used, as most of the pores in charcoal are present as single tubes passing throughout the entire grain, so they do not have any defined start or stop position in the growth direction of the wood. This is as expected, as these pores are the main water transportation channels in the plant. Because of these material related limitations of the data, a custom alignment procedure was designed in the form of an ImageJ macro.

The principle of the custom alignment procedure is based on 2D registration of orthogonal projections, with the aid of a modified version of the TurboReg plugin for ImageJ [6]. This plugin calculates the rigid transform matrix required for the best fit of two images. It can work without landmarks. (The modifications done included exporting the transform matrix as a table, and the plugin code was called directly bypassing its GUI. No changes were made to the fitting algorithm.) The calculated offsets and rotation angles in the *X*- *Y*- and *Z*-directions are then used to shift and rotate the post-experiment dataset to make it match the pre-experiment dataset. The data transformation is done using the GPU-accelerated CLIJ2 plugin for ImageJ [7] to reduce processing time. Since the interpolation methods of CLIJ2 are optimized for speed rather than quality, the final result data was made by applying the calculated affine transform matrix on the original 3D data using the TransformJ plugin for ImageJ [8]. This plugin uses high-quality interpolation.

The registration method for the carbon grains is iterative, outlined stepwise as follows:

1. Three projections of the average attenuation values are calculated for the *X*- *Y*- and *Z*-directions separately, for the pre- and post-experiment samples. (Maximum value projections can also be used, depending on the material)
2. The offset, rotation and scaling values are calculated using TurboReg for each of the pre- and post-pairs of the *X*, *Y* and *Z* projection images, and the equivalent affine transformation matrix is saved
3. The post-dataset is rotated (using CLIJ2) according to the calculated offset and rotation from step 2
4. The process is repeated from step 1, two or three times (depending on the material)
5. The final transform matrix is calculated by multiplying the affine matrices obtained from step 2 and applied to the post-experiment dataset
6. The transformed data is exported both as a 3D volume and coloured overlay images for easy visual presentation.

## High Temperature Experiments

Various high temperature experiments were done in different projects to investigate the behaviour of bio-carbon under experimental conditions replicating the conditions under the relevant industrial processes. μCT was used to characterize the samples before and after these experiments, to observe and measure the changes in the carbon structure. The following describes the type of experiments that were done:

### *Heating in Argon and $CO_2$ Atmospheres*

In both the Si and FeSi furnaces carbon is used as a reductant. The solid carbon will not react with the ore until temperatures are well above 1000 °C. However, the carbon reductant will undergo devolatization and can react with any $CO_2$ through the Boudouard reaction before it starts to react with the ore. To better understand how the carbon material behaves in the furnace it is important to understand the effects of the devolatilization and Boudouard reaction on the physical structure of the carbon reductant. To study the effects of heating on the carbon structure in both inert and $CO/CO_2$ atmospheres, samples were heated in a TGA at 10 °C a minute to 1100 °C under argon. For the tests with only argon, the samples were held for 30 min after which the furnace was shut off and allowed to cool naturally to room temperature under Ar. For the experiments with $CO/CO_2$, the gas was switched to $CO/CO_2$ until a weight loss corresponding to 20% of the fixed C was measured. The samples were then cooled to room temperature under Ar.

### *Potassium-Impregnation*

The effects of potassium on the Boudouard reaction are important for the FeMn industry as in the industrial furnace potassium becomes trapped and is recirculated resulting in concentrations up to 5%. To study the effects of potassium on the carbon reductant potassium was first deposited on the sample. To do this the gas impregnation method was used where the charcoal was placed above an alumina crucible that contained a mixture of carbon and potassium carbide inside a sealed Kanthal crucible. The Kanthal crucible was heated to 1000 °C inside a resistance heated furnace with a flow of 3 lpm of Ar as a carrier gas. Above 1000 °C the potassium carbide and carbon will react to form potassium gas; this potassium gas travels upwards where it is subsequently deposited on the charcoal sample. The samples are held for times between 1 and 3 h depending on the desired potassium content. More details on the experimental setup and results can be found in the work of Kaffash and Tangstad [9] Potassium is an important contaminate in the FeMn furnace as it will catalyse the Boudouard reaction thus its effect is important. The charcoals investigated are commercial products received from the industry.

### *Carbon-Impregnation from $CH_4$*

Increasing the carbon content and mechanical strength is important to increase the usage of charcoal in the Si and FeMn industries. One possible method to achieve this is to deposit carbon from the cracking of $CH_4$. To deposit the carbon the sample is heated in a Kanthal crucible to above 900 °C in a resistance heated furnace and $CH_4$ gas is introduced. The $CH_4$ will subsequently crack and deposit carbon on the surface and pores. The deposition of carbon is found to increase with a longer residence time of $CH_4$ in proximity to the carbon and higher temperatures. This has been shown to increase the mechanical strength and reduce $CO_2$ reactivity. More details on this experimental setup can be found in the work of Kaffash et al. [10]. The charcoals investigated are industrial products delivered from a local silicon producer.

### *Macro TGA Reactivity Test*

This test has the same purpose as the test with heating in argon and $CO_2$ described above, but was done on charcoals intended as reductants in Mn production. The charcoal will undergo devolatization and reactions with surrounding gases at different temperatures, and finally reach the bottom of the furnace to react with the ore. For the Macro TGA reactivity test, the temperature was held at 1100 °C with purging of Ar for 30 min, and then the purging gas subsequently shifted to a mixture of $CO/CO_2$. This work was published by Wang et al. [11]. Two types of bio-carbon were studied with the Macro TGA. The first type of charcoal was produced by running a laboratory scale fixed bed carbonizer on Norwegian birch wood chips under atmospheric pressure without purging of $N_2$. The second type of bio-carbon was produced from the same feedstock, but with the running of a flash carbonizer under pressure. In this treatment, wood chips were ignited by an electrical heating coil at the bottom of a reactor vessel pressurized with air at 7.9 bar. After 6 min of ignition time, compressed air was fed into the top of the reactor and flowed downwards through the feedstock bed. The ignition caused a flash fire of the feedstock and a flame front moved upward, trigging the carbonization of feedstock into charcoal. More details about the reactor and operational procedures for producing bio-carbon through the flash carbonizer can be found in Van Wesenbeeck et al. [12].

## *SINTEF SiO Reactivity Test for Carbon Reductants*

A SiO reactivity test has been developed at SINTEF [13]. It is used to measure the reactivity of carbon materials towards SiO gas, thereby choosing a suitable reductant for Si and FeSi production. The principle of this test is basically a gas analysis technique, where a gas with a known concentration of SiO(*g*) is passed through a packed bed of carbon materials to be tested at a temperature of 1650 °C and $pSiO <$ 0.01 bar. The carbon material will then react with SiO(*g*) to form SiC according to reaction Eq. (2)

$$SiO(g) + 2C = SiC + CO(g) \tag{2}$$

The experimental setup for the SiO reactivity test comprises of three separate alumina chambers: SiO(*g*) generation chamber, reaction chamber and condensation chamber. These are represented schematically in Fig. 3 [13].

The main reactions and distribution of gas flow taking place in each chamber are also shown in Fig. 3.

During the test, the CO in the off gas is continuously monitored. From this, the amount of SiO(*g*) that passes through the carbon bed unreacted can be calculated, and is used as the measure of reactivity. During the start of the test, when the SiO(*g*) starts to react with carbon materials, the off gas contains 18% CO. The test is stopped when CO in the off gas is 6%, indicating that all the SiO(*g*) passes unreacted through the carbon bed. This indicates that the carbon materials in the reaction chamber are completely converted to SiC. The time taken for the concentration of CO to decrease from 18 to 6% depends on the reactivity of the carbon materials, which again depends on the physiochemical properties and structure of the carbon. A $\mu$CT measurement of

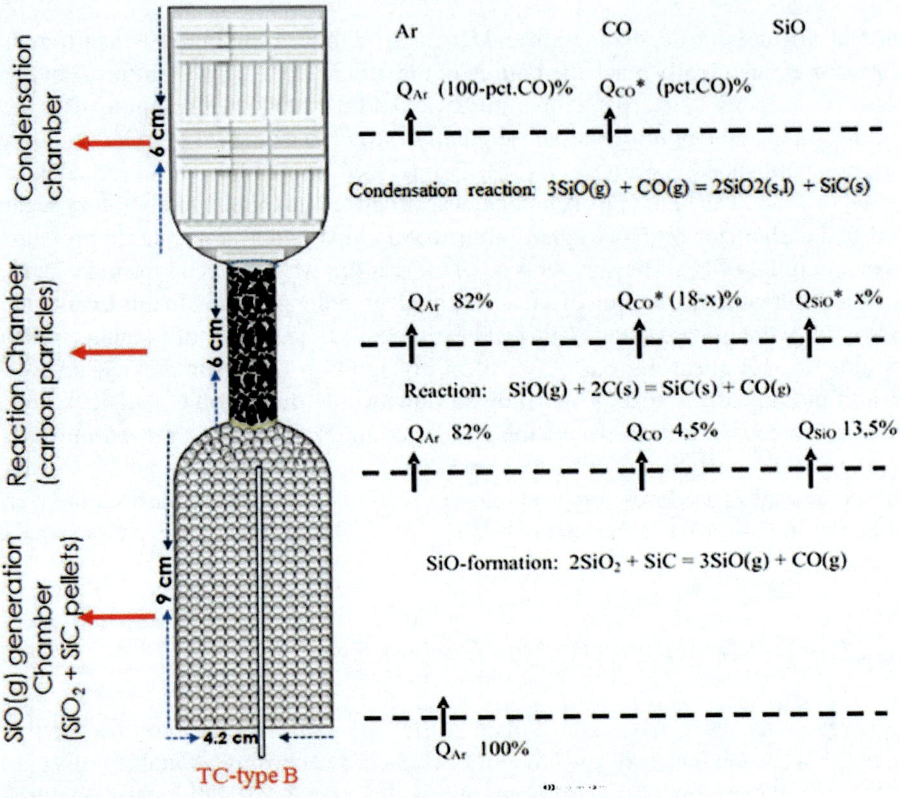


**Fig. 3** SiO reactivity experimental furnace and the reactions involved

the reaction chamber holding C materials (below the top condensation chamber) was done before and after the SiO test. The test was stopped at 15% CO for only a partial conversion in the carbon bed in the reaction chamber. A partial conversion can show if there is a difference in reactivity according to the location in the reaction chamber, or grain-to-grain variation of the carbon structure. Also in this experiment, the charcoals investigated are industrial products delivered from a local silicon producer.

## Qualitative Results

Each of the figures in the following sections consists of three parts: The left part is a cross-section of the μCT data acquired before the experiment. The middle part is a cross-section of the matching plane from the μCT data of the sample acquired after the experiment. The right part is an overlay of these two images where the pre-experiment image has a red colour, and the post-experiment image has a cyan colour. (The sum of red and cyan is grey, so identical parts will appear grey.)

## *Heating in Argon and $CO_2$ Atmospheres*

Figure 4 shows a comparison of pre- and post-experiment charcoal from a hardwood, heated in argon at 1100 °C. The carbon structure is unchanged, expect for an overall uniform shrinkage.

Figure 5 shows a comparison of pre- and post-experiment charcoal from the same type of tropical hardwood, heated for in $CO_2$. This sample shows the same degree of shrinkage as the argon sample, but there is also some significant removal of material from the surface. The pores inside the sample are of the same size after the experiment. This shows that the $CO_2$ reaction predominantly occurs at the sample surface, and not inside the pores.

Figure 6 shows a comparison of pre- and post-experiment charcoal from a softwood, heated for argon. Figure 7 shows the same, using a $CO_2$ atmosphere. The result is similar to the tropical hardwood sample: There is overall shrinkage, but the internal structure is otherwise unchanged after heating in argon atmosphere. In $CO_2$ the sample both shrinks and there is removal of material from the surface. For the softwood, the border of the surface is more diffuse, showing that the gas penetrates somewhat deeper into the pores compared to the hardwood.

Figure 8 shows (for comparison purposes) the same experiment done for a metallurgical coke from fossil sources heated argon, and Fig. 9 shows the same for $CO_2$. There is no significant shrinkage for the coke sample, neither in argon nor $CO_2$ atmosphere. However, there is selective erosion of particular sub-grains after the $CO_2$ experiment. These sub-grains seem to have a different structure or chemistry making them highly reactive to $CO_2$.

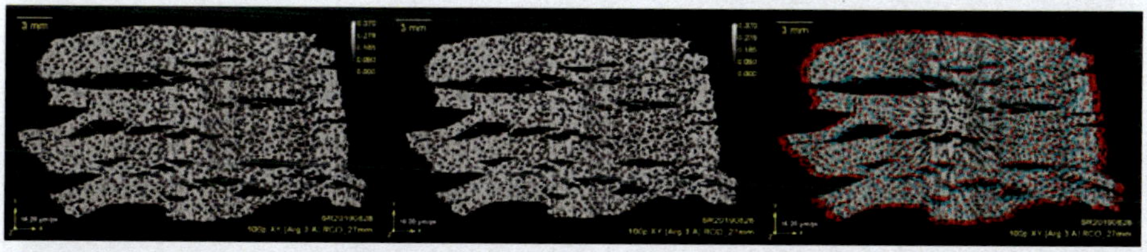

**Fig. 4** Charcoal from hardwood heated in argon at 1100 °C; before (left), after (middle) and comparison (right)

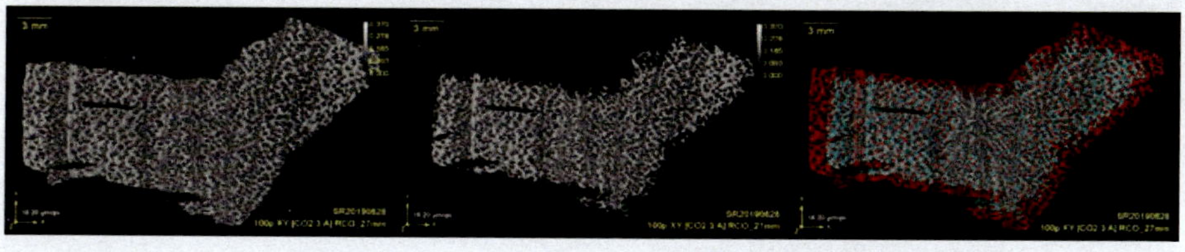

**Fig. 5** Charcoal from hardwood heated in $CO_2$ at 1100 °C

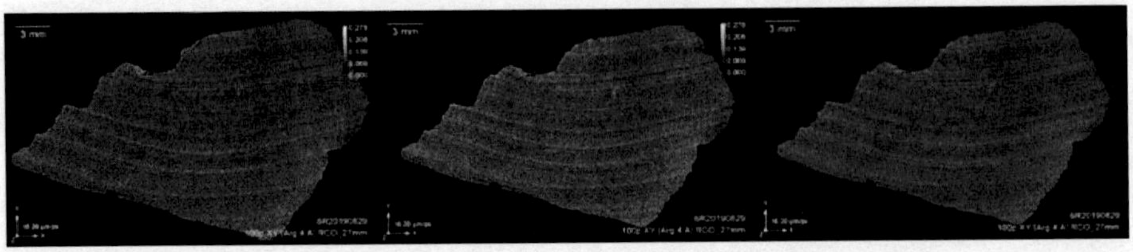

**Fig. 6** Charcoal from softwood heated in argon at 1100 °C

**Fig. 7** Charcoal from softwood heated in $CO_2$ at 1100 °C

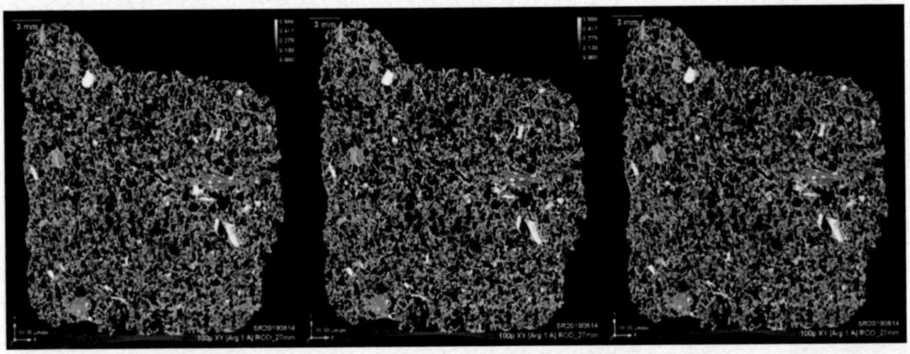

**Fig. 8** Metallurgical coke from fossil sources heated in argon at 1100 °C

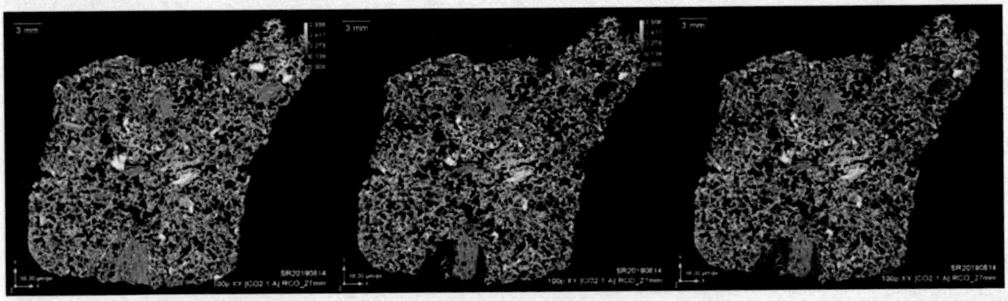

**Fig. 9** Metallurgical coke from fossil sources heated in $CO_2$ at 1100 °C

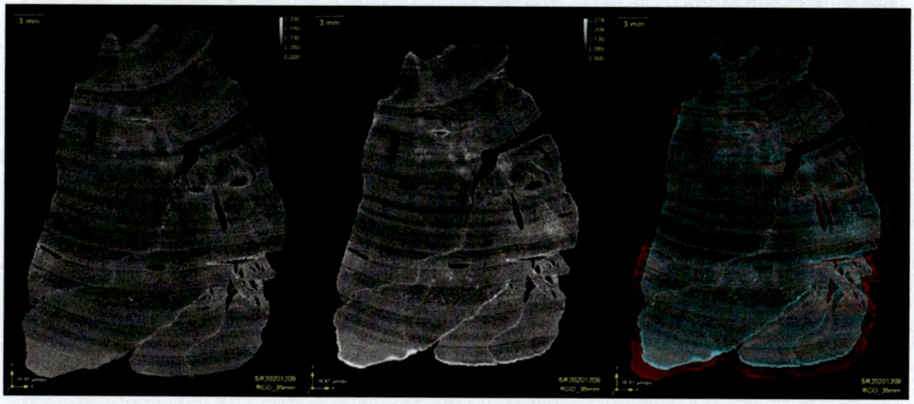

**Fig. 10** Charcoal from softwood impregnated with potassium; before (left), after (middle) and comparison (right)

For all these samples, the carbon structure (hardwood vs softwood, fossil vs bio-coke) seems to matter more than the pore structure. There are no obvious visible differences in reactivity related to the local pore structure of the grain.

## *Potassium-Impregnation*

Figure 10 shows a softwood charcoal from an experiment where the grains were impregnated with potassium. The overall structure has not changed besides some overall shrinkage. The potassium can be seen as a brighter shade of grey. There is an intrusion throughout the sample, but more along the cracks and at the bottom-facing surface of the sample.

## *Carbon-Impregnation from $CH_4$*

Figure 11 shows a charcoal from an experiment where the grains were impregnated by carbon deposited from $CH_4$ gas. There is a large overall shrinkage of the grain, but no obvious density change of the carbon when comparing the post- and pre-experiment images, and no visible density gradient from the surface to the middle of the grain. The carbon deposits are just a few μm thick [10] and thus too small to be seen in the μCT data.

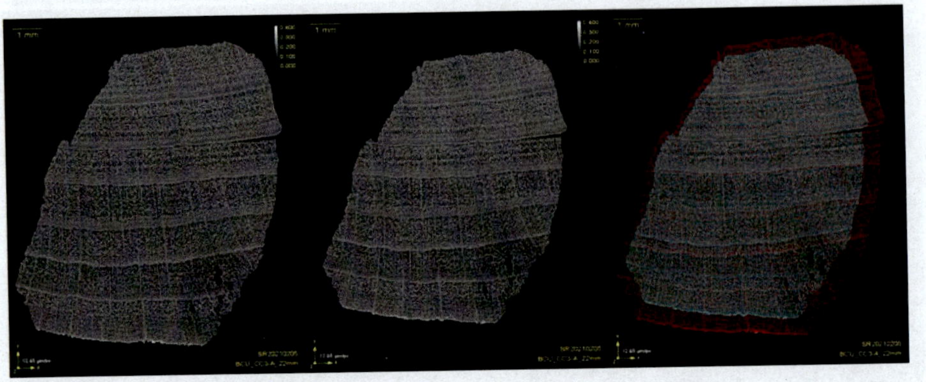

**Fig. 11** Charcoal with carbon deposited from $CH_4$; before (left), after (middle) and comparison (right)

## *SiO Gas Reactivity*

Figure 12 shows a crucible containing charcoal grains before and after the SiO reactivity test. The SiO reactivity causes carbon to convert into SiC, which has higher X-ray attenuation (brighter shade of grey in the μCT images). The left part (vertical cut) and the upper middle part (horizontal cut) is of the crucible before the test. The right part (vertical cut) and the lower middle part is (horizontal cut) is of the crucible after the test. Unfortunately, the grains moved during handling of the crucible, so it is not possible to provide an exact match of the pre- and post-situation of the grains. However, a clear trend can be seen in that the grains in the lower part of the crucible are fully reacted and the middle part is partially reacted. The top part of the grains is also partially reacted, but to a lower degree than the middle part. The reaction progresses as a uniform front from the surface of the grains and inwards, apparently independent of the local porosity.

Figure 13 shows SiO reactivity pre- and post-experiment images of a single charcoal grain from the lower part of the crucible. The grain is fully reacted, and no structural changes nor significant shrinkage can be observed. (Since SiC has a relatively high X-ray attenuation compared to carbon, the grey level scale in the SiO reaction images has a wider range than the previous sections in this paper. The unreacted carbon is therefore relatively dark in these images.)

Figure 14 shows SiO reactivity pre- and post-experiment images of a single charcoal grain from the middle part of the crucible. This grain is also fully reacted. There is no significant shrinkage, but the outermost 0.5–1 mm is removed. There are also many small SiC crystals on the walls of the largest pores.

Figure 15 shows SiO reactivity pre- and post-experiment images of a single charcoal grain from the upper part of the crucible. This grain only reacted at the surface. There is no significant shrinkage or structural changes. There are also here some small SiC crystals on the walls of the largest pores. These crystals are present deeper

**Fig. 12** Crucible before (left + upper middle) and after (right + lower middle) SiO reactivity test

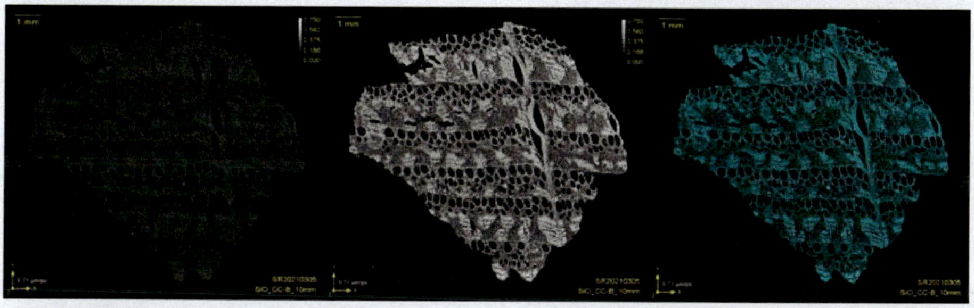

**Fig. 13** Charcoal from SiO reactivity test, bottom of crucible; before (left), after (middle) and comparison (right)

into the charcoal grain than the fully reacted layer, perhaps indicating that these crystals form earlier than the main conversion of the carbon.

For comparison purposes, Fig. 16 shows a fossil based coal grain pre- and post SiO reactivity test, from the middle region of the crucible. The grain is partially converted; the outermost parts are fully converted while the core is not converted. The conversion seems to follow the carbon structure rather than the pores. This figure illustrates the large difference in pore structure of the bio-carbon (Fig. 13) compared to fossil carbon (Fig. 16). The open pore network structure of bio-carbon seems to be beneficial to the carbon/SiO gas reaction.

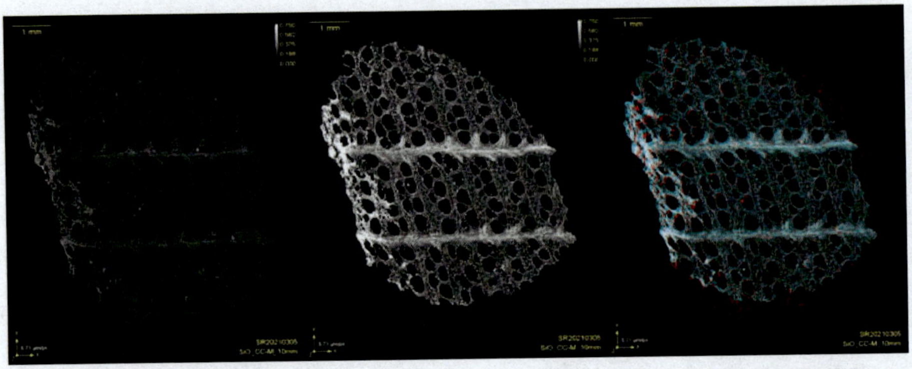

**Fig. 14** Charcoal from SiO reactivity test, middle of crucible

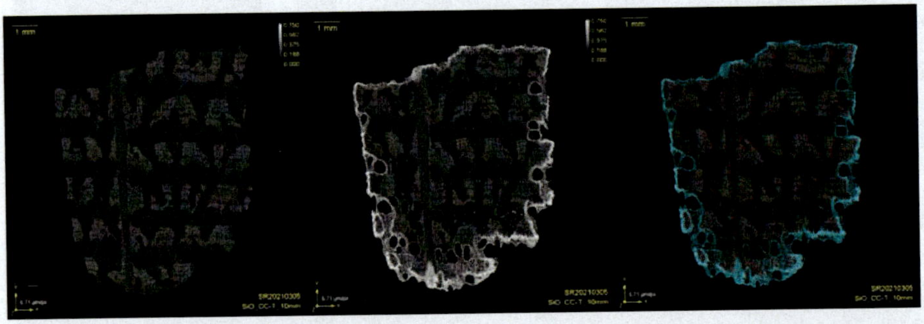

**Fig. 15** Charcoal from SiO reactivity test, top of crucible

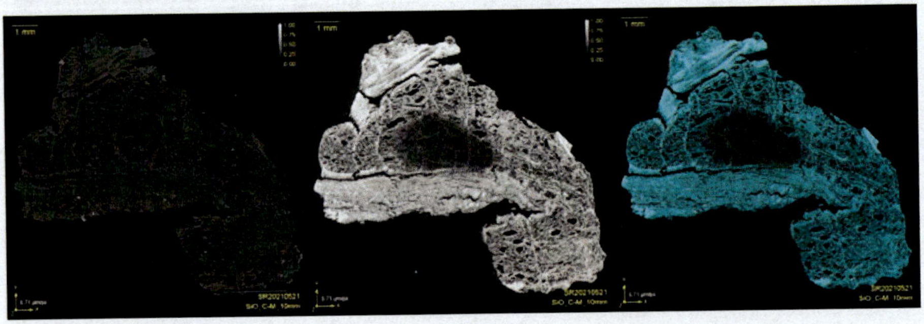

**Fig. 16** Fossil coal from SiO reactivity test, middle of crucible

## *Macro TGA Reactivity Test*

Figure 17 shows a reference charcoal made without purging gas during pyrolysis stage before and after the macro TGA test. The grain has shrunk in the horizontal direction post-testing, but there are no other structural changes.

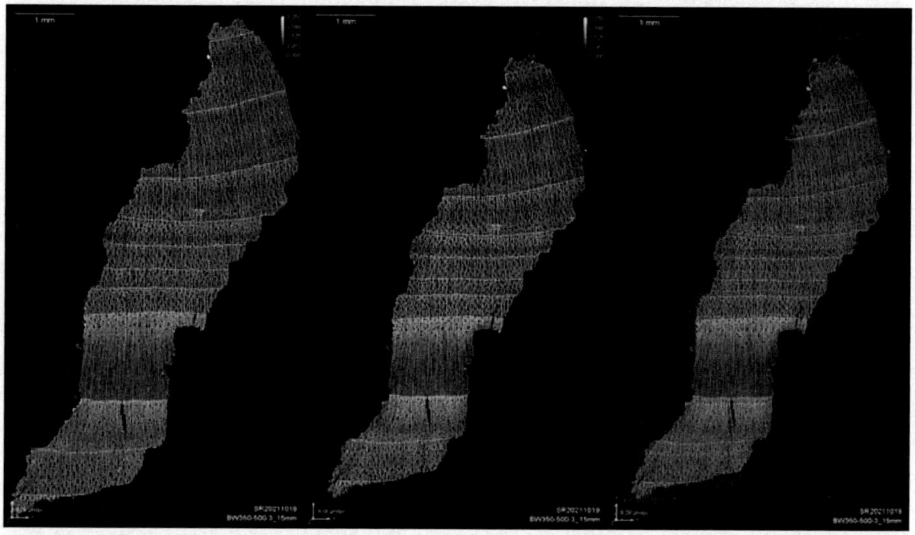

**Fig. 17** Charcoal BW before (left) and after (middle) macro TGA reactivity test; comparison overlay (right)

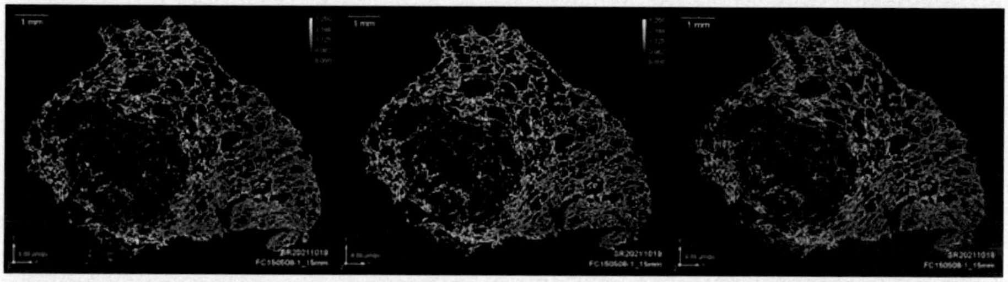

**Fig. 18** Charcoal FC from a pressurized carbonizer before and after macro TGA reactivity test

Figure 18 shows a charcoal from a pressurized carbonizer before and after macro TGA test. The overall structure is different from the reference charcoal. The carbon has a more fused appearance. Post-test, there is some general shrinkage of the grain, and some of the outermost pieces of carbon are lost. The structure of the inner parts of the grain is intact. The shrinkage of the charcoal from a pressurized carbonizer is much less than the normal pyrolysis charcoal.

## Quantitative Results

Table 1 shows an example of image analysis results for all pre- and post-experiment µCT images shown in this paper. "Material shrinkage" is calculated from the scaling factor needed to align the pre- and post-experiment images. The value represents one-dimensional shrinkage (scaling factor); the shrinkage across the sample. The volume shrinkage is larger; the scaling factor is then raised to the power of 3. "Material loss" is calculated from the integrated attenuation values in the post-experiment versus pre-experiment images. The material loss corresponds to total loss, independent of the volume. If the reaction product has higher X-ray attenuation than the carbon (higher average atomic number), the "material loss" value will then be negative. The shrinkage is based on the rigid scaling factor required to make features in the post-experiment sample match the pre-experiment sample, and is thus independent of the degree of material removal. "Density Increase" is based on the sum attenuation, divided by the sample volume. This means that if the grain shrinks but material is not lost, the density will increase. Vice versa, if material is lost but the sample does not change its overall size, the density increase is negative.

Comparing the heating in argon versus $CO_2$ experiments, there is little material loss in the argon atmosphere. The charcoals shrink, thus increasing the density. For $CO_2$ there is a high material loss, higher in the charcoals than the coke. The charcoals have a higher shrinkage than material loss, so there is a net density increase for the charcoals.

The charcoal sample from $CH_4$ carbon deposition has both a significant shrinkage and material loss, due to loss of volatiles (incomplete pyrolysis). There is a small net density increase. The reported increase in strength [10] can both be a result of the observed shrinkage, the loss of volatiles or the carbon deposition (density increase). The coke grains from the carbon deposition experiments showed a large grain-to-grain variation in both shrinkage, material loss and density, so it is difficult to conclude which of these effects are most important. This paper should be viewed more as an example of what can be assessed from μCT measurements than a definitive explanation of the behaviour of the bio-materials.

The charcoal sample from the potassium-impregnation experiment shows some shrinkage, and a large negative material loss and density increase. This is as expected, as K has diffused throughout the structure, increasing the X-ray attenuation.

The charcoal samples from the SiO reactivity test show no shrinkage. The middle position sample appears to have expanded a bit, but upon inspection of the data, the auto-alignment has calculated a slightly wrong rotation around the Y axis, so the intersection plane is a bit larger than it should have been.

The charcoal sample from the macro TGA test shows both shrinkage and material loss. The macro TGA results indicated 15–19 wt% volatile matter in the charcoal samples left after pyrolysis. This volatile matter leaves the sample during the heat treatment in the test, and will thus be part of the measured material loss. The pressurized carbonizer charcoal sample from the macro TGA reactivity test shows a small shrinkage and increased density (negative material loss).

The charcoal produced under pressurized conditions has a more compact structure and a rather different chemical order of carbon matrix than those produced at atmospheric pressure. The charcoal surface is covered by secondary carbon with an ordered structure, which might be less reactive, explaining a smaller shrinkage of

**Table 1** Material shrinkage and material loss of samples post-experiment versus pre-experiment

| Material type | Experiment | Temperature (°C) | Material shrinkage (%) | Material loss (%) | Density increase (%) |
|---|---|---|---|---|---|
| Metallurgical coke | Heating in argon | 1100 | 0.20 | 2.46 | −1.41 |
| Hardwood charcoal | Heating in argon | 1100 | 5.61 | 5.32 | 5.81 |
| Softwood charcoal | Heating in argon | 1100 | 8.13 | 2.48 | 16.20 |
| Metallurgical coke | Heating in $CO_2$ | 1100 | 0.28 | 15.05 | −9.39 |
| Hardwood charcoal | Heating in $CO_2$ | 1100 | 9.83 | 25.85 | 1.72 |
| Softwood charcoal | Heating in $CO_2$ | 1100 | 21.19 | 34.96 | 6.63 |
| Charcoal | Carbon deposition from $CH_4$ | 900 | 11.37 | 18.55 | 2.10 |
| Charcoal | Potassium-impregnation | 1000 | 7.79 | −43.57 | 68.57 |
| Charcoal, lower crucible position | SiO reactivity | 1650 | 0.09 | −248.55 | 209.39 |
| Charcoal, middle crucible position | SiO reactivity | 1650 | −1.86 | −248.98 | 187.17 |
| Charcoal, upper crucible position | SiO reactivity | 1650 | 0.01 | −46.64 | 36.73 |
| Coal, middle crucible position | SiO reactivity | 1650 | −0.20 | −267.46 | 231.66 |
| Charcoal | Macro TGA reactivity test | 1100 | 11.27 | 17.22 | 4.17 |
| Charcoal, pressurized carbonizer | Macro TGA reactivity test | 1100 | 3.02 | −12.00 | 11.01 |

the charcoal grain. In addition, the volatiles remaining in the charcoal after pressurized carbonization might release again during the heating stage with the purging of Ar. With the compact structure, the volatiles in the close pores might react with hot carbon matrix and decompose back to the surface of solid carbon. Cracking and condensation of volatiles on hot carbon matrix might block pores and can thus be observed as an increase of density. But the material loss still occurred for the charcoal produced in pressurized conditions, which are less than those produced at atmospheric conditions with a more open structure and high porosity.

## Summary

This paper shows how μCT can successfully be used as a method for characterizing the structure and behaviour of bio-carbons in metallurgical processes. μCT is non-destructive, and can therefore be applied to the very same sample before and after a laboratory experiment, which examines the combined effect of thermal treatment and gas reactivity on the sample. Automated image analysis can be applied to the μCT 3D data, measuring dimensional changes (shrinkage or expansion) as well as density changes (material removal or deposition). Since the structure of bio-carbon is quite different from fossil carbon, μCT is very useful for observing the structural differences' impact on the metallurgical process of interest.

**Acknowledgements** The authors would like to acknowledge the use of the μCT laboratory of the X-ray Physics Group at NTNU, part of the National Infrastructure NEXT, funded by NTNU, the Norwegian Research Council, and European program INFRAIA (90659800).

## References

1. Rasband WS (1997–2002) ImageJ, U. S. National Institutes of Health, Bethesda, Maryland, USA, https://imagej.nih.gov/ij/
2. Schneider CA, Rasband WS, Eliceiri KW (2012) NIH Image to ImageJ: 25 years of image analysis. Nat Methods 9:671–675
3. Rørvik S, Lossius LP (2017) Characterization of prebake anodes by micro X-ray computed tomography. In: Light metals 2017. The minerals, metals & materials series, pp 1237–1245
4. Berger MJ, Hubbell JH, Seltzer SM, Chang J, Coursey JS, Sukumar R, Zucker DS, Olsen K. XCOM: photon cross sections database. NIST Standard Reference Database 8 (XGAM), NIST, PML, Radiation Physics Division
5. X-ray form factor, attenuation, and scattering tables (NIST Standard Reference Database 66). https://www.nist.gov/pml/x-ray-form-factor-attenuation-and-scattering-tables
6. Thévenaz P, Ruttimann UE, Unser M (1998) A pyramid approach to subpixel registration based on intensity. IEEE Trans Image Process 7(1):27–41
7. Haase R, Royer LA, Steinbach P, Schmidt D, Dibrov A, Schmidt U, Weigert M, Maghelli N, Tomancak P, Jug F, Myers EW (2020) CLIJ: GPU-accelerated image processing for everyone. Nat Methods 17:5–6. https://doi.org/10.1038/s41592-019-0650-1
8. Meijering EHW, Niessen WJ, Viergever MA (2001) Quantitative Evaluation of convolution-based methods for medical image interpolation. Med Image Anal 5(2):111–126
9. Kaffash H, Tangstad M (2022) $CO_2$ gasification of densified biomass: the influence of K on reaction rate. JOM 74:1900–1907. https://doi.org/10.1007/s11837-021-05150-7
10. Kaffash H, Surup GR, Tangsad M (2021) Densification of biocarbon and its effect on $CO_2$ reactivity. Processes 9:193. https://doi.org/10.3390/pr9020193
11. Wang L, Skreiberg Ø, Smith-Hanssen N, Jayakumari S, Jahrsengene G, Rørvik S, Turn S (2022) Investigation of the properties and reactivity of biocarbons at high temperature in a mixture of $CO/CO_2$. Chem Eng Trans 92:697–702. https://doi.org/10.3303/CET2292117

12. Van Wesenbeeck S, Wang L, Ronsse F, Prins W, Skreiberg Ø, Antal MJ (2016) Charcoal "Mines" in the Norwegian woods. Energy Fuels 30:7959–7970
13. Jayakumari S, Ringdalen E (2022) Effect of varying SiO contents on Si and FeSi production. Silicon for the chemical and solar industry XVI, Trondheim, 14–16 June 2022. https://papers.ssrn.com/sol3/papers.cfm?abstract_id=4118642

# Part IV
# Energy Valorization in Metallurgy

# The Use of Concentrating Solar Energy for Thermal Decomposition in Oxide and Carbonate Minerals

Lina Hockaday and Quinn Reynolds

**Abstract** Concentrating solar energy can deliver high temperature process heat to metallurgical processes. An overview of mineral resources in South Africa and Australia and the possibilities of using concentrating solar thermal energy for thermal decomposition as novel low-carbon pre-treatment processes are investigated. The paper will consider the thermodynamics of oxide and carbonate minerals and evaluate the potential carbon emission reductions as well as changes in energy demand of such processes. The state of the art of concentrating solar technologies for materials treatment as applied to the thermal decomposition reactions are reviewed.

**Keywords** Pyrometallurgy · Process technology · Extraction and processing · Solar materials processing · Decarbonisation

## Introduction

Thermal decomposition reactions take place when compounds disassociate into products without the need for additional reactants. A well-known example of a thermal decomposition reaction is the calcination of limestone or calcite to lime. Calcination is the process of heating materials to high temperatures in the presence of oxygen or air, enabling the thermal decomposition reactions. The required high temperatures range from 600 to 1400 °C, depending on the application, but are generally in the order of 800–1000 °C. In this paper, the reaction enthalpy, $\Delta H$, at standard conditions of 25 °C and pressure of 1 atmosphere was determined using the HSC Reaction database [1, 2].

$$CaCO_3 \rightarrow CaO + CO_2(g);\ \Delta H = +178\,\text{kJ} \tag{1}$$

L. Hockaday (✉)
Senior Engineer, Minerals, Energy and Chemical Engineering, Curtin University, Western Australia School of Mines, Kent street, Bentley, WA 6102, Australia
e-mail: lina.hockaday@curtin.edu.au

Q. Reynolds
MINTEK, Pyrometallurgy Division, Mintek, 200 Malibongwe Drive, Strijdom Park, Randburg, South Africa



C. Fleuriault et al. (eds.), *Advances in Pyrometallurgy*, The Minerals, Metals & Materials Series, https://doi.org/10.1007/978-3-031-22634-2_18

Thermal decomposition is also often used to remove bound water from hydrated minerals. The dehydration reaction may be seen in reaction (2) using goethite as an example.

$$2FeO.OH \rightarrow Fe_2O_3 + H_2O(g);\ \Delta H = +56\,kJ \tag{2}$$

Another use of thermal decomposition reactions is in reduction reactions such as the generic reaction (3).

$$M_xO_y \rightarrow M_xO_{y-1} + \frac{1}{2}O_2 \tag{3}$$

Reduction reactions may be used as a pre-treatment step to reduce ores before smelting, lowering the reductant requirement of subsequent treatment. These reactions are also often the first step in thermochemical water-splitting. An in-depth review of metal oxides applied to thermochemical water-splitting for hydrogen production using concentrated solar energy may be found in literature [3] and will not be considered in this paper. The use of thermal decomposition to achieve reduction of metal oxides is generally not applied in mineral processing as these reactions require higher temperatures than reduction reactions. It may be time to rethink this, as the use of fossil-fuel based reductants becomes more undesirable while concentrating solar technologies mature and become more readily available.

## Overview of Minerals Resources in South Africa and Australia

South Africa and Australia are countries that are both known for their mineral resources. Both countries also receive high levels of annual solar irradiation, as shown in and. The economics for concentrating solar thermal energy is very dependent on the solar resource and for this paper we will consider regions with annual solar irradiance of above 2500 $kWh/m^2$ as possible locations for concentrating solar facilities (Figs. 1 and 2).

In South Africa, this region includes ore deposits containing copper, lead, zinc, iron, manganese, andalusite, baryte, diamonds, gypsum, limestone, rare earths, tantalum, silver, tin, and uranium [5, 6]. Of these resources, the following are actively being mined [7]: lead, copper, zinc, gypsum, diamonds, iron ore, manganese ore, and limestone for cement production. The mineral resources of South Africa are shown in Fig. 3, with the high resolution map being available *online* (https://www.geoscience.org.za/images/Maps/rsadeposits.gif) [5].

The mineral resources for Australia are shown in Fig. 4 together with mining and minerals processing operations and their energy intensity [8, 9]. In the area with very high solar insolation, the mining operations for the commodities of iron, gold, copper, uranium, nickel, lead, zinc, and silver are in operation.

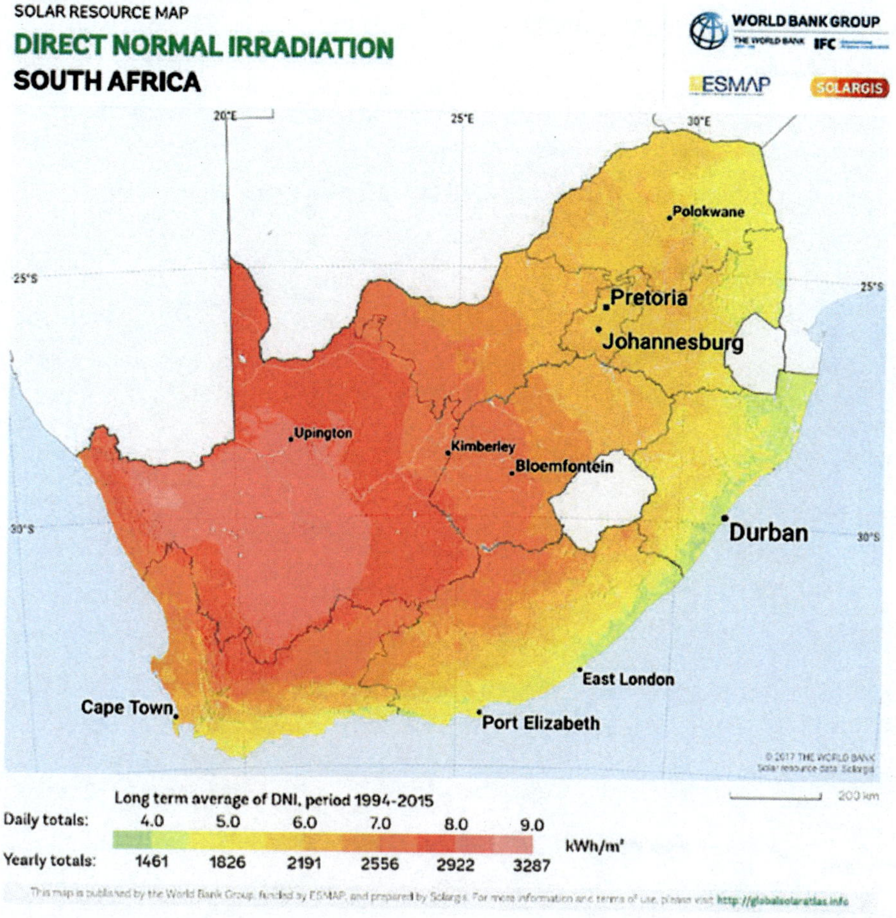


Fig. 1 Annual direct solar irradiation for South Africa [4]

## Possible Uses of Concentrating Solar Thermal Energy in Thermal Decomposition

Concentrating solar technologies, CST, can be used for heating, electricity production, and steam generation. Articles and reviews of the advances in solar thermal technologies, for process heat, describes the solar collectors and receivers as well as their integration into industrial processes [10–14].

Eglington et al. [9] discussed potential applications of concentrated solar thermal technologies in the Australian minerals processing and extractive metallurgical industry. They identified thermal decomposition CST applications in alumina calcining, nickel drying and calcining, cement production, and magnesia production, see Table 1. Thermal decomposition reactions do not feature significantly in the

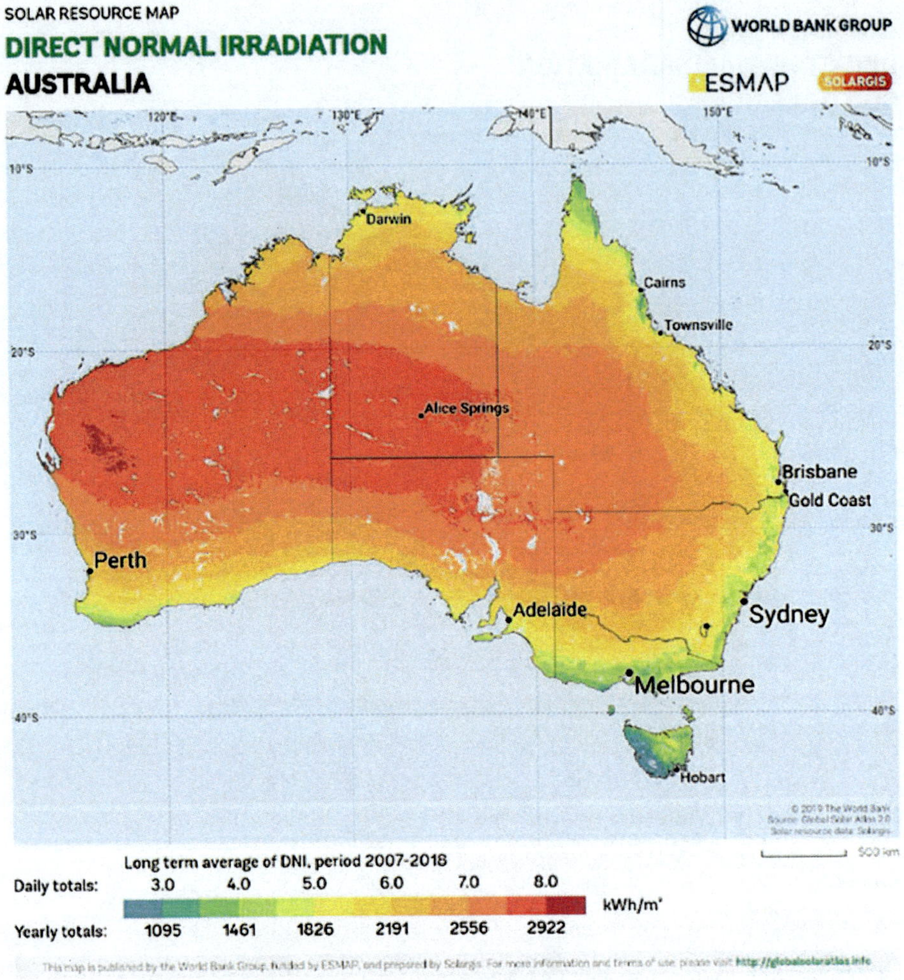


**Fig. 2** Annual direct solar irradiation for Australia [4]

processing of the other commodities or energy requirements are met by alternative sources such as to oxidation of sulphide ores in copper production.

## *Alumina Calcination*

The calcination of alumina is a dehydration process proceeding as follows:

$$2Al(OH)_3 \rightarrow \alpha Al_2O_3 + 3H_2O, \Delta H = +185.2\,kJ \quad (4)$$

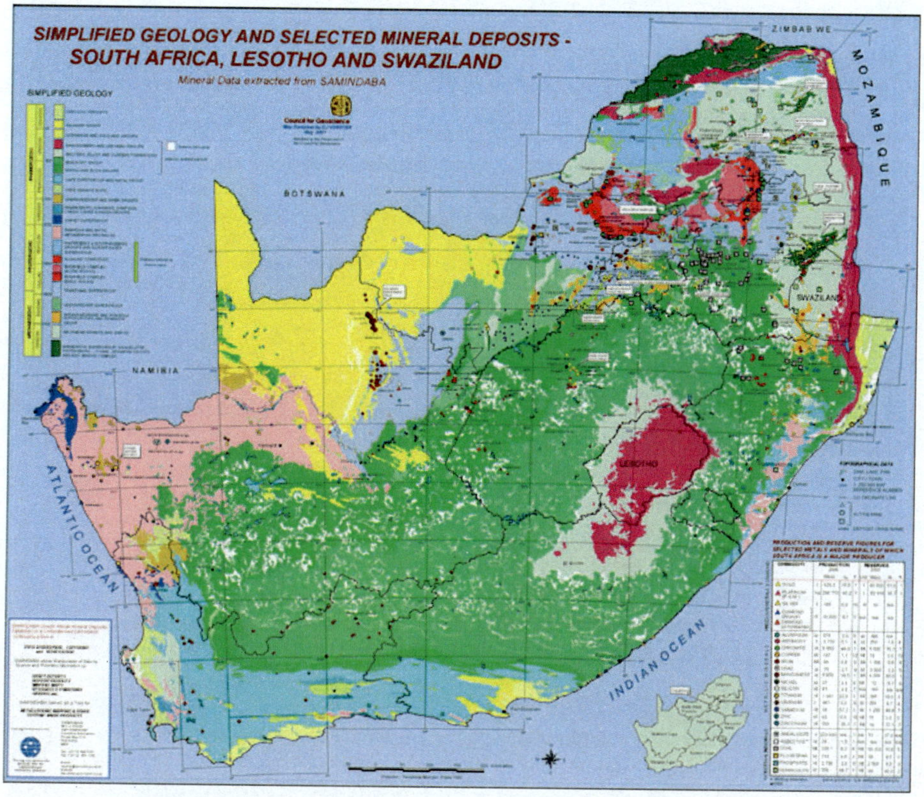


**Fig. 3** South African mineral resources [5]

The use of a solar-driven reactor to calcine alumina has been reported [15] to yield chemical conversions of up to 95.8% for nominal residence times of approximately 3 s. This technology could avoid the use of combustion derived $CO_2$ emissions for the calcination stage of the Bayer process if successfully scaled up.

## *Cement Calcination*

In addition to the carbon dioxide released upon calcination of lime for cement manufacturing, the combustion of fossil fuels for heating the calcination kiln generates about 35% of the greenhouse gas emissions of the Australian cement industry [9]. The use of CST for the production of lime has been extensively documented [16, 17] and proved technically feasible. The described type of solar kiln could be used together with conventional burner technology and fossil or renewable fuel for continuous operation.

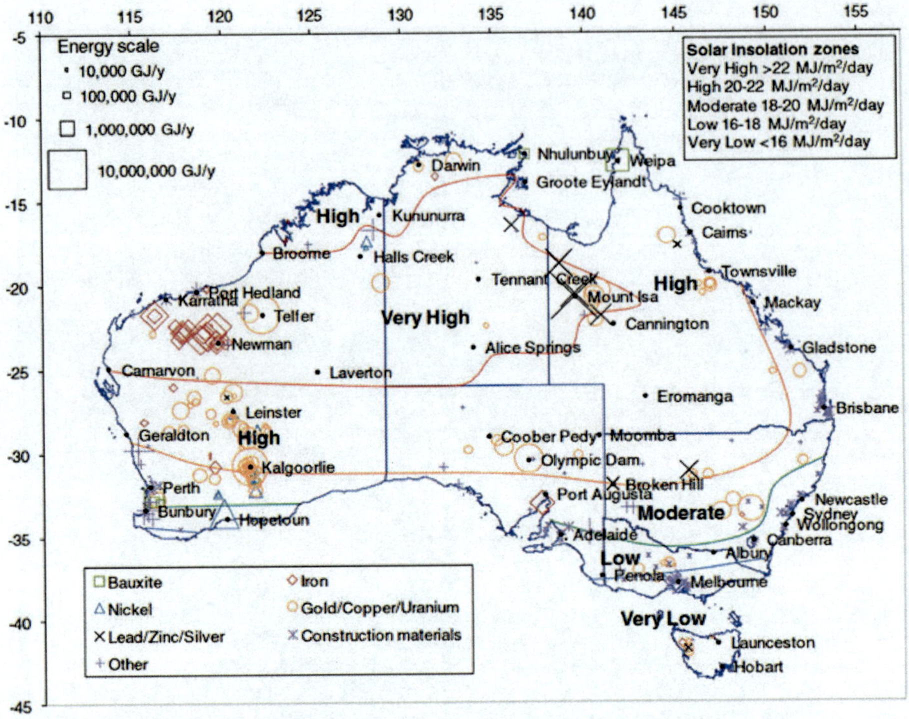


**Fig. 4** Approximate daily insolation levels averaged over a year for Australia compared with the location and energy intensity of mining and minerals processing operations from [9] after [8]

**Table 1** Summary of potential high temperature thermal decomposition applications of CST in minerals processing and metallurgical industries in Australia and South Africa

| Metal/Mineral | Characteristic process operation | Characteristic temperature (°C) | Fuels reactants |
|---|---|---|---|
| Alumina | Calcining | 1000 → 1100 | Natural gas |
| Cement | Calcining | 1000 → 1450 | Natural gas/coal/biomass |
| Nickel | Calcining | ≈1100 | Coal |
| Magnesia | Calcining | ≥1000 | Natural gas |
| Manganese | Pre-reduction | ≈600 | Furnace off-gas/coke |
| | Calcination | >950 | Coke |
| | Sintering | 1200 → 1300 | Coke |

## *Nickel Calcination*

According to Pickles [18], nickel oxide ores can contain up to about 45% water, both as free moisture and as chemically combined water in hydrated minerals. These ores are generally dried and calcined if processed by the pyrometallurgical ferronickel process or the Caron process, but not for the hydrometallurgical high pressure acid leach process. The predominant mineral in nickel laterite ores is goethite (FeO.OH) and reaction (2) is therefore the applicable thermal decomposition reaction. For nickel saprolite ores the predominant mineral is serpentine, ($Mg_3Si_2O_5(OH)_4$), and reaction (4) is applicable. If the decomposition takes place in the presence of solid reductant coal, pre-reduction is also achieved with $Fe^{3+}$ to $Fe^{2+}$ and some $Ni^{2+}$ to metallic nickel [19, 20]. The drying kinetics of nickeliferous limonitic laterite ores are described in detail by Pickles [18]. Drying is concluded to be boundary layer controlled below 100 °C and diffusion controlled at higher temperatures.

$$Mg_3Si_2O_5(OH)_4 \rightarrow 3MgO + 2SiO_2 + 2H_2O(g);\ \Delta H = +251\,\text{kJ} \quad (5)$$

According to thermodynamic simulations [20], roasting of nickel laterite and saprolite ores can achieve significant Ni recoveries to a metallic phase at roasting temperatures below 1000 °C, but slow reaction kinetics may require higher operating temperatures in practise.

## *Magnesia*

Magnesia is produced in a similar calcination process to lime from the thermal decomposition of magnesite in the reaction:

$$MgCO_3 \rightarrow MgO + CO_2(g),\ \Delta H = +101\,\text{kJ} \quad (6)$$

Magnesia is manufactured in three different grades defined by the calcination temperature. The 3 grades are:

1. dead burned MgO, treated at temperatures above 1400 °C, sometimes up to 2000 °C
2. hard burned MgO, calcined at temperatures of 1000–1400 °C
3. light burned MgO, also called caustic-calcined magnesite, calcined at 700–1000 °C.

Of these products, light burned magnesia is the preferred product due to its high reactivity and use in cement production. It requires longer residence times due to the lower reaction rates at lower temperatures in order to achieve the high specific surface area required [21]. At higher temperatures the magnesia crystals grow and eventually sinter, leading to reduced specific surface area as particles agglomerate and change in morphology.

Although Australia's magnesia production is located in Rockhampton, Queensland with average daily insolation levels of only 5–5.6 kWh/m$^2$, the solar thermal calcination of magnesite could be achieved in solar receivers described by Meier [17, 22] according to Eglington et al. [9].

## *Manganese Pre-reduction and Calcination*

The Kalahari manganese fields in South Africa and the Groote Eyland Manganese ores can vary in composition according to grade and mineralogy [23–26]. The Groote Eyland manganese ores are predominantly pyrolusite ores, ($MnO_2$), with some cryptomelane. Pyrolusite decomposes to lower oxides upon heating in air according to the following reactions:

$$2MnO_2 \rightarrow Mn_2O_3 + \frac{1}{2}O_2(g);\ \Delta H = +81\,\text{kJ} \tag{7}$$

$$3Mn_2O_3 \rightarrow 2Mn_3O_4 + \frac{1}{2}O_2(g),\ \Delta H = +110\,\text{kJ} \tag{8}$$

$$Mn_3O_4 \rightarrow 3MnO + \frac{1}{2}O_2(g),\ \Delta H = +228\,\text{kJ} \tag{9}$$

The reaction kinetics of reactions (7) and (8) have been reported by Terayama et al. [27], while reaction kinetics of reactions (8) and (9) have been published by Alonso et al. [28] and Botas et al. [29] as observed in conventional non-isothermal thermogravimetric set-ups. Reaction (7) occurs between 550 and 650 °C while reaction (8) only proceeds at temperatures above 800 °C and reaction (9) proceeds at temperatures above 1400 °C. It should be noted that in the presence of solid carbon or carbon monoxide, the reduction reactions are more stable (have a lower Gibbs free energy) and will take place preferentially. In general, it appears that reduction of manganese oxides to $Mn_3O_4$ is possible at temperatures below 1000 °C through thermal decomposition. Such a pre-reduction may lower the carbon demand for subsequent processing of $MnO_2$ ores (for reduction reactions) by 50%. Braunite minerals are however more stable than pyrolusite and bixbyite and only partially decompose. It should be remembered that the reduction in carbon demand will be offset by higher overall process energy demand as most of the manganese oxide reduction reactions with solid carbon or carbon monoxide are exothermic.

The manganese ores from the Kalahari manganese field contains mainly bixbyite ($Mn_2O_3$), braunite ($Mn_7SiO_{12}$), and hausmannite ($Mn_3O_4$), with calcite minerals [23, 30–32] as impurities. The merit of thermal decomposition will therefore depend on the ore mineralogy. Several on-sun studies on manganese ore samples of 200–500 g have also been published [33–35] with calcination levels above 80% reported for ores containing 15–25% calcite. This pre-treatment will lower the energy requirement for subsequent smelting as these endothermic reactions will no longer occur in the ferromanganese smelter.

## Conclusion

The benefit of using CST for thermal decomposition reactions are highly dependent on the location of implementation, the ore mineralogy of the materials treated and the scale of implementation. The potential exists to replace fossil fuels usage in calcination applications and in selected pre-reduction applications, but the challenge is to scale up the promising research already done to commercial readiness.

Calcination of gibbsite, nickel laterites, calcite, and manganese ores appear as promising applications in South Africa and Australia, given the location of the deposits, thermodynamics of the reactions and the existing state of the art of CST. The main challenge remains to scale technologies to enable demonstration of larger through-put.

## References


1. Outotec (2019) HSC Chemistry. https://www.outotec.com/products/digital-solutions/hsc-chemistry/. Accessed 18 May 2019
2. Roine A et al (2002) Outokumpu HSC chemistry for windows. Report for Outotec Research Oy, Finland, 02103-ORC-T
3. Abanades S (2019) Metal oxides applied to thermochemical water-splitting for hydrogen production using concentrated solar energy. ChemEngineering 3(3). https://doi.org/10.3390/chemengineering3030063.
4. SolarGIS (2019) iMaps. Solar resource maps and GIS data for 200+ countries. https://solargis.com/maps-and-gis-data/overview/. Accessed 27 Feb 2019
5. Council for Geoscience (2022) Council for Geoscience Interactive Web Map. https://maps.geoscience.org.za/portal/apps/sites/#/council-for-geoscience-interactive-web-map-1. Accessed 7 Sept 2022
6. Voster CJ et al (2005). Simplified geology, selected mines and mineral deposits South Africa, Lesotho and Swaziland. Council for Geoscience, Pretoria, South Africa. http://www.geoscience.org.za/images/DownloadableMaterial/RSA_Mineral_Map.pdf. Accessed 7 Sept 2022
7. Department of Mineral Resources and Energy (2022) Operating mines. https://www.dmr.gov.za/mineral-policy-promotion/operating-mines. Accessed 7 Sept 2022
8. Beath AC (2012) Industrial energy usage in Australia and the potential for implementation of solar thermal heat and power. Energy 43(1):261–272. https://doi.org/10.1016/j.energy.2012.04.031.
9. Eglinton T et al (2013) Potential applications of concentrated solar thermal technologies in the Australian minerals processing and extractive metallurgical industry. JOM 65(12):1710–1720. https://doi.org/10.1007/s11837-013-0707-z
10. Amsbeck L et al (2016) Particle tower technology applied to metallurgic plants and peak-time boosting of steam power plants. In: Paper presented at SolarPACES2016 Cape Town, South Africa. https://doi.org/10.1063/1.4949148
11. Ho CK (2017) Advances in central receivers for concentrating solar applications. Sol Energy 152:38–56. https://doi.org/10.1016/j.solener.2017.03.048
12. Kodama T et al (2017) Particle reactors for solar thermochemical processes. Sol Energy 156:113–132. https://doi.org/10.1016/j.solener.2017.05.084
13. Kumar L et al (2019) Global advancement of solar thermal energy technologies for industrial process heat and its future prospects: A review. Energy Convers Manage 195:885–908. https://doi.org/10.1016/j.enconman.2019.05.081

14. Moumin G et al (2019) Solar treatment of cohesive particles in a directly irradiated rotary kiln. Sol Energy 182:480–490. https://doi.org/10.1016/j.solener.2019.01.093
15. Davis D et al (2017) Solar-driven alumina calcination for $CO_2$ mitigation and improved product quality. Green Chem 19(13):2992–3005. https://doi.org/10.1039/C7GC00585G
16. Koepf E et al (2016) A review of high temperature solar driven reactor technology: 25years of experience in research and development at the Paul Scherrer Institute. Appl Energy 188:620–651. https://doi.org/10.1016/j.apenergy.2016.11.088
17. Meier A et al (2006) Solar chemical reactor technology for industrial production of lime. Sol Energy 80(10):1355–1362. https://doi.org/10.1016/j.solener.2005.05.017
18. Pickles CA (2003) Drying kinetics of nickeliferous limonitic laterite ores. Miner Eng 16(12):1327–1338. https://doi.org/10.1016/S0892-6875(03)00206-1
19. Wang F et al (2023) An integrated process of $CO_2$ mineralization and selective nickel and cobalt recovery from olivine and laterites. Chem Eng J 451. https://doi.org/10.1016/j.cej.2022.139002
20. Elliott R et al (2016) Thermodynamics of the reduction roasting of nickeliferous laterite ores. JMMCE 04(06):320–346. https://doi.org/10.4236/jmmce.2016.46028
21. Ebrahimi-Nasrabadi K et al (2013) Time-temperature-transformation (TTT) diagram of caustic calcined magnesia. CIM J 6(1):42–50
22. Ebert et al (2019) Operational experience of a centrifugal particle receiver prototype. In: Paper presented at SolarPACES2018, Casablanca, Morocco, 2019. https://doi.org/10.1063/1.5117530.
23. Sorensen B et al (2010) Properties of manganese ores and their change in the process of calcination. In: Paper presented at the twelve international ferroalloys conference, Helsinki, Finland, 6–9 June 2010
24. Tangstad M et al (1995) The ferromanganese process- material and energy balance. In: Paper presented at InfaconVII, Trondheim, Norway, 11–14 June 1995
25. Ringdalen E et al (2010) Ore melting and reduction in silicomanganese production. Metall and Mater Trans B 41(6):1220–1229. https://doi.org/10.1007/s11663-010-9350-z
26. Chetty D (2008) A geometallurgical evaluation of the ores of the northern Kalahari manganese deposit, South Africa. Ph.D. thesis, University of Johannesburg
27. Terayama K et al (1983) Study on thermal decomposition of $MnO_2$ and $Mn_2O_3$ by thermal analysis. Trans JIM 24(11):754–758. https://doi.org/10.2320/matertrans1960.24.754
28. Alonso E et al (2013) Kinetics of $Mn_2O_3$–$Mn_3O_4$ and $Mn_3O_4$–MnO redox reactions performed under concentrated thermal radiative flux. Energy Fuels 27(8):4884–4890. https://doi.org/10.1021/ef400892j
29. Botas JA et al (2012) Kinetic modelling of the first step of $Mn_2O_3$/MnO thermochemical cycle for solar hydrogen production. Int J Hydrogen Energy 37(24):18661–18671. https://doi.org/10.1016/j.ijhydene.2012.09.114
30. McRae LB (1979) The prereduction of Mamatwan-type ores. Report for National Institute for Metallurgy, Randburg, South Africa
31. Eric RH et al (1992) The mechanism and kinetics of the carbothermic reduction of mamatwan manganese ore fines. Miner Eng 5(7):795–815. https://doi.org/10.1016/0892-6875(92)90247-7
32. Steenkamp JD et al (2015) Characterisation of two UMK manganese ore samples and one briquette sample. Report for MINTEK, Randburg
33. Hockaday L et al. Solar thermal treatment of manganese ores. In: Paper in AIP conference proceedings 2033, Santiago, Chile. https://doi.org/10.1063/1.5067152
34. Hockaday L et al (2019) The solar thermal treatment of manganese ore pellets using closed-loop forced convection of air. In: Paper presented at SolarPACES2018, Casablanca, Morocco. https://doi.org/10.1063/1.5117659
35. Hockaday L et al (2021) A comparison of direct concentrating solar thermal treatment of manganese ores to fossil fuel based thermal treatments. In: Paper presented at the sixteenth international ferro-alloys congress, Trondheim, Norway, 27–29 Sept 2021. https://doi.org/10.2139/ssrn.3926254

# Linde's Industrial Gas Technology in Nonferrous Processing: Combining CFD with Partial Experimental Verification and Validation

**William Mahoney, Adrian Deneys, Jiaye Gan, and Ahmed Abdelwahab**

**Abstract** In primary and secondary nonferrous sectors, industrial gases and associated injection equipment are applied to oxyfuel and pneumatic injection technologies. The former is applied to decarbonization through the improved utilization of alternative fuels including a growing interest in hydrogen. The latter is concerned with gas jetting for metal mixing and refining. CFD is used to support the design, installation and start-up of oxyfuel burners and gas injection equipment. Verification and validation (V&V) in CFD output can be enhanced via experimental efforts; however, the extraordinary difficulty in taking measurements in production furnaces under hot conditions is fully appreciated. We present an approach on a partial V&V method relying on basic assumptions in the computational and experimental domains, wherein posteriori error estimation is partially fulfilled through experiment. We draw upon industrial installations in the following reactors: fluidized bed roaster (g-g), cylindrical-horizontal furnaces (g-l) top-blown and submerged) and straight grate/rotary kilns (g-s).

**Keywords** Modeling and simulation · Sustainability · Recycling and secondary recovery · CFD verification and validation

## Extended Abstract

Linde's gas applications technology personnel are fully concerned with industrial trials and commercial development programs within the high-temperature industries. In the primary (smelting) and secondary (recycling) nonferrous sectors, Linde's industrial gases and associated injection equipment are applied to oxyfuel burners and pneumatic injection technologies. The former is today often applied to decarbonization efforts through the increased use of alternative fuels (of all forms), including a growing interest in industrial hydrogen trials. The latter is generally concerned with

W. Mahoney (✉) · A. Deneys · J. Gan · A. Abdelwahab
Linde Technology Center, Tonawanda, NY 14150, USA
e-mail: William.j.mahoney@linde.com



C. Fleuriault et al. (eds.), *Advances in Pyrometallurgy*, The Minerals, Metals & Materials Series, https://doi.org/10.1007/978-3-031-22634-2_19

controlled gas jetting into industrial furnaces for refining, mixing power, oxidation, and other purposes.

Linde utilizes computational fluid dynamics (CFD) to support the design, installation, and operation of oxyfuel burners and pneumatic injection equipment. These studies further contribute towards risk minimization and the reduction of time required towards process optimization. Verification and validation (V&V) in CFD output can be enhanced via physical/experimental efforts; however, the extraordinary difficulty in taking direct measurements in production furnaces under hot conditions is fully appreciated. We present an approach on a partial V&V method relying on basic assumptions in the computational and experimental domains, wherein *posteriori error estimation* is partially fulfilled through laboratory methods.

Examples illustrating gas-into-gas (g-g), gas-into-liquid (g-l) and gas-into-solid (g-s) physical fields are described. We draw upon industrial examples in the following reactors: fluidized bed roaster (g-g), cylindrical-horizontal furnaces (g-l) (top-blown and submerged) and straight grate/rotary kilns (g-s). An industrial trial concerning the optimization of alternative fuel utilization is also reviewed (oxyfuel).

(g-g): Zinc calcine (ZnO) can be produced from zinc sulfide (ZnS) concentrates by oxidative roasting in fluidized bed roasters (FBR). Air flow through the FBR nozzle plate serves to fluidize the particulate bed, which also provides the oxygen required for conversion of the concentrate to calcine. The conical section below the nozzle plate from where the air is supplied to the particulate bed is known as the wind-box.

It is well known that oxygen gas can be admixed with the wind-box air to realize several process benefits. Oxygen enriched air can be utilized to increase production rate through the roaster while maintaining or even reducing residual sulfur in the calcine. The use of oxygen also affords the operator the ability to control the superficial velocity through the bed by throttling down air flow and using technical oxygen to balance the oxygen coefficient requirement of the concentrate. This is particularly effective when the concentrate contains high fraction of fines, which are prone to being carried out of the roaster as carry-over.

The fresh concentrate is injected (thrown) into the FBR via slinger feeder which then falls as a particulate stream into the roaster where it is subsequently mixed with the reacting bed. The region where the concentrate stream enters the bed is known as the feed-zone, illustrated schematically in Fig. 1 (top-right). The feed-zone is known to contain an oxygen deficiency relative to the rest of the particulate bed, which can cause issues such as accretion/build-up. It is desirable to increase the rate of oxygen supplied to the feed-zone to obviate this local deficiency in the oxygen coefficient requirement.

Controllable amounts of oxygen can be directed into the feed-zone by employing oxygen jets formed from injectors (or lances) mounted onto and through the wind-box wall. In this way, the oxygen is locally concentrated in the air underneath the nozzle plate, where it is carried into the feed-zone as oxygen enriched air through the nozzle plate.

The technical problems associated with injector design, i.e., sizing the oxygen injectors as well as their physical arrangement is summarized in the following steps. Firstly, CFD is used to analyze the flow field of the wind-box air. Figure 1 left (top and

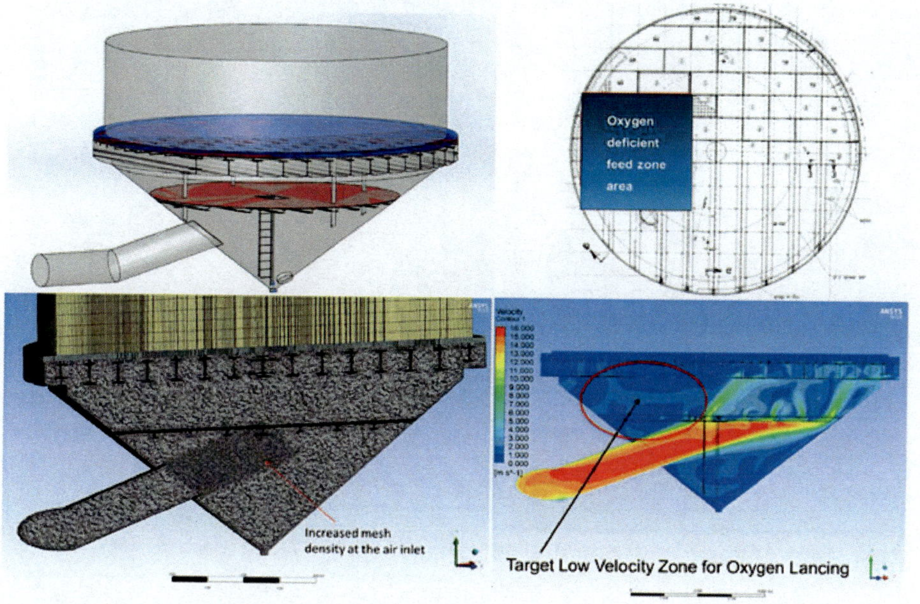


**Fig. 1** Left (top and bottom) wind-box CFD model; Right-top: feed zone illustration; Right-bottom: wind-box CFD

bottom) shows an example CFD model set-up. The model is a full-scale 3D wind-box. This baseline analysis gives information about the velocity field of the wind-box air flow—Fig. 1 (right-bottom). Secondly, armed with this knowledge, injectors can be designed to deliver oxygen jets to directly target the feed-zone from underneath the nozzle plate. As the oxygen jets traverse the distance between injection point and the target feed-zone, they partially mix with the wind-box air. This reduces the concentration of oxygen in the jet so that a jet of oxygen enriched air (of controllable composition) contacts the nozzle plate.

The following method is used to partially verify and validate the CFD work concerning the oxygen jetting behavior in the wind-box. Firstly, the injector is tested in the laboratory in the open air using a pitot tube from which information about the axial jet velocity and oxygen concentration profile can be derived—Fig. 2 (left-top). Secondly, CFD is subsequently used to model the laboratory experimental set-up. Thirdly, the oxygen velocity and oxygen concentration profiles are then compared in Fig. 2 (left-bottom). When acceptable agreement between the experimental and CFD data is achieved for the laboratory model, the oxygen injectors on the wind-box are then subsequently modeled using CFD with the goal to predict how well the oxygen jets target the prescribed feed-zone area. In this way, we have developed confidence in the prediction of the oxygen jet trajectory and its mixedness with the wind-box air prior to contacting the nozzle plate in the feed-zone—Fig. 2 (right—top and bottom).

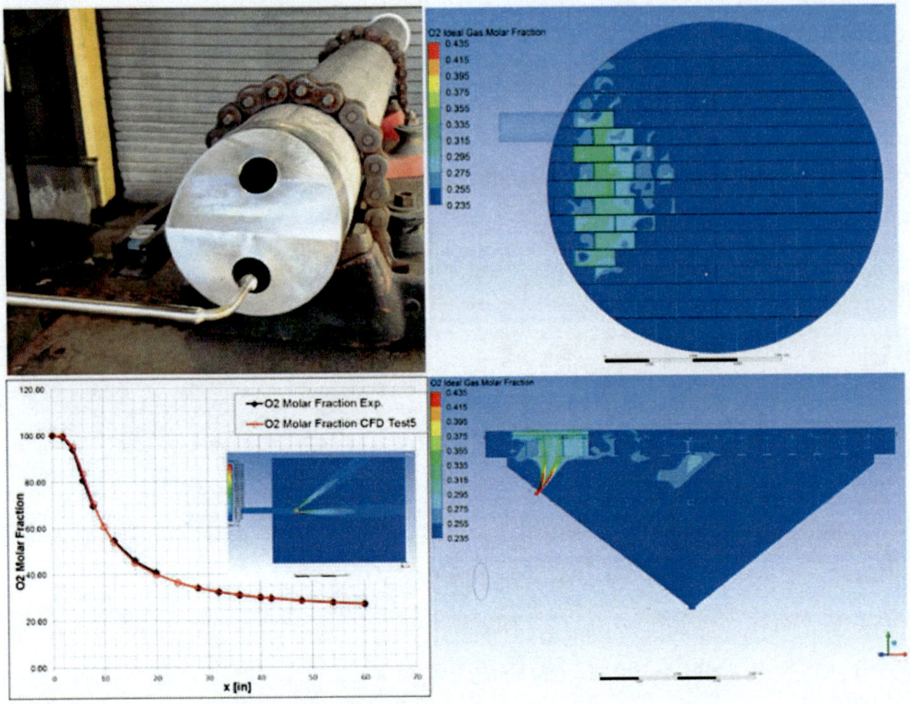


**Fig. 2** Left-top laboratory oxygen injector with probe; Left-bottom oxygen concentration by experiment and CFD: Right-top CFD showing increased oxygen concentration in the feed zone; Right-bottom CFD showing oxygen concentration wind-box cross section

# Sulphuric Acid Plants in Metallurgical Facilities: Options for Energy Optimization

**Shailesh Sampat, Joseph Kelly, Maria de Campos, and Sina Mostaghel**

**Abstract** Highly exothermic reactions in a Sulphuric Acid Plant are sources of energy that can be used effectively in metallurgical facilities. Unfortunately, this aspect does not receive the attention that it deserves while designing smelting and/or hydrometallurgical plants. High temperatures in gas circuits allow heat recovery as steam which is a very valuable source of energy. Over the last two decades, development of higher quality materials of construction has allowed the operation of acid circuits at high temperatures, resulting in increased opportunity for heat recovery as steam. This paper discusses various options for recovering heat and utilizing it effectively in metallurgical facilities.

**Keywords** Sulphuric acid plants · Pyrometallurgical facilities · Heat recovery · Energy optimization

## Introduction

Most sulphide smelting metallurgical complexes have a Sulphuric Acid plant. Pyrometallurgical plants processing sulphide concentrates use $SO_2$ gas generated in smelting processes to produce Sulphuric Acid and avoid pollution. Since acid production from this gas stream is unavoidable and, in some cases unnecessary, it is commonly referred as "fatal" acid. This "fatal" acid is generally used either to leach oxides of metal or to produce phosphatic fertilizers. Hydrometallurgical complexes

S. Sampat · J. Kelly · M. de Campos · S. Mostaghel (✉)
SNC-Lavalin, 191 the West Mall, Toronto, ON M9C 5K1, Canada
e-mail: sina.mostaghel@snclgroup.com

S. Sampat
e-mail: shailesh.sampat@snclgroup.com

J. Kelly
e-mail: joseph.kelly@snclgroup.com

M. de Campos
e-mail: maria.decampos@snclgroup.com



C. Fleuriault et al. (eds.), *Advances in Pyrometallurgy*, The Minerals, Metals & Materials Series, https://doi.org/10.1007/978-3-031-22634-2_20

use Sulphur as a raw material to produce acid and use it to leach oxide concentrates. If both sulphide and oxide concentrates are available in the same region, acid produced by the pyrometallurgical process could be used to leach oxide ore in the hydrometallurgical operation. Shortfall of acid is either procured from other sources or produced internally by burning Sulphur. In case acid production from pyrometallurgical process is too high, acid is sold to others, often at considerably lower prices.

In any case, Sulphuric acid plant (SAP) can produce plenty of energy in different forms due to highly exothermic reactions during the process. Heat can be generated in the form of high-pressure super-heated steam, intermediate-pressure steam, hot air or hot water depending on the selected technology. In many cases, this ends up being a major source of energy in the complex and reduces plant's dependence on external energy sources.

After a few years of operation, characteristics of ore/concentrate may change significantly, which may have a significant impact on metal production process parameters and even technologies employed. Several strategies are adopted to cope up with varying compositions/characteristics of the feed material. Acid plants also need to be modified to match the new strategy and the modified process parameters/technologies. This paper suggests a few options for modifying acid plants efficiently to match the updated operating strategies.

## Metallurgical Gas-Based Acid Plants

Metallurgical gas plants are essentially pollution control measures for $SO_2$ capture from the gas stream originating from pyrometallurgical units such as roasters, smelting or converting furnaces. Depending on the feed composition, metal produced, and technologies adopted, $SO_2$ generating reactions will be defined. Some of the most common and general reactions for copper and zinc production processes, and the resulting heat generation per mole of SO2, are listed below.

### *Copper Smelting and Converting Processes*

In copper smelting processes, the iron-copper-sulphide concentrate is combusted by oxygen blow, to remove sulphur as $SO_2$ in the off-gas and produce an intermediate sulphidic phase (matte). Silica is added as a flux to produce a fayalitic slag and remove iron. The simplified overall reaction, as shown in Eq. 1, is exothermic and produces 86.85 kcal/mole $SO_2$ (all heat of reactions were calculated using HSC-Sim thermodynamic package).

$$3CuFeS_2 + SiO_2 + 4.5O_2 = Fe_2SiO_4 + 3.5SO_2 + 1.5Cu_2S + FeS + \left(86.85\frac{\text{kcal}}{\text{moleSO}_2}\right) \tag{1}$$

The generated matte in the smelting process will be further oxidized in the converting process to produce metallic copper, more $SO_2$ gas and fayalitic slag, as shown in Eq. 2.

$$2Cu_2S + 2FeS + SiO_2 + 5O_2 = 4Cu + 4SO_2 + Fe_2SiO_4 + \left(83.26\frac{\text{kcal}}{\text{moleSO}_2}\right) \tag{2}$$

## *Zinc Roasting*

In the zinc sulphide smelting or roast-leach processes, the first step is typically roasting and preparing the feed material for the downstream pyro- or hydrometallurgical operations. The simplified and general reaction involved in the roasting process is shown in Eq. 3. As can be seen, the combustion reaction is strongly exothermic and can maintain the temperature.

$$2ZnS + 3O_2 = 2ZnO + 2SO_2 + \left(105.66\frac{\text{kcal}}{\text{moleSO}_2}\right) \tag{3}$$

## *Acid Plant*

The main reactions in an acid plant are shown in Eqs. 4 and 5.

$$\text{Heat of conversion } SO_2 + 0.5O_2 = SO_3 + \left(23.45\frac{\text{kcal}}{\text{kmole}}\text{at } 25\,^\circ\text{C}\right) \tag{4}$$

$$\text{Heat of formation } SO_3 + H_2O = H_2SO_4 + \left(54\frac{\text{kcal}}{\text{kmole}}\text{at } 25\,^\circ\text{C}\right) \tag{5}$$

## Heat Recovery

Figures 1 and 2 show material and energy block flow diagrams of a typical double contact and single contact metallurgical gas acid plant, respectively. $SO_2$ bearing gas stream leaves smelter / roaster at temperatures exceeding 1000 °C. This gas is cooled to about 350 °C in a waste heat boiler to recover heat as high-pressure steam up to 66 barg. Typically for a 300,000 tpa copper plant, 42 tph of steam can be generated by cooling the process gas in a metallurgical plant. Partially cooled gas stream then passes through a hot electrostatic precipitator to remove most of the particulate matter from the gas stream.

This partially cleaned gas then goes through a series of scrubbers and wet ESPs to remove other impurities to make gas suitable for Sulphuric Acid Plant. This process also cools the gas essentially to the wet bulb temperature and saturates the gas stream with water. This gas is then dried to remove all the moisture content and then reheated to 400 °C to convert $SO_2$ to $SO_3$ in multiple catalyst beds. Part of the heat generated in this reaction is used for reheating the feed gas. As a result, excess heat available in metallurgical acid plant is lower than that of a Sulphur burning plant, where gas leaving the waste heat boiler can be fed directly to the catalyst beds. In a double contact plant, cold gas leaving inter-pass absorption tower also needs to be reheated using the heat of conversion; hence, heat available for recovery is further reduced. This disadvantage is more than compensated by reduction of $SO_2$ emissions from > 1000 ppm up to < 100 ppm.

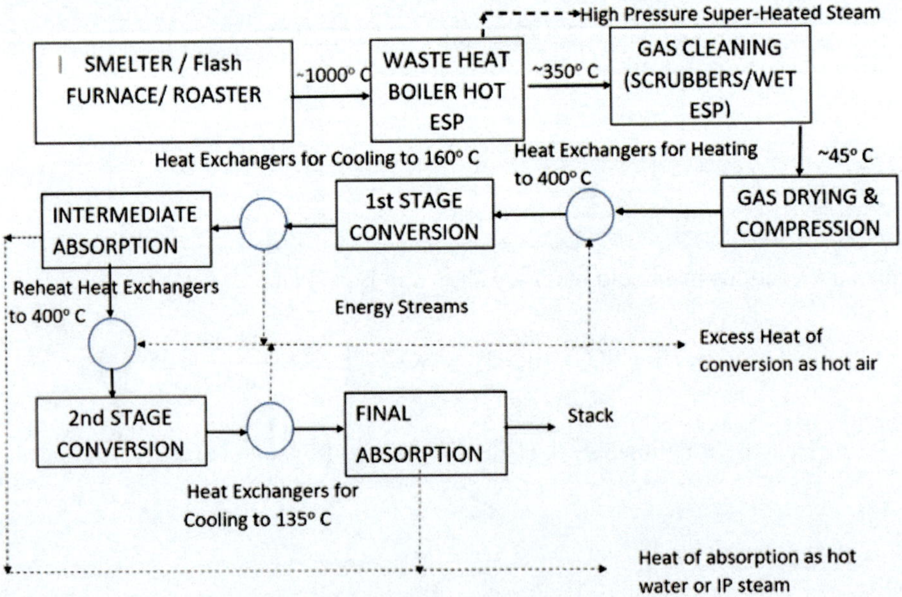


**Fig. 1** Double contact (DCDA) metallurgical gas plants block diagram

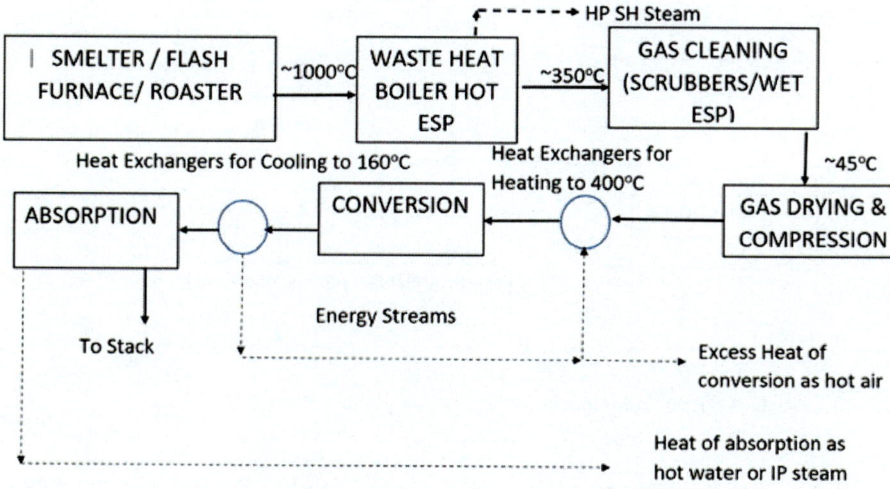


**Fig. 2** Heat recovery single contact metallurgical gas plants block diagram

Flash furnaces and fluidized bed roasters operate under fairly stable conditions, resulting in steady gas flow with high $SO_2$ content. This results in a steady operation of the acid plant with better heat recovery. However, copper converters (typically Peirce-Smith Converters) frequently stop operating to skim slag, add cold recycling materials, etc. This means the gas stream flow and $SO_2$ concentration vary widely, which results in an unstable operation of SAP. It is important to note that excess heat available for recovery increases with increasing $SO_2$ concentration. Standard acid plant allows operation of up to about 13% $SO_2$ in the inlet gas stream due to limitation of the maximum allowable temperature of the catalyst. This allows some production of high-pressure steam. Higher $SO_2$ concentration also reduces the power consumption of the acid plant due to lower gas volumes.

In most of the metallurgical gas plants, excess heat from the acid plant is taken out in air-cooled heat exchangers. In a typical metallurgical gas-based acid plant of 2000 tons per day capacity, total heat lost through hot air vented to atmosphere is $8.9 \times 10^6$ kcal//h, although there are practical solutions for recovering heat from these air streams. Hot air can be used in equipment such as dryers or even as feed to flash furnace or roaster. Steam or hot water can also be generated using hot air. If the flow or temperature of the hot air varies or is inadequate for final use, a supplementary heat source can be added to stabilize process conditions. Refer to Fig. 3 for a potential block diagram for using hot air. While finalizing the layout of a green field metallurgical complex, it is important to ensure that units receiving hot air generated in the acid plant are located near the acid plant to minimize duct lengths and pressure losses.

Alternatively, air-cooled heat exchangers can be replaced by boilers to directly generate steam from the process gas. Typically, 0.17 tons of intermediate-pressure steam can be produced per ton of acid produced.

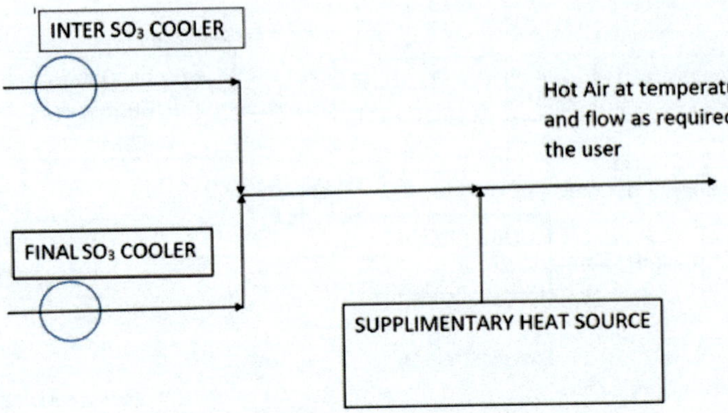


**Fig. 3** Utilization of hot air from metallurgical gas acid plants

If the $SO_2$ content of the gas at the inlet is higher than 13.5%, a pre-converter can be installed in the large acid plants. This scheme can handle up to 18% $SO_2$ concentration, reduce power consumption and increase generation of HP super-heated steam. Table 1, provided by our technology licensor MECS, compares critical parameters of conventional plant with pre-converter design with or without Heat Recovery System (HRS) from the acid circuit.

**Table 1** Critical parameters of design showing the effect of heat recovery system (HRS) system [1]

| | Conventional design | | Pre-converter design | |
|---|---|---|---|---|
| | Without HRS | With HRS | Without HRS | With HRS |
| Capital cost | 100% | 131% | 111% | 142% |
| Power consumption | 100% | 97% | 76% | 82% |
| Steam export | | | | |
| High pressure | 100% | 99% | 175% | 175% |
| Medium pressure | 0 | 0.45 t/t of acid | 0 | 0.44 t/t of acid |
| Cooling water usage | 100% | 58% | 85% | 40% |

## Sulphur Burning Plants

Sulphur burning plants in a metallurgical complex produce the acid required for leaching oxide ores. The major reactions involved are shown in Eqs. 6–8. Total heat of 148.29 kcal/kmole can be generated at 25 °C.

$$\text{Heat of combution } S + O_2 = SO_2 + \left(70.94\frac{\text{kcal}}{\text{kmole}} \text{at } 25\,^\circ\text{C}\right) \tag{6}$$

$$\text{Heat of conversion } SO_2 + 0.5O_2 = SO_3 + \left(23.45\frac{\text{kcal}}{\text{kmole}} \text{at } 25\,^\circ\text{C}\right) \tag{7}$$

$$\text{Heat of formation } SO_3 + H_2O = H_2SO_4 + \left(54\frac{\text{kcal}}{\text{kmole}} \text{at } 25\,^\circ\text{C}\right) \tag{8}$$

$$\text{Total Heat Generated} = \left(148.29\frac{\text{kcal}}{\text{kmole}} \text{at } 25\,^\circ\text{C}\right) \tag{9}$$

Figures 4 and 5 show block diagrams of Single Contact and Double Contact Sulphur burning acid plants. Most of the Sulphur burning plants are double contact types designed to minimize $SO_2$ emissions. Gas stream from third converter bed outlet is taken through an intermediate absorption tower to remove $SO_3$ formed to ensure better conversion in final catalyst bed. Cold gas leaving the intermediate absorption tower is at ~ 80 °C and needs to be reheated to 400 °C before it enters final catalyst bed. Part of the heat of conversion is generally used to reheat this gas. In a single absorption plant, entire heat of conversion is available for heat recovery. For this reason, High-Pressure (HP) steam generated in a double contact plant is lower than the steam generated in a single contact plant. This disadvantage is offset by lower emissions from double contact plant, which makes it the preferred option in most of the cases.

Heat can be recovered as HP super-heated steam as well as IP steam from Sulphur burning plants. Part of the steam is utilized within the plant in Sulphur section and in deaerator. Rest of the steam is used in condensing turbines to generate power. Power generated by this steam is enough to run the entire complex; in many cases reducing dependence on external power sources.

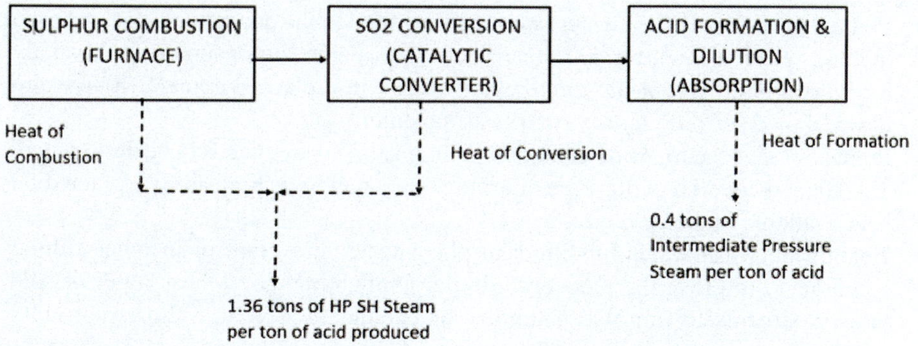


**Fig. 4** Heat recovery in single contact sulphur burning plants

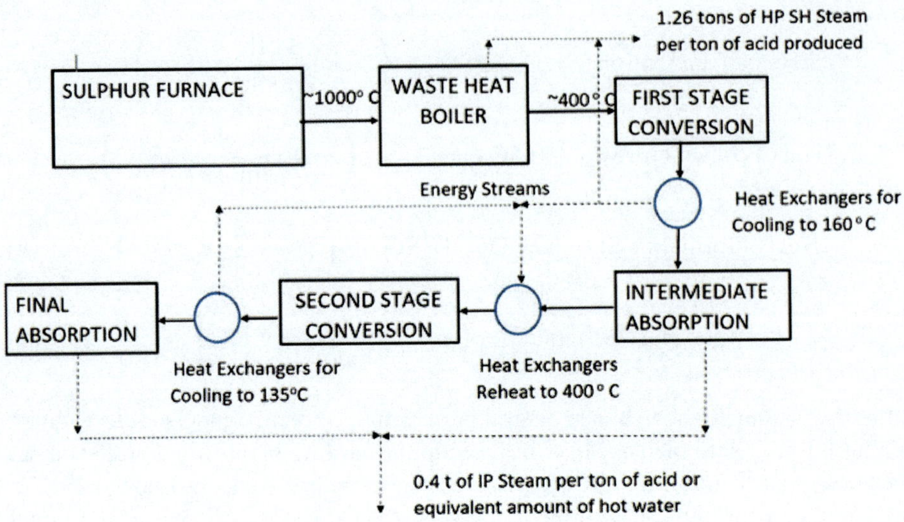


**Fig. 5** Heat recovery in DCDA sulphur burning plants

## Options for Low-Grade Heat Recovery in Acid Plant

Low-grade heat is produced in acid circuits. Most of the older plants operated acid circuits at temperatures below 120 °C, due to lack of suitable materials of construction. This resulted in most of the heat being wasted in cooling towers. After development of HRS technology by MECS, it is possible to recover heat from the acid circuit in form of intermediate-pressure steam. However, incorporating HRS technology increases the capital cost of the plant by ~ 30%.

If heat recovered as hot water can be utilized within the complex, additional capital costs could be reduced to much lower percentages. A few less-expensive options for low-grade heat recovery from the acid circuits are listed below:

- Preheat the boiler to feed water or condensate feed to the deaerator. This proposed method reduces low pressure steam consumption in the deaerator and usually recovers around 7% of the total heat available in the acid circuit. SNC-Lavalin has designed several plants with this arrangement.
- Preheat water feed to the desalination plant if saline water needs to be desalinated. Total heat recovered in this type of arrangement can be as high as 40% of the total heat available in the acid circuit.
- Heating up wash water for filters in phosphoric acid plant or in other similar locations to improve the recovery. Similar applications to preheat water or cold process streams to improve solubility of chemicals. SNC-Lavalin is currently working on a similar scheme for a Lithium producer, where heat recovery as water would reduce the LP steam consumption in the complex, equivalent to 0.27 tons per ton of acid produced having a heat value equivalent to about 40 MW.

A detailed description of SNC-Lavalin's experience in heat recovery and energy integration in different types of chemical and metallurgical complexes has been published before [2].

## Modifying Acid Plant to Suit Ore/Concentrate Characteristics

Depletion of ore bodies and variation in composition of the concentrates can result in the following problems:

- Lower metal and acid production from the plant.
- Concentration of $SO_2$ reduces in the gas streams entering the acid plant. This results in thermal imbalance in the acid plant, which requires external heat sources.
- Lower concentration of $SO_2$ also results in moisture imbalance which may force production of acid with lower concentration.
- Owner may prefer to increase the metal production from the oxide concentrate, resulting in higher acid requirement if the oxide concentrate is available nearby.
- Similarly, if oxide concentrate has lower metal content, owner may try to increase metal production by treatment of sulphide concentrates.

These scenarios can be managed by the following adjustments in the acid plant:

1. Converting a metallurgical gas-based acid plant to a hybrid plant. Hybrid plant works with a dual feed arrangement with liquid Sulphur as well as Sulphur-containing metallurgical gas. This essentially allows operation of the plant with 100% Sulphur feed as well. This kind of revamp has several advantages over just burning a fuel to maintain the thermal balance:
    - Acid production increases
    - Fluctuations in metallurgical gas flow and $SO_2$ concentration can be compensated by adjusting Sulphur feed, which can stabilize the operation of the acid plant.
    - In addition to maintaining thermal balance, moisture balance is also maintained which results in a steady product acid concentration.
2. Converting Sulphur Burning Plant to metallurgical gas-based plant if the oxide-based ore is too lean and owner decides to abandon leaching to focus completely on pyrometallurgical operation. Gas cleaning section needs to be added, Sulphur section needs to be removed and other equipment need to be modified.

SNC-Lavalin has experience in both types of proposed revamps.

## Conclusions

Energy recovery from acid plants located in Pyro-Metallurgical complexes is often neglected. Industry needs to pay more attention to this and explore possibilities of recovering more energy from various exothermic reactions that take place in the Sulphuric Acid Plant located in the complex. Utilizing the hot air that is vented from the gas coolers or revamping the plant to recover heat as LP steam should be considered as potential options. Incorporating new technologies such as pre-converter and HRS to increase steam generation can be helpful.

If quality and/or grade of ore/concentrate has changed after several years of operation, modifications in the acid plant should be considered to match the revised acid requirements/production capacity and energy recovery alternatives.

## References

1. McLaws DJ (2014) High $SO_2$ feed gas Sulphuric Acid Plant Design Option. In: 53rd Annual conference of metallurgists (COM2014), Vancouver, BC, Canada, Sep 28–Oct 01, 2014–09
2. Sampat S, Kelly J, DeSilva N (2022) Sulphuric Acid Plant integration in a chemical complex. Sulphur Mag 399:1–5, March–April 2022

# Experimental Analysis of Zinc Melting Using CSP

**Pieter J. A. Bezuidenhout, Willem G. Le Roux, and Joalet D. Steenkamp**

**Abstract** The industrial sector is expected to be a large greenhouse gas producer in the future, with the majority of emissions being attributed to fossil fuel-based heat generation. By directly applying solar thermal energy to a high-temperature industrial process, the reliance of the industrial processes on fossil fuel-based energy sources can be lessened or completely removed. Preliminary investigations into the production and beneficiation of zinc using concentrated solar power (CSP), in a South African environment, show promise and even more so with a focus only on the cathode casting operations. To evaluate the practical implications of directly applying CSP to a high-temperature materials processing application, a study was launched into the direct solar melting of zinc metal. In this study, the setup and performance of a novel cavity receiver for zinc melting using CSP are discussed.

**Keywords** Concentrated solar power · Zinc · Solar melting

## Introduction and Background

The industrial sector is predicted to be by far the biggest greenhouse gas producer by 2060 [1]. Process heat represents about two-thirds of the industrial energy demand, of which 90% is currently supplied by burning fossil fuels [1]. Countries with a good solar resource have the potential to make use of concentrated solar power (CSP) to revolutionize the industrial heat supply by making use of this sustainable alternative.

P. J. A. Bezuidenhout (✉)
MINTEK, 200 Malibongwe Drive, Strijdom Park, Randburg, South Africa
e-mail: pjabezuidenhout@gmail.com

W. G. Le Roux
Department of Mechanical and Aeronautical Engineering, University of Pretoria, Pretoria, South Africa

J. D. Steenkamp
Department of Chemical and Metallurgical Engineering, University of the Witwatersrand, Johannesburg, South Africa



C. Fleuriault et al. (eds.), *Advances in Pyrometallurgy*, The Minerals, Metals & Materials Series, https://doi.org/10.1007/978-3-031-22634-2_21

Concentrating solar power (CSP) technologies have been shown to achieve temperatures in excess of 1000 °C [2, 3], which makes this an ideal candidate for meeting the industrial heat demand.

Most existing renewable energy solutions in the industrial sector are aimed at electrification, using photovoltaic (PV) technologies, or for heating applications below 200 °C which is typically earmarked for agri-processing and textile industries [4, 5]. Approximately half of all industrial heat applications are classified as high-temperature applications (> 400 °C) [1], which is why a focus should be placed on finding cleaner alternatives for these applications if the large carbon footprint of the industry sector is to be addressed. The melting and re-melting of low-melting-temperature metals have been identified as a possible starting point for the research into using renewable solutions as a heat source for high-temperature industrial applications. This not only allows the carbon footprint of these processes to be addressed but also improves the circular economy of these metals. For the study discussed in this paper, zinc (Zn) metal was selected based on its availability, material properties, importance in modern society, and the potential impact on other industrial applications such as hot-dip galvanizing, casting applications, and recycling.

To melt Zn metal, an operating temperature of above 420 °C is required. Of the most common CSP technologies, only two meet this high operating temperature requirement, with both being classified as focal point technologies [6, 7]. These technologies are heliostat field reflectors (HFR) and parabolic dish reflectors (PDR). Both technologies employ an enlarged reflective surface to focus incoming solar irradiation to a single focal point, at which a receiver can be placed to make use of the collected heat. An HFR system employs a field of individual mirrors that track and direct the sunlight to a fixed central tower receiver, achieving concentration ratios of up to 1500 and temperatures of close to 2000 °C [7]. A PDR makes use of a parabolically shaped dish, lined with a reflective surface, to collect and focus incoming sunlight to a receiver positioned at the focal point. By fitting the entire dish structure and receiver to a two-axis tracking system, concentration ratios of up to 2000 and temperatures of close to 1500 °C are attainable [6]. This allowed PDR systems to be classified as some of the most efficient solar collector technologies [8]. The complexity and cost of the technology have, however, limited its uptake to proof of concept and specialty applications [8].

To investigate the concept of solar melting of zinc, a multi-faceted parabolic dish reflector (PDR) will be used in this work. This solar collector was selected based on the scale of the concept test work and due to the availability of this technology at the University of Pretoria, in South Africa. The method and results of the initial experimental test work will be discussed in detail.

## Thermal Analysis

This study will make use of an existing multi-facet parabolic dish reflector developed by the University of Pretoria [9, 10]. A cavity receiver is positioned at the focal point of the multi-facet parabolic dish to make use of the concentrated heat. Cavity receivers offer good thermal efficiencies in combination with a parabolic dish reflector system [3]. To determine the thermal efficiency of the cavity receiver as well as the amount of heat available for melting the zinc metal, some governing heat transfer equations were considered. The thermodynamic model was developed in the Python coding language and is based on the heat loss mechanisms from the cavity receiver, as shown in Fig. 1. These heat losses include conduction losses ($Q\_conduction\_loss$), convection losses ($Q\_convection\_ loss$), and radiation losses ($Q\_radiation\_ loss$). The net heat transfer, $Q_{net}$, can be calculated using Eq. 1, with $Q_{in}$ the solar thermal input as calculated in Eq. 2.

$$Q_{net} = Q_{in} - (Q\_conduction\_\ loss + Q\_convection\_\ loss + Q\_radiation\_\ loss) \quad (1)$$

$$Q_{in} = I \cdot A_D \cdot \Gamma \cdot f_s \cdot \rho_R \quad (2)$$

Note that $I$ is the solar irradiance for a specific location, $A_D$ the dish aperture area, $\Gamma$ the intercept factor, $f_s$ the shading factor, and $\rho_R$ the reflectivity of the mirror facet.

The developed numerical model evaluates each of the heat loss mechanisms separately, and the experimental data will be used for the validation of this model. But to estimate the overall thermal efficiency of the system, the heat required to melt down the batch of zinc was calculated as shown in Eqs. 3–5. $Q_{solid\_heating}$ represents the energy required to bring the solid zinc from ambient temperature to melting temperature and $Q_{latent}$ the energy required for the phase change from solid to liquid. $Q_{required}$ is the total amount of energy required to melt down the charged batch of zinc.

$$Q_{solid\_\ heating} = m_{zn} \cdot c_p \cdot (T_{melt} - T_{amb}) \quad (3)$$

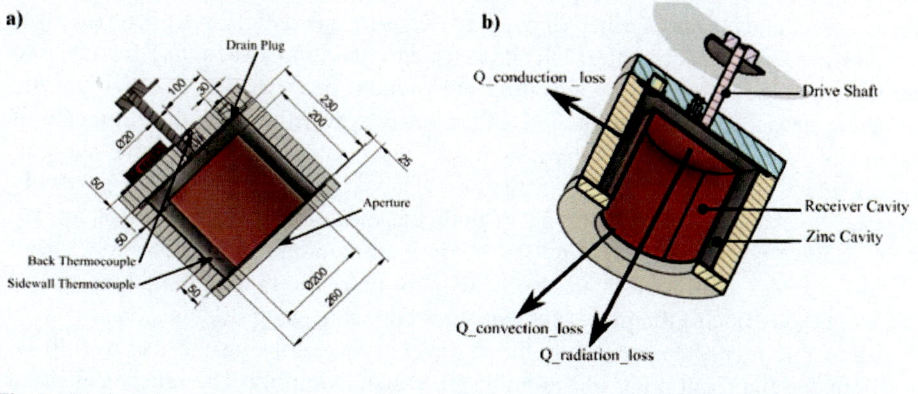


**Fig. 1** Section view of cavity receiver with **a** showing dimensions and **b** showing heat losses

$$Q_{\text{latent}} = m_{\text{zn}} \cdot L_f \quad (4)$$

$$Q_{\text{required}} = Q_{\text{solid\_ heating}} + Q_{\text{latent}} \quad (5)$$

Note that $m_{\text{zn}}$ is the mass of the fed-batch of solid zinc, $c_p$ the specific heat capacity of zinc in the solid phase, $L_f$ the latent heat of fusion of zinc, $T_{\text{melt}}$ the melting temperature of zinc assumed as 420 °C, and $T_{\text{amb}}$ the as-fed ambient temperature assumed as 20 °C.

With the energy input from the sun calculated and the total amount of energy required to melt down the fed-batch, the overall thermal efficiency ($\eta_{\text{th}}$) of the system can be calculated using Eq. 6.

$$\eta_{\text{th}} = \frac{Q_{\text{required}}}{Q_{\text{in}}} \cdot 100 \quad (6)$$

## Methodology

The details relating to the experimental setup will be discussed in detail to provide an in-depth understanding of the equipment used for the test work together with the experimental procedure followed to collect the experimental data.

### *Design*

The multi-faceted parabolic dish approach was adapted from previous work done at the University of Pretoria [9, 10]. This assembly relies on six vacuum-formed membrane dish facets to work independently to collect incoming solar irradiance and focus it to a collective focal point as shown in Fig. 2. The calibration method for the individual facets is described in detail by Roosendaal et al. [9]. A pneumatic ball valve is fitted to the back of the dish and is used to maintain a vacuum drawn behind the reflective membrane. By controlling the vacuum pressure, the concavity of the reflective surface can be manipulated. This concavity together with the alignment of the mirror facets on the facet mounting base is used to focus the incoming sunlight into the receiver aperture. A linear actuator and a bearing in the base of the assembly are used to manually track the sun in both the elevation and azimuth plane. By maintaining a tracking error below 1°, the optimum amount of heat can be focused into the receiver aperture and heat spillage is limited. The two-axis tracking makes this technology one of the most efficient solar collector technologies.

Adjustable receiver arms allow the receiver frame, housing the receiver, to be positioned at the focal point of the multi-facet dish assembly. The receiver is fitted

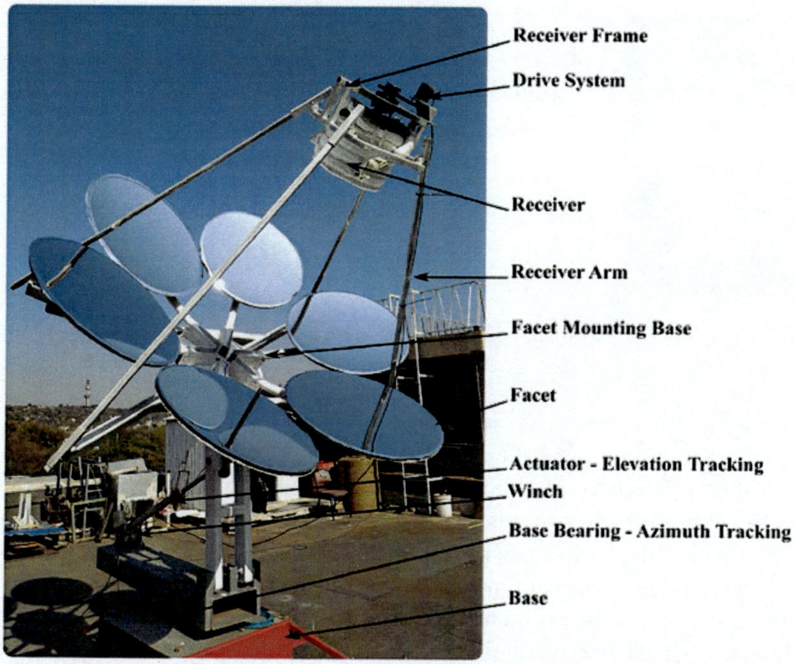


**Fig. 2** Shows the experimental setup with the multi-facet dish and the receiver installed

to the receiver frame by means of a drive shaft that is kept in alignment using two flanged pillow block bearings, spaced 20 mm apart. This arrangement enables the receiver to be supported in a cantilever approach while limiting the rotational friction of the rotating system. A sprocket and chain drive system is employed together with a NEMA 23 stepper motor in order to rotate the whole receiver assembly. The rotational speed is controlled using a stepper motor driver and an Arduino microcontroller.

The receiver assembly, as shown in Fig. 3a), consists of a receiver cavity fabricated from 350 WA mild steel and insulated using two layers of 25 mm thick insulating ceramic fibre blanket. The receiver design consists of a primary cavity that faces the parabolic dish and receives the concentrated solar thermal energy, and a secondary enclosed cavity that contains the zinc metal (shown in Fig. 1). This arrangement allows the zinc metal to be contained while in direct contact with the heated cavity exposed to the concentrated sunlight. The blanking flange, to which the drive shaft is also welded, contains a steel plug of 30 mm in diameter, that is removed after the zinc is melted, to drain the molten material from the zinc container. In the case of adding the solid zinc feedstock, the blanking flange can be removed in order to expose the secondary cavity and allow for easy filling of the cavity with a new batch of zinc metal. The blanking flange can then be bolted back into place and a high-temperature sodium silicate-based sealant is used as gasket material to ensure a liquid-tight seal on the flange mating faces.

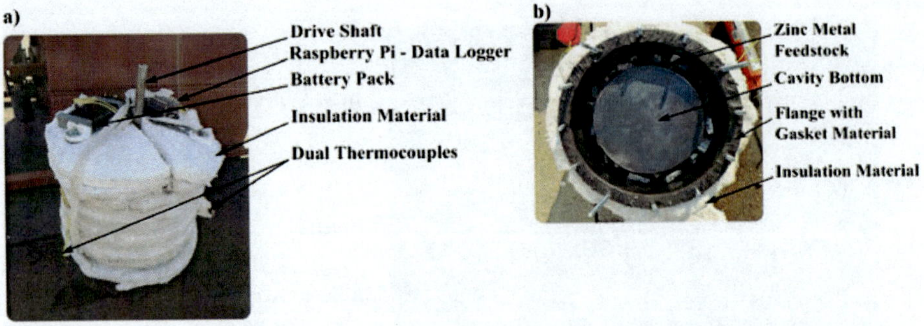


**Fig. 3** **a** Shows the receiver assembly and **b** the zinc cavity with the zinc feedstock

The entire receiver assembly is rotated to aid in the mixing and heat transfer in the zinc metal and to avoid the potential for hot spot formation in the cavity as a result of the individually focused mirror facets. This aids in the efficiency of the heating process but adds significant complexity to the control and measurement of the system. As no electrical wires can be connected to the receiver, alternative methods had to be developed to remotely monitor the temperature of the cavity and the zinc metal. The solution to this problem was found by making use of a Raspberry Pi 3 [11], which is a single-board computer that can be powered by a battery pack. This solution made it possible to use Virtual Network Computing (VNC) and local Wi-Fi to monitor the temperatures of the system remotely [12]. The Raspberry Pi logs the temperatures locally to avoid the situation where an internet connection failure can corrupt the data being collected, but the logging method still allows the operator to view the temperatures remotely. After the successful completion of an experimental test, the data can be downloaded from the Raspberry Pi for analysis.

In order to measure the temperature, four dual-junction K-type thermocouples were used, resulting in 8 temperature measurements. Each thermocouple had a junction at the tip and a second junction at 30 mm back from the tip. This arrangement allowed for two temperature measurements from a single access port, which reduced the potential for molten zinc to leak from an opening in the receiver side wall. The tip of the thermocouple was embedded in the primary cavity sidewall, through the zinc cavity, allowing for accurate measurement of the cavity wall temperature as it was exposed to sunlight. The second junction was positioned just inside the zinc cavity, allowing for accurate measurement of the zinc metal temperature. Three of the dual-junction thermocouples were positioned 50 mm up from the aperture, in the sidewall of the receiver, and spaced 120° apart, while the last one was positioned in the blanking flange from the back. The reason for the three thermocouples on the sidewall was to confirm temperature uniformity in the receiver and to serve as redundancy. If one of the thermocouples failed, two more would continue to record an accurate representation of the cavity wall temperature and that of the zinc metal. The side wall thermocouples were positioned low enough to always be in contact with the zinc, in both the solid en molten state. Due to the orientation of the receiver throughout an experiment, the thermocouple in the back would for most of the test measure the air-gap temperature inside the receiver and not the zinc temperature. This is because the zinc will drain forward as it melts down and collect around the face side of the receiver, in contact with the side wall thermocouples.

### *Material Preparation*

The feed material for the experiment was Special High Grade (SHG) zinc cast anode offcuts. SHG zinc has a purity of 99.995%. The anode offcuts were supplied in strips of 250 mm with a thickness of approximately 16 mm and varying widths. These strips were then cut into smaller stock utilizing a hydraulic guillotine, in order for the feed material to fit into the zinc cavity of the receiver. The zinc feedstock can be seen in Fig. 3b).

### *Measurements and Data Collection*

A Raspberry Pi 3 microcomputer was used for the remote logging of the process temperatures [11]. Two MCC134 thermocouple data acquisition (DAQ) HAT (hardware attached on top) boards for Raspberry Pi by Measurement Computing [13] were used to interpret the readings from the thermocouples. These HAT boards are "plug-and-play" additions to the Raspberry Pi microcomputer that expand the capability of this device. The HAT boards are stackable to add more thermocouple inputs to the Raspberry Pi—for this test work, two of these HAT boards were used, each capable of having four thermocouples connected to them. The two MCC 134 DAQ HAT boards allowed the 8 thermocouple readings to be remotely recorded on the Raspberry Pi. The fastest recording interval of 1 s was used to enable high-resolution recordings of any temperature changes that took place in the system. These readings together with a time step were saved in a CSV (comma-separated values) file for post-processing of the data.

### *Experimental Method*

The typical experimental method is described in a step-by-step approach:

1. Charge zinc feedstock into the zinc cavity, with thermocouples already secured in place.
2. Apply high-temperature sealant on the back flange and secure the blanking flange in place using 12 bolts. Also, fit two layers of insulating ceramic fibre blanket to the back blanking flange.

3. Fit the Raspberry Pi unit and battery back to the back of the receiver, on top of the 2 layers of insulation. The assembly is secured in place by having four of the flange bolts extend through the fibre blanket. Once secured in place, the thermocouples are connected to the Raspberry Pi and all the cabling is secured to the receiver by high-temperature glass tape.
4. The Raspberry Pi is powered on and a connection check between the device and the monitoring laptop is done to verify that temperature readings can be observed remotely.
5. The cable winch attached to the base of the solar dish is used to lower the receiver frame to ground level for easy access.
6. The receiver is then fitted to the receiver frame and secured in place by tightening the drive shaft in place.
7. The cable winch is used again to raise the assembly into place and the linear actuator is connected to the assembly for solar tracking purposes.
8. The dish is then pointed towards the sun and aligned in both axes using the shadows cast by the structure as a reference. A vacuum is then drawn behind each reflective membrane to concentrate the incoming solar irradiation into the cavity of the receiver.
9. The drive system is activated and the cavity is rotated at approximately 30 revolutions per minute (RPM), and the test is then started. Every minute the dish tracking is adjusted in both the elevation and azimuth axis using the structure shadows as a reference. At the same time, the temperature increase in the receiver is noted.
10. The temperature will steadily increase to 420 °C, the melting point of zinc, after which it will level out as the material undergoes the phase change. Once the temperature starts to rise again, it serves as an indication that the phase change is complete and that the entire batch of zinc is melted down.
11. The solar dish is swiveled out of alignment with the sun and the cable winch is used again to lower the receiver to ground level. Once at ground level, the steel plug at the back of the receiver is opened and the molten zinc is allowed to drain out of the receiver into a steel ladle.
12. Once all the zinc is drained from the receiver, the receiver is removed from the receiver frame and is allowed to cool down. The solar dish is then also returned to its stow position to be used again at a later stage and the vacuum behind the membranes is released.
13. The Raspberry Pi and battery back are removed from the receiver and the data can now be retrieved from the device for analysis. The casted ingot is also removed from the steel ladle to be weighed and used for the mass balance.

## Results

The Direct Normal Irradiance (DNI) readings for a test done on the 21st of August 2022 are shown in Fig. 4 [14]. The average DNI input for the test period was 906 W/m$^2$.

Figure 5 shows the temperature measurements obtained from the experiment on the 21st of August 2022. Eight thermocouples were connected to the Raspberry Pi, but two on the shell of the receiver were damaged during the installation of the receiver. Fortunately, there were two back-ups and these temperatures corresponded well with one another to provide confidence in these measurements. "Receiver_1" and "Receiver_2" recorded the cavity wall temperature on the sidewall and "Receiver_back" the cavity roof temperature. "Zinc_1" and "Zinc_2" recorded the zinc temperature and "Zinc_back" recorded the back of the zinc cavity. From the measurements, it can be observed, as expected, that Zinc_1 and Zinc_2 were in constant contact with the zinc feedstock and that the temperature increased to the melting point of 420 °C, where it then leveled out. Due to the orientation of the receiver during operation, Zinc_back was not in constant contact with the zinc metal and was therefore measuring the zinc cavity air temperature, but as the melting temperature was reached, molten material started to move around in the receiver (due to the rotation) and as a result came in contact with the back thermocouple periodically, resulting in the uneven readings. The batch is estimated to be fully melted after 114 min, which is when the temperature started to rise again for the first time after the latent phase. The receiver was then drained according to the procedure explained previously and casted into a metal ladle and a steel cupcake pan to produce the ingots as shown in Fig. 6.

If a conservative intercept factor of 80%, a shading factor of 1, a facet reflectivity of 97% [9], and a mirror surface area of 2.7 m$^2$ [10] are assumed, an efficiency of 28.05% is achieved with this receiver design. This means that approximately 28% of the solar input was absorbed by the zinc metal for the duration of the test. Of the

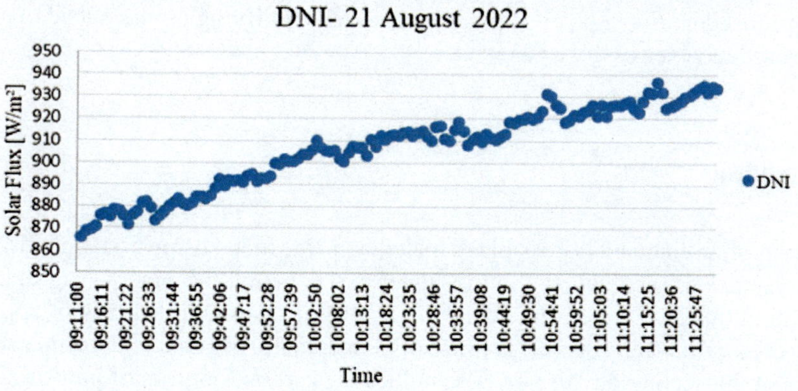


Fig. 4 Direct Normal Irradiance (DNI) at the time of the experiment. [14]

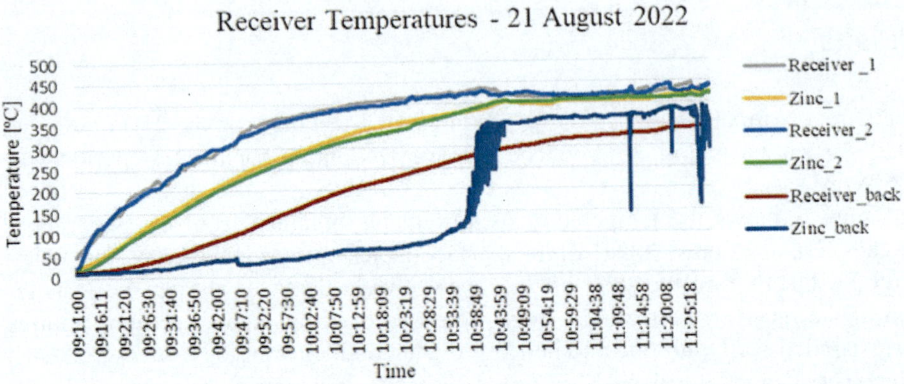


**Fig. 5** Receiver temperatures

**Fig. 6** Casted zinc metal ingot

14.5 kg of zinc feedstock charged for the batch, 10.3 kg were drained from the receiver in the molten state, resulting in approximately 29% of the inventory remaining in the receiver. This is largely due to the heat losses from the receiver cavity during the tapping procedure and a layer of process material solidifying on the outside face of the exposed receiver cavity as a result.

## Conclusion

This study showed that it is possible to melt down zinc metal using CSP only. 14.5 kg of zinc metal was melted down in just under two hours on the 21st of August 2022, resulting in this receiver design achieving an overall thermal efficiency of approximately 28%. This study serves as proof of concept and will motivate further development of this technology in a drive to reduce the carbon footprint of industrial heat applications. The results from this experimental study will also be used for the validation of a python-based heat-transfer model developed in support of this work. The python model can be used for scale-up and feasibility studies of this technology.

**Acknowledgements** This paper is published with the permission of Mintek.

## References


1. Philibert C (2017) Renewable energy for industry. Report from Internasional Energy Agency Insight Series. https://www.iea.org/reports/renewable-energy-for-industry. Accessed 7 Apr 2022
2. Rosa LG (2019). Solar heat for materials processing: a review on recent achievements and a prospect on future trends. ChemEngineering 3(4):83. https://doi.org/10.3390/chemengineering3040083
3. Kasaeian A, Kouravand A, Rad MAV, Maniee S, Pourfayaz F (2021) Cavity receivers in solar dish collectors: a geometric overview. Renewable Energy 169:53–79. https://doi.org/10.1016/j.renene.2020.12.106
4. WWF (2017) Industrial scale solar heat in South Africa, Report by WWF Netbank Green Trust. https://www.greencape.co.za/assets/Uploads/Industrial-Scale-Solar-Heat-in-South-Africa-opportunities-in-agri-processing-and-textiles.pdf. Accessed 10 April 2022
5. Joubert EC, Hess S, Van Niekerk JL (2016) Large-scale solar water heating in South Africa: status, barriers and recommendations. Renewable Energy 97:809–822. https://doi.org/10.1016/j.renene.2016.06.029
6. Kalogirou SA (2004) Solar thermal collectors and applications. Progr Energy Combus Sci 30(3):231–295. https://doi.org/10.1016/j.pecs.2004.02.001
7. Kumar L, Hasanuzzaman M, Rahim NA (2019) Global advancement of solar thermal energy technologies for industrial process heat and its future prospects: a review. Energy Convers Manage 195:885–908. https://doi.org/10.1016/j.enconman.2019.05.081
8. Coventry J, Andraka C (2017) Dish systems for CSP. Sol Energy 152:140–170. https://doi.org/10.1016/j.solener.2017.02.056
9. Roosendaal C, Swanepoel JK, le Roux WG (2020). Performance analysis of a novel solar concentrator using lunar flux mapping techniques. Sol Energy 206:200–215. https://doi.org/10.1016/j.solener.2020.05.050
10. Swanepoel JK, le Roux WG, Lexmond AS, Meyer JP (2021) Helically coiled solar cavity receiver for micro-scale direct steam generation. Appl Therm Eng 185:116427. https://doi.org/10.1016/j.applthermaleng.2020.116427
11. Raspberry Pi Ltd. (2022) Raspberry Pi. https://www.raspberrypi.com/. Accessed 9 Apr 2022
12. RealVNC (2022) RealVNC®—Remote access software for desktop and mobile | RealVNC. RealVNC®. https://www.realvnc.com/en/. Accessed 9 Apr 2022
13. Measurement Computing Corporation (2022) Thermocouple measurement HAT for Raspberry Pi® from measurement computing. MC Measurement Computing—MCC 134. https://www.mccdaq.com/DAQ-HAT/MCC-134.aspx. Accessed 7 Apr 2022
14. Brooks MJ, du Clou S, Van Niekerk JL, Gauche P, Leonard C, Mouzuoris MJ, Meyer AJ, van der Westhuizen N, van Dyk EE, Voster F (2015) SAURAN: a new resource for solar radiometric data in Southern Africa. J Energy Southern Africa 26

# Effect of Ore Pre-heating on Furnace Operation in High Carbon Ferromanganese Production—Lessons Learnt from Pilot-Scale Test Work

**Matale Samuel Moholwa, Sello Peter Tsebe, Derek Alan Hayman, Pieter Johannes Andries Bezuidenhout, Martin Bongani Sitefane, and Joalet Dalene Steenkamp**

**Abstract** The EU-funded PreMa project investigated the potential for a preheating stage to reduce the electrical energy requirement and $CO_2$ emissions produced during the production of high-carbon ferromanganese in a submerged arc furnace. Pilot-scale test work was conducted at Mintek in South Africa to demonstrate the potential effect of preheating on furnace operation. The A 300 kVA AC-furnace facility at Mintek was upgraded extensively for this purpose. The paper reports on the lessons learnt from the pilot-scale test work.

**Keywords** Furnace operation · Pre-heater · Ferromanganese production

## Introduction

High-carbon ferromanganese (HCFeMn) is utilised in the production of mild steel [1]. It contains between 74 and 82% Mn, 7.5% C, 1.2% Si, and between 9.3% and 17.3% Fe by mass [2]. The alloy is produced primarily in submerged arc furnaces (SAFs) and the process is electrical energy intensive requiring 2200–3900 kWh/ton alloy [1]. Calculations by Tangstad et al. [3] indicated that by heating the raw materials utilised in HCFeMn production to 873 K and evaporating the water, the energy savings could be as much as 20%. In the PreMa project [4], the hypothesis is tested that a two-stage process will result in 25% reduction in electrical energy consumption and 15% reduction in $CO_2$ emissions. As part of the project, a pilot-scale campaign was conducted at Mintek in South Africa to test this hypothesis and to demonstrate the potential effect of preheating on furnace operation.

The pilot-scale facility included an existing 120 kW rotary kiln linked to a newly designed AC furnace powered by an existing 300 kVA power supply. The paper presented here reports on the furnace containment system design, installation, and

M. S. Moholwa (✉) · S. P. Tsebe · D. A. Hayman · P. J. A. Bezuidenhout · M. B. Sitefane
MINTEK, Randburg, South Africa
e-mail: sammo@mintek.co.za

J. D. Steenkamp
University of the Witwatersrand, Johannesburg, South Africa


C. Fleuriault et al. (eds.), *Advances in Pyrometallurgy*, The Minerals, Metals & Materials Series, https://doi.org/10.1007/978-3-031-22634-2_22

**Table 1** Campaign schedule

| Campaign name | Start date | Schedule end date | Actual end date and reason for early end |
|---|---|---|---|
| Campaign 1 | 14th Feb 2022 | 24 February 2022 | 20th February 2022, furnace roof burn through |
| Campaign 2a | 23rd May 2022 | 6th June 2022 | 25th May 2022, oil leak from power supply tap changer |
| Campaign 2b | To be confirmed | – | – |

evaluation during the pilot-scale campaign. Design assumptions included a furnace process power factor of 0.7, which meant that the energy available to the process was 210 kW, and the assumption that the energy losses to the environment would be 50%, based on previous experience [5].

## Campaign Schedules

Table 1 shows the campaign schedule. The schedule underwent changes a few times during the past year. This was due to a number of equipment challenges experienced that necessitated premature termination. The original plan was to have two campaigns, namely Campaign 1 (cold feed) and Campaign 2 (hot feed). The duration for each of the two campaigns was planned for 10 days, termination of the campaign was to be followed by a systematic excavation of the furnace. Unfortunately, campaign 1 was terminated prematurely before enough data can be collected due to a roof burn through. In Campaign 2, the plan was revised to run one campaign in which cold and hot material would be processed sequentially. The campaign was planned for a duration of 15 days whereby cold material would be processed for 8 days and hot material for 7 days. Campaign 2 had to be terminated prematurely due to an oil leak from the power supply tap changer. This campaign was then labelled Campaign 2a. The current plan is to conduct one campaign, namely, Campaign 2b, in which cold and hot material will be processed in 10 days.

## Lessons in Furnace Design

This section of the paper focuses on the lessons learned about furnace design from the two short campaigns executed to date. Highlights and lowlights experienced during the campaigns provided insight on whether the assumptions made during the furnace design were valid or not. The knowledge gained was used to provide technical guidance with respect to improvements that need to be implemented to the furnace facility for the next campaign.

## *Feed System*

As and when material was required at the plant, bulk bags were transported by forklift from the storage area to the pilot facility area (Bay 1). The material was then charged into an allocated day-bin, of which the volume matched the feed bin. The day-bin was then lifted by an overhead crane and emptied into one of the four feed bins. Each day-bin was painted the same color as a feed bin, ensuring that the correct material was fed to the corresponding feed bin. There were four day-bins, one per feed material. The general arrangement of the feed system, kiln, furnace, and ladle is indicated in Fig. 1

The feed system had four raw material hoppers grouped into two streams.

Stream A delivered ore to the rotary kiln feeder. The ore was blended by controlling the feed ratio into the rotary kiln feeder using Schenk loss-in-weight (LIW) controllers, variable speed belt feeders, and the hot feed conveyor belt. The Schenk controllers were operated in gravimetric mode, which implied that the speed of the variable speed belt feeder was a function of the weight loss from the raw material hopper. This weight loss was measured and controlled in real time. The feed-rate set point of each feeder was calculated by the chief investigator (CI) as a function of the ratio of the final ore blend.

The ore blend was fed via the rotary kiln into the furnace whether it was preheated or not. The rotary kiln was fed by a vibrating tube feeder (see Fig. 1), which extended approximately 150 mm into the kiln to avoid back feeding. This feeder operated in volumetric mode with the aim to maintain a constant level of ore in the feeder box.

The rotary kiln was fitted with an electric variable speed drive, as well as mechanical variable speed gearboxes, to control the speed of rotation. The angle of the rotary kiln tube was also adjustable—this, combined with the speed of rotation, controlled the residence time of the material in the rotary kiln. The rotary kiln was equipped with

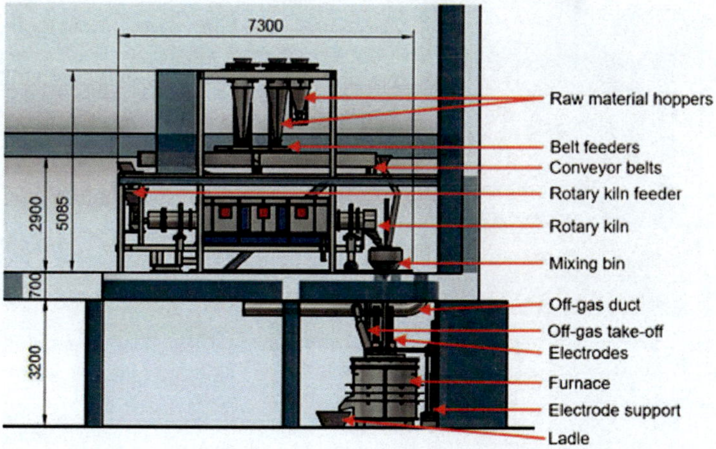


**Fig. 1** General arrangement of the feed system and furnace

three heating elements, each fitted with temperature controllers. The three control functions (i.e. speed of rotation, inclination of the tube, and operating temperature) controlled the temperature of the material exiting the kiln.

Stream B delivered cold coke and quartz from the bins into the mixing bin. Similar to the ore bins, Schenk LIW controllers, operated in gravimetric mode, controlled variable speed belt feeders and the cold feed conveyor belt. The feed-rate set point of each feeder was calculated by the CI as a function of the ratio of the coke or quarts to the ore blend.

Stream A and B were combined in the mixing bin and fed into the center of the furnace via a single feed pipe. This feed arrangement ensured adequate mixing of the feed materials and equal distribution of material inside the furnace. The design intent was to choke feed the furnace during operation in order for feed material to provide a gas seal in the feed pipe. After the hazard and operability (HAZOP) study, a slide gate valve was installed in the feed pipe just below the mixing bin to allow the furnace to be operated without choke feeding.

The two campaigns completed to date demonstrated that the design of the feed system was fit for purpose. None of the challenges experienced during the campaigns were due to the design and arrangement of the feed system. Therefore, no changes were made to the feed system.

### *Furnace Dimensions*

The shell of the furnace was manufactured from mild steel as per the dimensions indicated in Fig. 2.

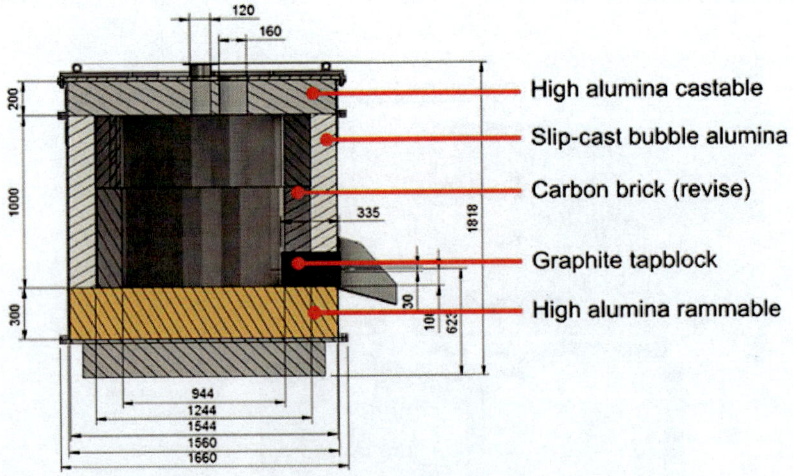


**Fig. 2** Furnace shell and refractory design for campaign 1 (dimensions in mm)

The furnace shell dimensions were based on the refractory design calculations. The steel shell, together with the refractory material, made up the furnace body and created a process cavity size of approximately 0.7 $m^3$. The shell design consisted of 3 major sections: the base, roof, and sidewall. These were also the sections in which the furnace body was fabricated, which simplified the transport and assembly of the furnace. The base and sidewall of the furnace were bolted together, after which the hearth was cast followed by the slip-cast bubble alumina refractory. The roof section was cast separately. The casting process was simplified by using steel formers to occupy the space for the main cavity while Styrofoam formers were used for all the auxiliary ports. The roof was fitted on the main body with the roof section slightly elevated and the refractory cured using a propane gas flame. Once the cast refractory material was cured, the roof was removed and the furnace cavity lined with carbon refractory bricks. Once the bricks were installed, the furnace roof was re-fitted to the main furnace body. Securing all the furnace sections together was achieved by bolting the flanges of the mating sections.

There were no issues experienced with the furnace hearth and sidewall refractory during operation. There were, however, issues encountered with the furnace roof design. The furnace roof in the first campaign consisted of a steel sheet with ports for electrodes and feed pipe. The steel sheet was supported by steel ribs to avoid warping. This design resulted in steel in between the electrodes, in an area known as the delta section. The presence of conductive matter in the delta section caused stray arcing on the roof. It is believed this phenomenon contributed to the roof refractory failure and eventually burn through in the roof. This necessitated the re-design of the roof for the second campaign.

Figures 3 and 4 illustrate the re-designed furnace roof with a fully cast delta section. Additional refractory anchors were added in this section to support the refractory material overhang. Not only was the steel in-between the electrodes removed, a 100 mm thick refractory disk was also cast to increase the thickness of the roof to 300 mm. This additional refractory material provides an extra layer of protection to the furnace operators in the event the roof overheats again. A spare disk was also made available to ensure timely replacement and to minimize furnace downtime should another failure occur in the delta section during Campaign 2.

## *Furnace Containment*

In HCFeMn production, modern refractory designs [6, 7] were typically based on a conductive lining design philosophy [8]. It was found that the reline-to-reline time significantly increased when process materials were frozen as a protective layer onto the hot face of the refractory system [6, 7]. For the PreMa campaign, quantifying the energy consumption of the process was one of the key objectives of the project, and given that energy losses from a small furnace were expected to be significant, it was decided to follow an insulating design philosophy typical of older refractory designs [6, 9]. Small furnace tends to freeze quickly since we are working with a

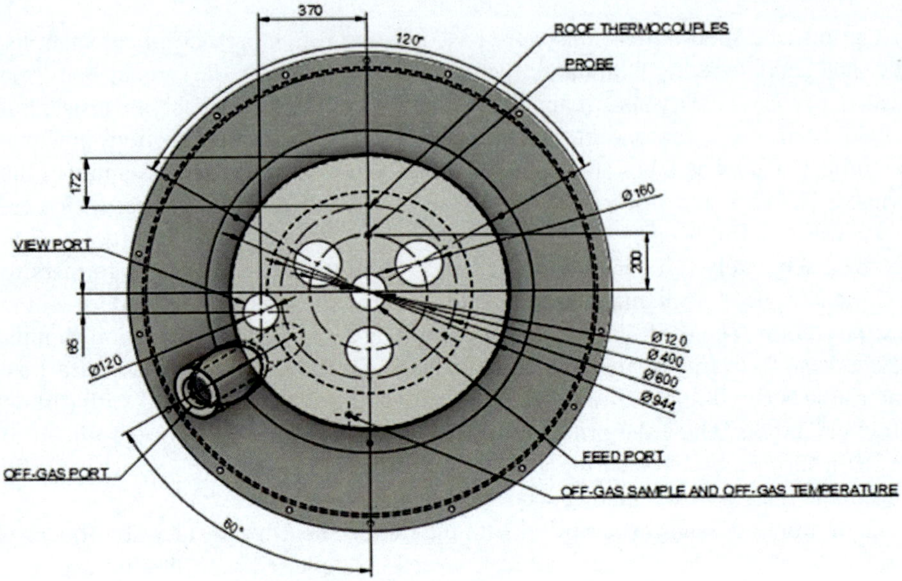


**Fig. 3** Roof design for campaign 2

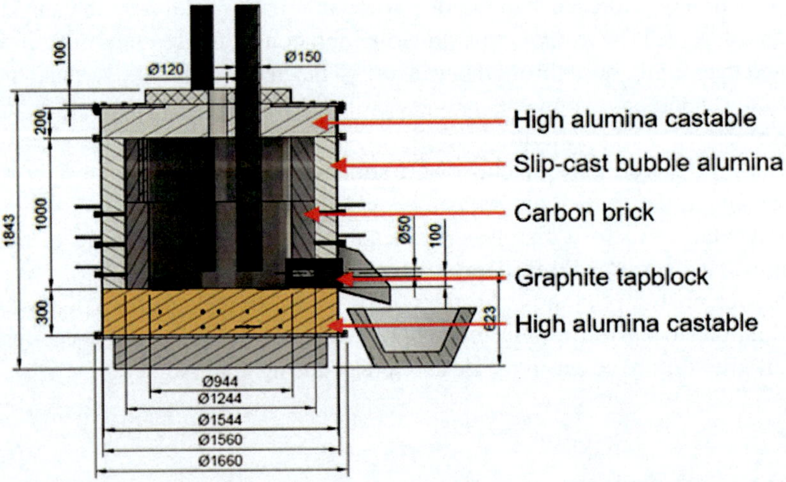


**Fig. 4** Furnace shell and refractory design for campaign 2

small quantity of molten material. Refractory design calculations were conducted in 1-D (1 Dimension) and for steady state conditions. For the refractory design, three sets of calculations were made: sidewall, hearth, and roof.

The heat losses for each zone and the total heat losses for the furnace are plotted in Fig. 5 over the duration of campaign 2a. The overall heat losses during campaign 2a were only 15.5% of the total power input. The heat losses are significantly lower

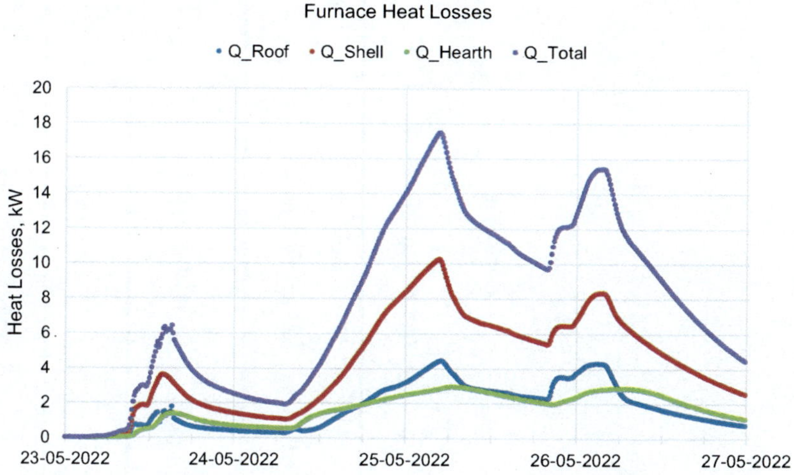


**Fig. 5** Heat losses from different zones of the furnace

than the 50% that was predicted based on previous experience. This is good because most of our energy input is utilized in the actual process. The heat losses are higher in the sidewall and lower in the hearth and the roof. This is because the available surface area of heat losses is bigger for the sidewall. There are more or less the same heat losses in the in the hearth and the roof because the available surface area is the same.

## Sidewall Refractory Design

It was decided to assess two types of refractory materials on the sidewalls. On the hot-face, material more compatible with high MnO slag was selected, i.e. C-based [6, 9] and MgO-based [10] refractory. The thermal conductivities of both refractory materials (C-based and MgO-based) were fairly high at the targeted operating temperatures. For the cold-face, materials with a very low thermal conductivity were selected for assessment, i.e. slip-cast bubble alumina (0.5 W/mK) and fireclay brick (1.4 W/mK). In industrial applications, C-based refractory materials were installed either as rammables [11] or as bricks [6, 7] and fireclay as bricks [9]. In pilot-scale applications, MgO-based refractory material was installed on the sidewall as bricks [10] and in the hearth as bricks or rammables, depending on the application. Bubble alumina is applied in laboratory-scale applications for insulation and sample support, typically in induction furnace experiments [12]. From an installation perspective in industrial application, when a rammable was installed as hot face refractory in the hearth and sidewalls, the back-lining had to be made of a pre-shaped material either as bricks or as a pre-cast shape with significant physical strength. If the back-lining was made of a rammable or pre-cast shape that was fairly fragile, the hot face refractory had to be made of bricks or a pre-cast shape. It was decided to slip-cast bubble

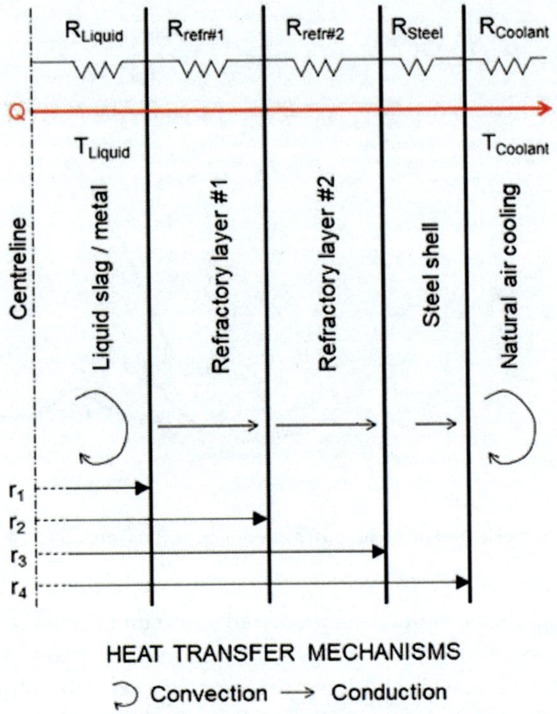


**Fig. 6** Variables applicable to sidewall refractory design

alumina in-situ using a pre-shaped polystyrene former followed by installation of pre-cut carbon bricks as hot-face refractory, using a carbon-based cement as sealant. The alumina slip-cast refractory was chosen to reduce heat losses to the surrounding through radiation and convection. The 1-D, steady state heat transfer calculations applied were described elsewhere [8]. The variables applicable were defined in Fig. 6 and quantified in Table 2.

## Hearth Refractory Design

For the hearth refractory, material compatibility with the sidewall refractory and having a low thermal conductivity were the important selection criteria as chemical wear of the refractory by alloy was not a concern. Alumina-based castable or rammable material was, therefore, selected. The heat transfer calculations applied were similar to the ones described elsewhere [8]. The variables applicable were defined in Fig. 7 and quantified in Table 3.

**Table 2** Quantified variables applicable to sidewall refractory design based on C-refractory where numbers in italic are inputs to calculations and numbers in black the results

| Var | Value | Unit | Description | Var | Value | Unit | Description |
|---|---|---|---|---|---|---|---|
| $r_1$ | *472* | mm | Liquid/refractory 1 interface | $h_{liquid}$ | 75 | $W/m^2$ | Coefficient of convection for liquid |
| $r_2 - r_1$ | *150* | mm | Refractory 1 thickness | $k_{refr1}$ | 10 | W/mK | Coefficient of conduction for C-based refractory |
| $r_2$ | 622 | mm | Refractory 1/refractory 2 interface | $k_{refr2}$ | 0.5 | W/mK | Coefficient of conduction for $Al_2O_3$-based refractory |
| $r_3 - r_2$ | *150* | mm | Refractory 2 thickness | $k_{steel}$ | 36 | W/mK | Coefficient of conduction for steel |
| $R_3$ | 772 | mm | Refractory 2/steel interface | $h_{air}$ | 300 | $W/m^2$ | Coefficient of convection for air |
| $r_4 - r_3$ | *8* | mm | Steel thickness | $R_{liquid}$ | 4.5E−03 | | Resistance to heat flow—liquid |
| $r_4$ | 780 | mm | Steel/air interface | $R_{refr1}$ | 4.4E−03 | | Resistance to heat flow—*C*-based refractory |
| $T_1$ | *1500* | °C | Hot liquid | $R_{refr2}$ | 6.9E−02 | | Resistance to heat flow—$Al_2O_3$-based refractory |
| $T_2$ | 1418 | °C | Refractory 1/refractory 2 interface | $R_{steel}$ | 4.6E−05 | | Resistance to heat flow—steel |
| $T_3$ | 128 | °C | Refractory 2/steel interface | $R_{coolant}$ | 6.8E−04 | | Resistance to heat flow—air |
| $T_4$ | 127 | °C | Steel/air interface | $h$ | *1* | m | Shell height |
| $T_5$ | *30* | °C | Air | $Q$ | 18.8 | kW | Energy loss through sidewall |

## Roof Refractory Design

When a furnace is operated in SAF mode, the roof refractory should not be exposed to excessive temperatures. This was the design basis for the roof refractory design. Mintek typically installed alumina-based castable refractory material in its furnace roofs t and it was decided to apply the same material. The variables applicable were defined in Fig. 8 and quantified in Table 4.

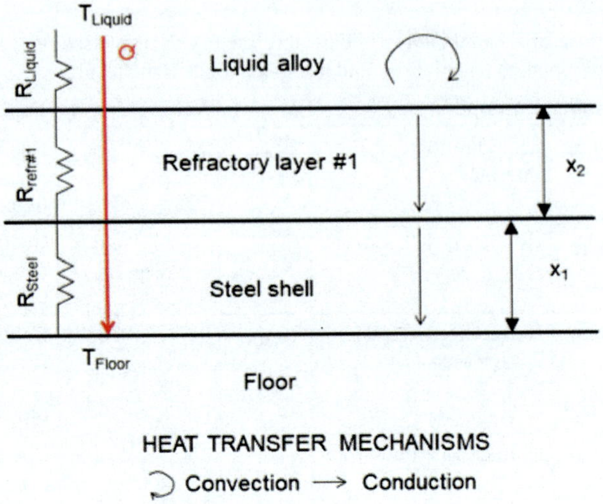


**Fig. 7** Variables applicable to hearth refractory design

**Table 3** Quantified variables applicable to hearth refractory designs where numbers in italic are inputs to calculations and numbers in black the results

| Var | Value | Unit | Description |
|---|---|---|---|
| $x_1$ | *300* | mm | Thickness of refractory 1 |
| $x_2$ | *20* | mm | Thickness of steel |
| $T_1$ | *1400* | °C | Temperature of liquid alloy |
| $T_2$ | 75 | °C | Temperature of refractory 1 cold face |
| $T_3$ | 69 | °C | Temperature of steel cold face |
| $T_4$ | *30* | °C | Temperature of floor |
| $h_{alloy}$ | *75* | W/m$^2$ | Coefficient of convection for liquid |
| $k_{refr1}$ | *2.4* | W/mK | Coefficient of conduction for $Al_2O_3$-based refractory |
| $k_{steel}$ | *36* | W/mK | Coefficient of conduction for steel |
| $R_{liquid\ alloy}$ | 7.0E−03 | – | Resistance to heat flow—liquid |
| $R_{refr1}$ | 2.4E−01 | – | Resistance to heat flow—$Al_2O_3$-based refractory |
| $R_{steel}$ | 1.1E−03 | – | Resistance to heat flow—steel |
| $A$ | 1.9 | m$^2$ | Area of hearth |
| $Q$ | 5.5 | kW | Energy loss through hearth |

## Lessons in Process Control

This part of the paper discusses the lessons learned related to the control of the process. There were a number of challenges during the two campaigns completed to date. These challenges provided insights and guidance about improvements

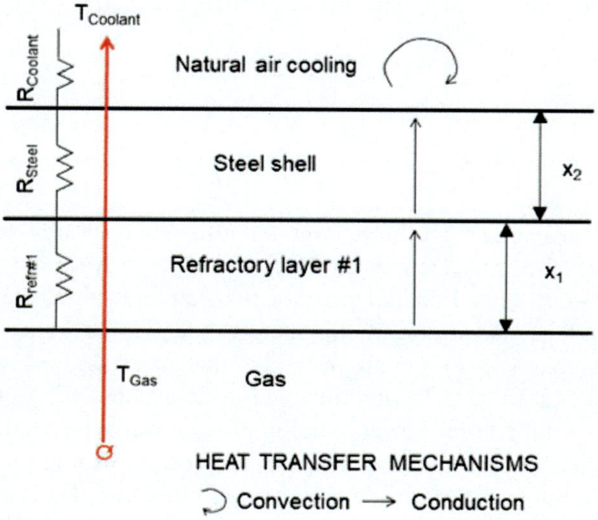


**Fig. 8** Variables applicable to roof refractory design

**Table 4** Quantified variables applicable to roof refractory designs where numbers in italic are inputs to calculations and numbers in black the results

| Variable | Value | Unit | Description |
|---|---|---|---|
| $x_1$ | *200* | mm | Thickness of refractory 1 |
| $x_2$ | *20* | mm | Thickness of steel |
| $T_1$ | *1500* | °C | Temperature of hot gas |
| $T_2$ | 71 | °C | Temperature of refractory 1 cold face |
| $T_3$ | 61 | °C | Temperature of steel cold face |
| $T_4$ | *30* | °C | Temperature of air |
| $h_{\text{hot gas}}$ | *300* | $W/m^2$ | Coefficient of convection for hot gas |
| $k_{\text{refr1}}$ | *2.4* | W/mK | Coefficient of conduction for refractory 1 |
| $k_{\text{steel}}$ | *36* | W/mK | Coefficient of conduction for steel |
| $h_{\text{air}}$ | *300* | $W/m^2$ | Coefficient of convection for air |
| $R_{\text{hot gas}}$ | 1.7E−03 | – | Resistance to heat flow—hot gas |
| $R_{\text{refr1}}$ | 1.6E−01 | – | Resistance to heat flow—refractory 1 |
| $R_{\text{steel}}$ | 1.1E−03 | – | Resistance to heat flow—steel |
| $R_{\text{coolant}}$ | 1.7E−03 | – | Coefficient of convection for air |
| $A$ | 1.9 | $m^2$ | Roof area |
| $Q$ | 9.0 | kW | Energy loss through sidewall |

that needed to be implemented. The major challenge experienced during the first campaign was coke bed management.

## *Coke Bed Management*

Good coke bed management is important for efficient operation of the SAF of this size. The coke bed management in the first campaign was poor. This resulted in numerous process-related challenges. For the first campaign, the start-up heel included coke and the coke in the feed recipe was added at 20% excess. During the warm-up period, there was a lot of arc instability that was stabilized by adding coke in 10 kg batches. All these coke additions led to an accumulation of a large coke bed which made it difficult for power to be supplied towards the hearth. Most of the power was supplied towards the roof instead of the hearth, leading to freezing of the material at the bottom of the furnace and tapping problems. The supply of power towards the roof also contributed to the roof refractory failure and the eventual burn through. Excavation of the furnace revealed that indeed the coke bed was too large and the electrode tips were located closer to the top of the furnace as shown in Fig. 9.

Because of the challenges experienced in the first campaign, several measures were implemented in the subsequent campaign to improve coke bed management. Firstly, there was no coke added in the start-up heel. The coke in the feed recipe was added at 100% stoichiometry. Arc instability was also experienced during warm up

**Fig. 9** Photograph showing electrode tip position determined during the excavation of the furnace

in the second campaign. The instability of the arc was stabilized by coke addition but this time it was only added in 1 kg batches. These measures had a positive impact on the operation and efficiency of the furnace. The furnace operator was able to keep the electrode tips down. This campaign managed to achieve a number of successful taps before the campaign was terminated prematurely due to failure of the power supply. The furnace excavation revealed that the electrode tips were located closer to the bottom of the furnace and the coke bed was only 170 mm in height.

## *Furnace Excavation*

After the premature termination of the second campaign, the furnace was systematically excavated to get the profile. Since the furnace was operating in submerged arc mode prior to termination, there was consensus among team members that the profile will be as good as that of a stabilized furnace. The excavation provided vital information about the operation of the furnace.

### Systematic Furnace Excavation Method

The roof of the furnace was removed without moving the electrodes. Six samples were taken from the top of the furnace burden at spots indicated in Fig. 10a. After collecting the first set of samples, a height of 10 cm of burden materials was removed from the furnace. The second set of 6 samples for the next level were collected. This was done for every 10 cm until the furnace hearth was reached as shown in Fig. 10b. The labelling of the samples was done by combing the level name and the sampling point name, e.g., the sample taken at L01 and sampling point 1 is labelled L01BI. Pictures were taken for each level before removing samples. The sample labelling key is shown in Table 5.

### Furnace Profile

Upon removing the roof, it was discovered that 30 cm of height in the furnace was empty. As such, sampling only began at level 7 (L07). The furnace appeared to be well mixed, the proportion of raw materials in a sample taken at any given point was similar. The burden was 47 cm in height as a result samples were collected at 5 levels. The percentage of fines present in the samples increased as the levels dropped towards the furnace hearth. After all the burden material was excavated, the wet coke bed was observed which was 23 cm above the furnace hearth. The coke bed was hard, therefore, a jack hammer was used for further excavations. Due to the difficulty to remove the coke bed, the decision was made to collect samples at four cardinal points with the tap hole area labelled North. The coke bed sample was also collected at the centre on the furnace. As the coke bed descended lower into the furnace, alloy

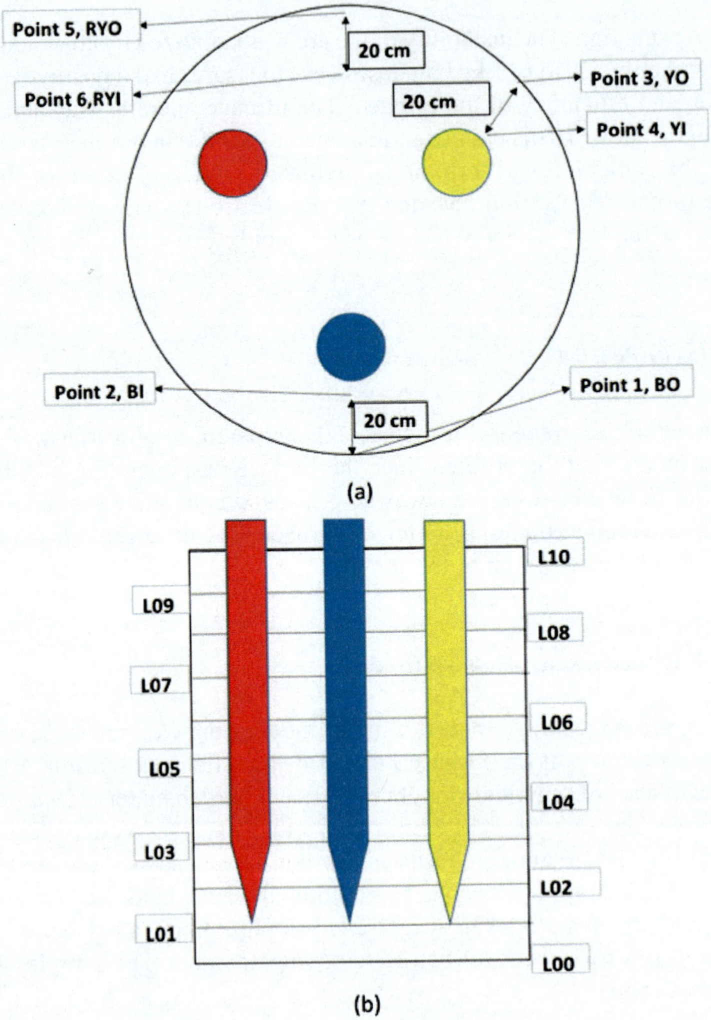


**Fig. 10** Showing sampling points from **a** top view and **b** side view

**Table 5** Sample labelling key

| Key | Definition |
|---|---|
| BI | Blue electrode, inner point |
| BO | Blue electrode, outer point |
| YI | Yellow electrode, inner point |
| YO | Yellow electrode, outer point |
| RYI | Mid-point of red &yellow electrode, inner point |
| RYO | Mid-point of red &yellow electrode, outer point |

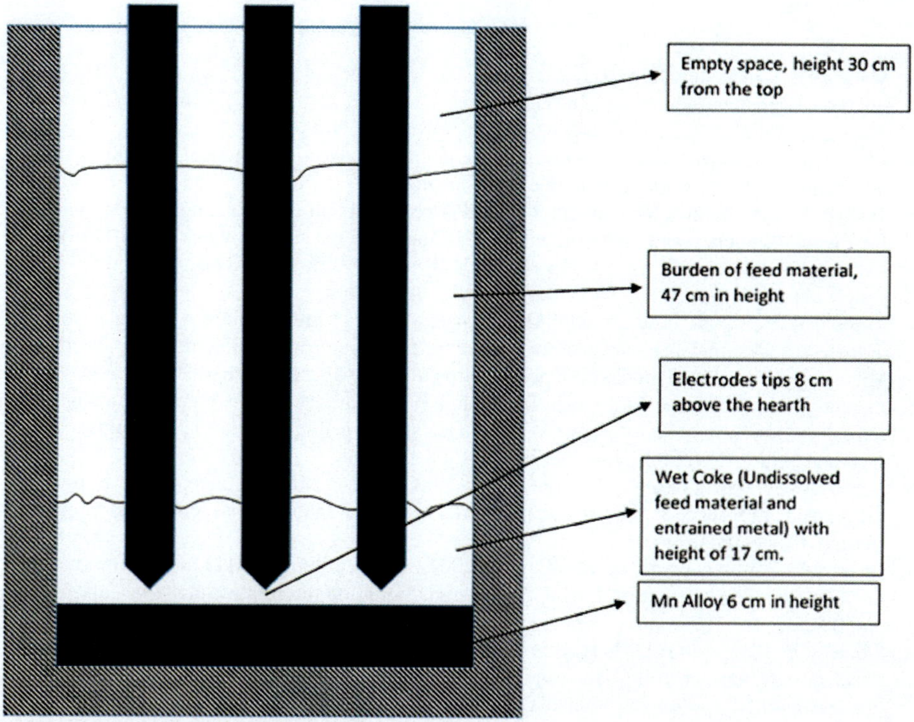


**Fig. 11** Furnace profile from the excavation

entrainments were observed. All three electrode tips were completely immersed into the coke bed at approximately 2 cm above the alloy and 8 cm above the hearth. Mn alloy was found at the bottom of the furnace at a height of 6 cm from the hearth. The cross-section of the furnace profile is depicted in Fig. 11.

## Next Steps

The next steps are to (i) receive delivery of the repaired power supply; (ii) installation and testing of the power supply; (iii) re-assembly of furnace and preparation of facility; and (iv) restart the furnace and continue with the campaign. Campaign 2b is expected to be executed in the third week of October 2022.

**Acknowledgements** The PreMa project is funded by the European Union's Horizon 2020 Research and Innovation Programme under Grant Agreement No 820561 and industry partners: Transalloys, Eramet, Ferroglobe, OFZ, and Outotec. The paper is published with permission from Mintek.

## References


1. Matinde E, Steenkamp JD (2021) Metallurgical overview and production of slags. In: Environmental geochemistry of metallurgical slags. The Royal Society of Chemistry, Cambridge (Online). Available: https://doi.org/10.1039/9781839164576-00014
2. ASTM-A99, Standard specification for ferromanganese. West Conshohocken, 2003: ASTM international, 2009 (Online). Available: www.astm.org
3. Tangstad M, Ichihara K, Ringdalen E (2015) Pretreatment unit in ferromanganese production. In: Energy efficiency and environmental friendliness are the future of the global Ferroalloy industry, Kiev, Ukraine, pp 99–106 (Online). Available: http://pyro.co.za/InfaconXIV/099-Tangstad.pdf
4. Ringdalen E, Gjøvik JE, Larssen TA, Tangstad M (2021) Pretreatment of manganese ores in different gas-atmospheres-a method to reduce energy consumption and $Co_2$ emissions in Mn-Alloy production, Trondheim Norway. https://doi.org/10.2139/ssrn.3930059
5. Lagendijk H, Xakalashe B, Ligege T, Ntikang P, Bisaka K (2010) Comparing manganese ferroalloy smelting in pilot-scale AC and DC submerged-arc furnaces. In: Infacon XII: sustainable future, Helsinki, Finland, pp 497–508
6. Hearn AM, Dzermejko AJ, Lamont PH (1998) 'Freeze' lining concepts for improving submerged arc furnace lining life and performance. In: 8th international ferroalloys congress, Beijing, China, pp 401–426
7. Hearn AM, Van Rensburg ASJ, Henning JR (2004) 'Freeze' lining on M12 furnace: motivation, installation and operation. In: INFACON X: transformation through technology, Cape Town, pp 500–507
8. Steenkamp JD, Denton GM, Hayman DA (2017) Insulating or conductive lining designs for electric furnace smelting? In: Applications of process engineering principles in materials processing, energy and environmental technologies: a symposium in honor of professor Ramana G. Reddy, San Diego, pp 209–220
9. Brun H (1982) Development of refractory linings for electric reduction furnaces producing Mn alloys at Elkem A/S-PEA plant, Porsgrunn, Norway. J Inst Refract Eng, pp 12–23
10. Muller J, Coetsee T, Groenewald JA, Du Toit AA, Zeelie D (2015) The AlloyStream process for HCFeMn production. In: INFACON XIV, Kiev, Ukraine, pp 91–98
11. Steenkamp JD, Pistorius PC, Tangstad M (2015) Chemical wear analysis of a tap-hole on a SiMn production furnace. J Southern Afr Inst Min Metall 115(3):199–208
12. Banda WK, Steenkamp JD, Matinde E (2020) An investigation into the wear mechanisms of carbon- and silicon carbide-based refractory materials by silicomanganese alloy. J S Afr Inst Min Metall 120(5)

# A Desktop Study on the Potential Use of South African Slags as Thermal Storage Medium

**Sifiso N. Sambo, Joalet D. Steenkamp, and Pieter J. A. Bezuidenhout**

**Abstract** Waste slag sources from South African pyrometallurgical smelters are evaluated to determine their potential for use as sensible heat storage media in Concentrated Solar Power applications. This offers the possibility of repurposing slag waste materials while tackling their environmental and handling problem. This work studies the key performance indicators of thermal energy storage systems to assess the suitability of pyrometallurgical slag for use as filling materials in packed bed thermal energy storage. The thermophysical properties of selected slags and their potential for long-term thermal stability is studied through the equilibrium module of FactSage 8.0 from 100 to 1000 °C. A validated model is employed to study the heat transport process in the fluid–solid interface with air as the heat transfer fluid. The maximum temperature of the studied thermal storage cycles is 750 °C. The overall system performance shows that pyrometallurgical slags have potential as energy storage materials.

**Keywords** South African slag · Crystal structure · Thermal stability · Numerical model

## Introduction

Renewable energy sources have a significant shortfall in that most of the sources are intermittent and the supply of energy fluctuates. To address this shortcoming, energy storage technologies can be employed. The International Energy Agency (IEA) has also recognized that energy storage technologies can support energy security and is

S. N. Sambo (✉)
MINTEK, 200 Malibongwe Drive, Strijdom Park, Randburg, South Africa
e-mail: Sifisos@mintek.co.za

J. D. Steenkamp
Department of Chemical and Metallurgical Engineering, University of the Witwatersrand, Johannesburg, South Africa

P. J. A. Bezuidenhout
Mechanical Engineer, MINTEK, Pyrometallurgy Division, Randburg, South Africa



C. Fleuriault et al. (eds.), *Advances in Pyrometallurgy*, The Minerals, Metals & Materials Series, https://doi.org/10.1007/978-3-031-22634-2_23

crucial to the decarburization of the energy sector [1]. Thermal energy storage (TES) is one of the most effective techniques for storing energy and its versatility has made it an ideal technology for improving the efficiency of different energy systems [2]. One of the more common industrial application of TES is in concentrated solar power (CSP) plants. This technology aids in addressing the mismatch between energy demand and supply on a CSP plant by storing excess solar energy in the form of heat to be used at a later period when demand arises.

There exist three basic techniques for TES. Sensible heat storage, latent heat storage, and thermochemical energy storage. Sensible heat storage employs a storage medium which can be in any state, as long as it remains in that state over the temperature range encountered during the TES process [3]. Latent heat storage relies on a medium that stores and releases energy through phase change. Latent heat storage generally has a much higher energy density than sensible heat storage and the heat extraction process can be designed to be nearly isothermal [2]. Thermochemical energy storage techniques store/release heat through reversible chemical reactions [2].

Sensible heat storage techniques have attracted the most interest for applications in CSP plants because they offer a low-cost solution with good thermal energy storage performance [3]. This technique uses a heat transfer fluid (HTF) to store energy (charging) in a solid medium and recover the energy (discharging) when the CSP plant energy supply is not available. Packed-bed systems have shown to achieve good thermal efficiencies and the simplicity of these systems makes it an ideal technique for application in TES [4, 5].

The choice of solid filler materials is limited by their thermophysical properties, range of working temperature, and compatibility with the HTF [6]. The filler materials are required to have high thermophysical properties and long-term thermal, chemical, and mechanical stability under thermal cycling. Chemical stability entails compatibility with the heat transfer fluid and the encompassing wall material of construction. Mechanical stability entails low thermal expansion and high compressive strength [7, 8].

Optimal operation of packed bed TES can be achieved by charging from the top and discharging from the bottom. This method takes advantage of the buoyancy stratification principle [9]. The ratio of the storage tank height and the length of the section above the thermocline (available useful energy) indicates the effectiveness of the TES system [9]. Improvements in stratification result from high aspect ratio of the storage tank. This is the column height to diameter ratio. The authors in reference [10] reported that an aspect ratio above 2 is enough to affect good stratification and higher ratios may result in no further significant improvements.

Charging time is dependent on the thermal properties of the filler materials. The energy requirements for charging the packed bed increase proportionally with filler materials' thermal capacity, the product of the solid materials' density, and heat capacity [11]. This reduces the rate of temperature increments during the charging cycle but leads to a high energy density resulting in reduced packing volume requirements. The thermal conductivity of the filler material also plays an important role on the charging and discharging rate of the packed bed [7, 9, 11].

Several filler materials such as natural rocks, concrete, and compound materials have been investigated for their potential as filler materials [9, 10, 12]. This work studies the potential of using South African pyrometallurgical slag as filler materials and air as the HTF. Thermal storage up to 750 °C of HTF on discharge is considered. The filler materials' thermophysical properties and their potential long-term stability is evaluated through FactSage 8.0 [13]. The thermal storage performance is studied through numerical modelling of thermal cycles.

The advantages of using air as the HTF is its availability, no toxicity, and its ability to maintain its form at any temperature [4]. However, the reduced density of air at high temperatures significantly increases its superficial flow velocity, consequently increasing pumping power requirements [9].

Pyrometallurgical slag is generally a waste source with limited applications in the current industrial market, and as a result is generally landfilled in large slag dumps. The chemical and thermal properties of pyrometallurgical slag are fairly well understood. The availability of slag in the processed form in landfills promises the benefit of cost reduction while mitigating their environmental impact. The ferroalloy and platinum group metals (PGM) industries are the major producers of slag in South Africa [4, 13].

Installed capacities are estimated at 5.4 million ton ferrochrome (FeCr), 0.99 million ton high carbon ferromanganese (HCFeMn), 0.16 million ton silicomanganese (SiMn), and 1.44 million ton PGM concentrate smelting [13–15]. All of these commodities are produced or processed using electric furnaces with slag to valuable product ratios ranging between 765 and 5740 kg/ton, see Table 1. The slags produced are either stored in landfill or sold as low-value products i.e. aggregates or shot for sandblasting.

**Table 1** Typical compositions (weight percent) of slags produced and slag/alloy or slag/matte[a] ratios (kg slag/ton alloy or matte)

| Slag type | | $Al_2O_3$ | CaO | $Cr_2O_3$ | FeO | MgO | MnO | $SiO_2$ | Slag/Alloy |
|---|---|---|---|---|---|---|---|---|---|
| HCFeMn | Average | 8.2 | 29.9 | – | 6.3 | 23.6 | 28.3 | 8.2 | 765 |
| | Std. dev | 7.6 | 2 | – | 0.6 | 6.3 | 3.8 | 7.6 | 70 |
| SiMn | Average | 4.7 | 25.9 | – | 0.4 | 5.4 | 14.4 | 46.3 | 1025 |
| | Std. dev | 0.1 | 1 | – | 0.1 | 0.5 | 2.3 | 0.1 | 35 |
| FeCr | Average | 23.3 | 3.4 | 11.4 | 7.5 | 26.8 | – | 31.5 | 1138 |
| | Std. dev | 2.3 | 0.6 | 4.5 | 8.8 | 5 | – | 4.5 | 48 |
| PGM | Average | 3.3 | 8.8 | 1.5 | 17.8 | 21.7 | – | 44.8 | 5740[a] |
| | Std. dev | 1.6 | 2.7 | 0.9 | 3.3 | 7.6 | – | 2.3 | 1905 |

[a] refers to slag/matte ratio not slag/alloy like the other slags, also specified on the label

## FactSage Simulations

The equilibrium module of the FactSage 8.0 thermochemical simulation software package was used to investigate thermal stability and the specific heat capacity of the different slags [13]. For the purpose of this study, the calculated average slag compositions presented in Table 1 were used. Temperatures between 100 and 1000 °C were considered. Indicators of potential long-term thermal stability, as well as the thermophysical properties required for the modeling of the heat transport of the thermal energy storage cycles, were obtained.

### *Liquid Phase Formation at Equilibrium*

The potential for liquid phase formation in the temperature range 500 and 1500 °C was calculated for each slag composition at equilibrium. Calculations were done using the FactPS and FToxid databases and normalized slag compositions. Figure 1 presents results for the liquid phase formation of the different slags. No liquid phase formed in the temperature range of 100–1000 °C. The solidus temperatures were calculated at 1052 °C for SiMn, 1180 °C for PGM, 1231 °C for FeCr, and 1255 °C for HCFeMn slag. The liquidus temperatures were calculated at 1302 °C for SiMn, 1500 °C for PGM, > 1500 °C for FeCr, and > 1500 °C for HCFeMn slag. The maximum temperature of this work (750 °C) is significantly lower than the liquid formation temperature for all slags.

### *Phase Distribution (Equilibrium)*

The phase composition in the temperature range 100 and 1000 °C under equilibrium conditions were calculated for each slag composition. The equilibrium phase

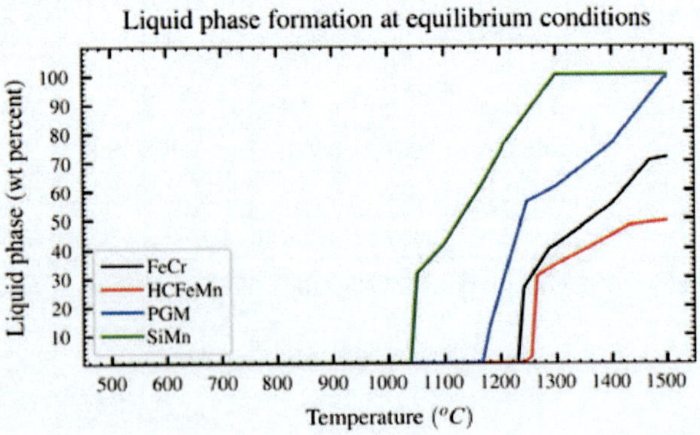


**Fig. 1** Liquid phase formation under equilibrium conditions, as calculated in FactSage 8.0

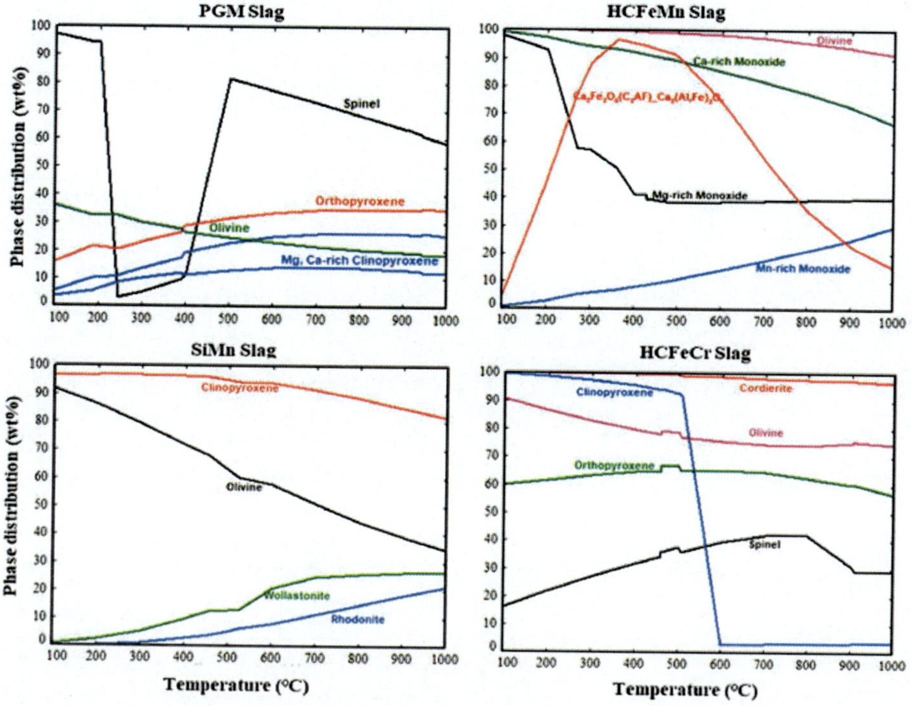


**Fig. 2** Equilibrium phase distribution as a function of temperature for the different slags

distributions as a function of temperature for the different slags are presented in Fig. 2.

Table 2 present the slags' mineral phases and their crystal structures at equilibrium. The slag with the least number of crystal structures is HCFeMn, at 3 with 4 different crystalline phases potentially forming. The other slags have 4 different crystal structures, but 7 different crystalline phases for the SiMn slag and 6 different crystalline phases for the HCFeCr and PGM slags. The number of crystal structures is of importance based on the assumption that a change in crystal structure during thermal cycles could lead to the decrepitation of the slag. Decrepitation results in fine particles, indicating compromised long-term thermal stability of the filler materials. This further increases the pressure drop requirements in the storage tank, consequently increasing HTF pumping power requirements.

## *Specific Heat Capacity and Thermal Conductivity*

The product of the materials' density and heat capacity, referred to as the thermal capacity, is used to assess the thermal storage capacity of the filler materials. The thermal conductivity was calculated using the model prepared by Mills et al. [16],

**Table 2** Crystal structures present in each of the typical slags at equilibrium conditions in the temperature range 100–1000 °C

| Slag type | Cubic | Monoclinic | Orthorhombic | Tetragonal | Triclinic |
|---|---|---|---|---|---|
| HCFeMn | Mn, Mg, Ca-rich monoxide | – | Olivine | $Ca_3MgAl_4O_{10}$ | – |
| SiMn | Grossularite, Spessartine | Clinopyroxene | Olivine | – | Wollastonite, Anorthite, Rhodonite |
| FeCr | Spinel | Sapphirine | Orthopyroxene, Olivine, Cordierite | – | Anorthite |
| PGM | Spinel | Mg, Ca-rich clinopyroxene, Mg-rich clinopyroxene | Orthopyroxene, Olivine | – | Anorthite |

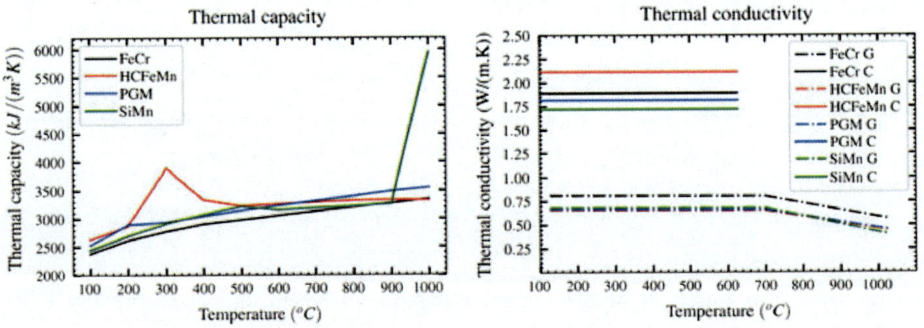


**Fig. 3** Thermal capacity and conductivity of the different slag types with temperature, G: glassy slag, C: crystalline slag

which produced comparable results to experimental measurements on slags of the same mineral composition as reported in literature [17]. The model is able to calculate the thermal conductivities for both glass and crystalline slags. Figure 3 presents the thermal capacity and thermal conductivity variation with temperature from 100 to 1000 °C. The increasing thermal capacity of SiMn may be due to sintering at it approaches its liquid phase formation temperature.

## Numerical Modelling of Thermal Energy Storage Cycles

A validated one-dimensional three-phase numerical model was used to simulate the thermal energy storage system [12]. The model was developed in previous work to study packed bed heat transfer between air and manganese ore in a shaft type column

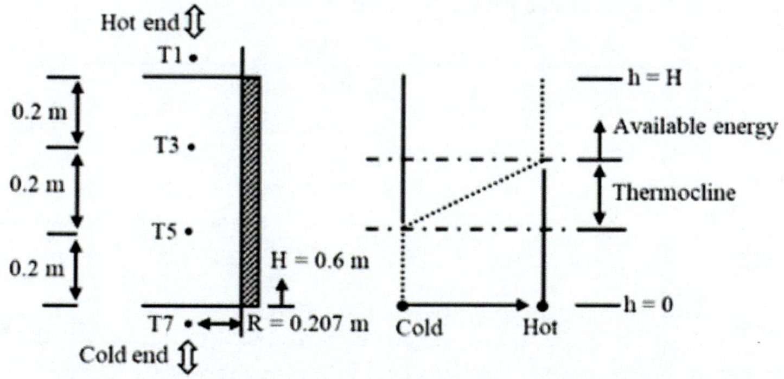


**Fig. 4** Schematic illustration of the simulated packed bed system with a depiction of a thermocline

[18]. The studied system in the previous work was on the preheating of manganese ore from room temperature to 600 °C with air fed at heat increments from room temperature to 750 °C. Details on model development are discussed in a Master's thesis [19].

The root mean square deviation of temperature profiles predicted by the model and experimental measurements quantified the deviation of the predicted temperature profiles to the measured temperature profiles. An average temperature error of 23 °C was reported across the packing temperature points for the entire process duration.

The packed bed dimensions and operating parameters used to simulate the thermal storage packing of this work were adopted from the manganese ore preheating work [18]. This was done to limit any discrepancies which may arise from applying the numerical model outside of its domain of validity. Only the differences in the slags' thermal capacity ($\rho C_p$) and their thermal conductivity ($k$) produce any differences in the results of the simulated heat transport of the thermal storage cycles.

The slags' thermal properties were fitted using sixth order polynomial functions for use in the heat transfer model. Table 3 presents the packed bed geometry and the operating parameters of the charging and discharging cycles of the slag-air thermal storage packing. Figure 4 shows a schematic diagram of the simulated packed bed thermal storage system with a depiction of the thermocline.

The thermal storage system was simulated for 275 min of process time, with four cycles comprising of two charging and two discharging cycles. 150 min of dynamic charging was afforded for the packed bed to reach equilibrium, which took place just before the 100 min point. A discharge cycle of 30 min was opted for, with a cut-off temperature just above 400 °C. A second charging cycle (constant temperature charging) was carried for 45 min, followed by a second discharge cycle. The cycle's progress temperature profiles are presented in Fig. 5.

Figure 6 shows the final temperature profile of the three phases (air, packed slag, and wall) at the end of the charging and discharge cycle along the packing height. The resulting fluid and solid temperature profiles are identical. The reported wall temperature is averaged between the inner and outer ends across its thickness. The

**Table 3** Dimensions of the simulated packed bed thermal storage system

| Operation parameters | Value | Geometry | Value |
|---|---|---|---|
| $G$ (Nl/min) | 1200 | Tank height (m) | 0.6 |
| Void fraction | 0.48 | Tank radius (m) | 0.207 |
| $T_{\text{charging, dynamic}}$ (°C) | 20–750 | Particle diameter (m) | 0.011 |
| $T_{\text{charging, constant}}$ (°C) | 750 | | |
| $T_{\text{Discharging, constant}}$ (°C) | 25 | | |
| $t_{\text{Charging, constant}}$ (min) | 150 | | |
| $t_{\text{Discharging}}$ (min) | 60 | | |
| $t_{\text{Charging}}$ (min) | 60 | | |

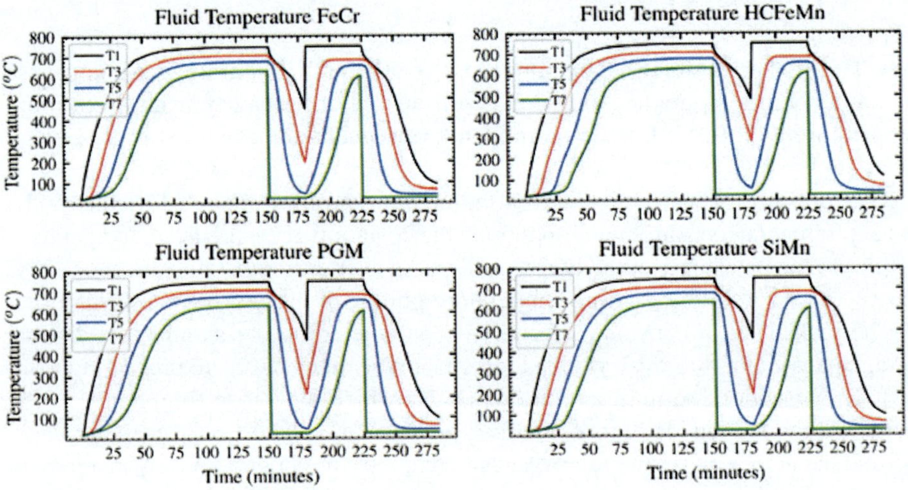


**Fig. 5** Temperature profiles of the thermal cycles for the different slags

reduced temperature at both ends is the result of the boundary conditions applied during model formulation.

The slag containing high energy due to its higher thermal capacity compared to other slags results in higher final temperature profile at the end of the 30 min discharge period. This further gives it a head start on the next charging cycle. The effect of thermal capacity during the constant discharge and charging cycles can be seen by comparing the cycle's progress on feed and exit fluid temperatures, see Fig. 7.

It can be seen that the equilibrium state of the charging cycle is reached at the same time for all slags and all final temperature profiles are identical. This is due to the fixed time for the transfer process between cycles. However, the performance of the different slags can be compared through the phase's energy balance for the duration of the thermal cycles. Theoretically, the packed bed can be charged to the

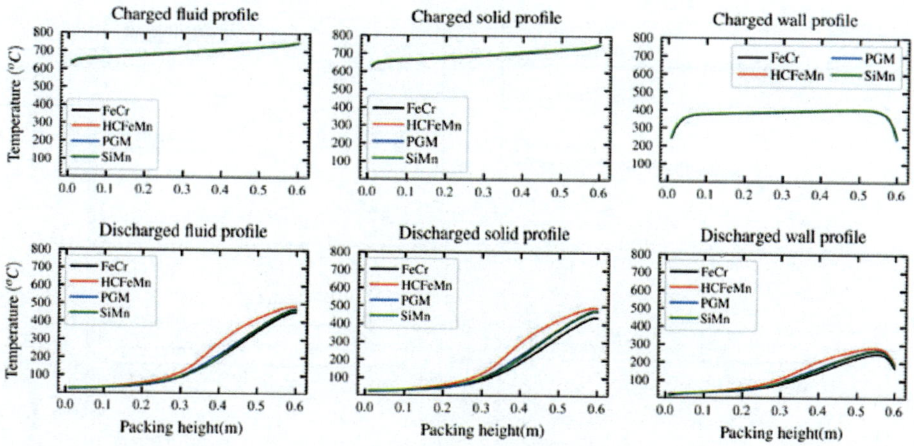


**Fig. 6** Phases charged and discharged final temperature profiles

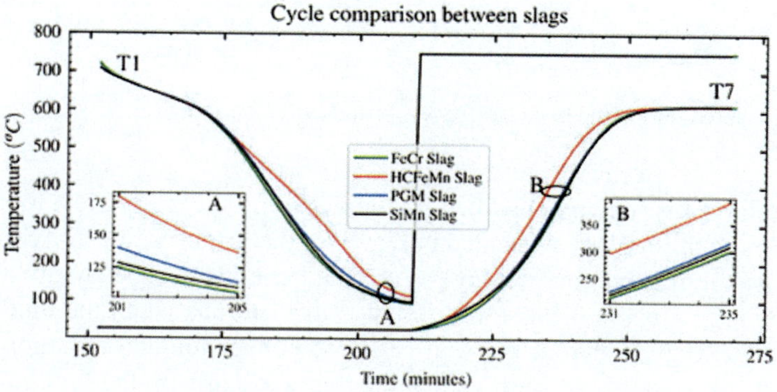


**Fig. 7** Fluid temperature at charging inlet ($T_7$) and exit ($T_1$) points of the packed bed during charging (210–270 min) and discharging (150–210 min) cycles for different slag type as packing materials

maximum temperature of 750 °C across its entire length, the resulting stored energy is referred to as the maximum theoretical stored energy.

The maximum theoretical stored energy further hint at the maximum storage capacity of the filling materials in a TES system. About 45% of the possible thermal storage capacity of the packed slag was exploited, as predicted by the numerical model. This is indicated by the capacity ratios of the different slags presented in Fig. 8. The capacity ratio can be improved by minimisation of energy lost to the environment.

The storage system's energy balance was calculated every time step during the simulation. Performance indicators were deduced from the energy balance as cumulative sums for the duration of the charging and discharging cycles. The net input

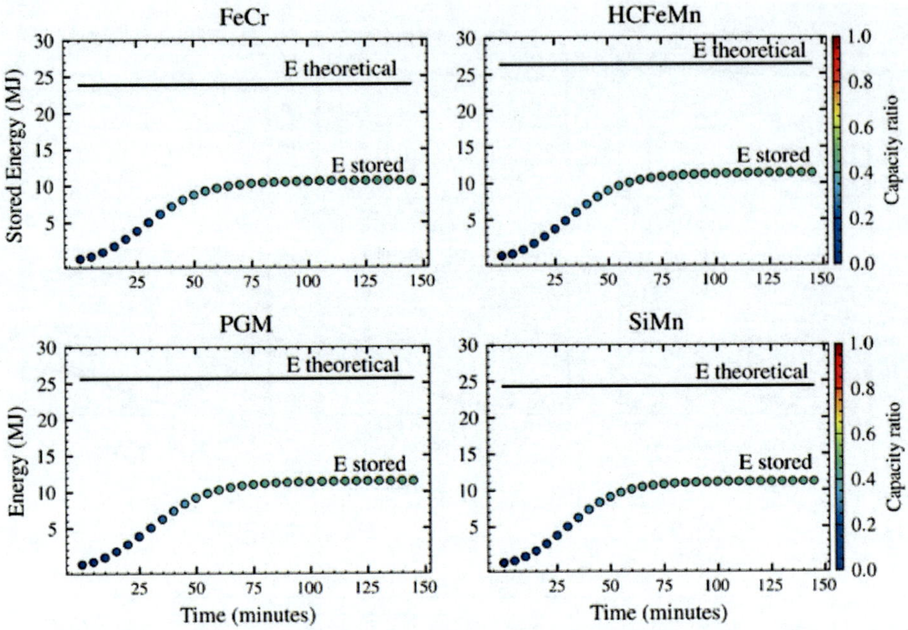


**Fig. 8** Energy stored during the first charging cycle and capacity ratio

energy, stored energy, energy absorbed by the wall, and the energy lost to the environment through the wall are presented in Fig. 9 for the first charging cycle (dynamic charging) and the second charging cycle (constant charging). The energy absorbed by the wall is the sensible heat stored in the wall materials when the unit is operating, and the energy lost from the wall is due to convection and radiation outside the vessel. It can be deduced from the results that the slag with the highest thermal capacity (HCFeMn) requires more energy to charge, as expected.

The second charging cycle benefit from the already present energy due to discharge cut-off temperature and less energy is required to return the packing to the same temperature profile as the initial charging cycle. Further reduction in the energy transferred to the wall, consequently the energy lost to the environment can be

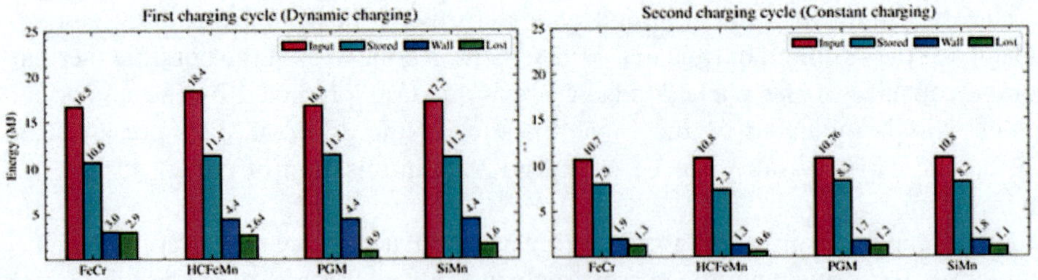


**Fig. 9** System energy at the end of the first and second charging cycle fed with fluid at ramping up temperature (dynamic charging) and constant temperature (constant charging) respectively

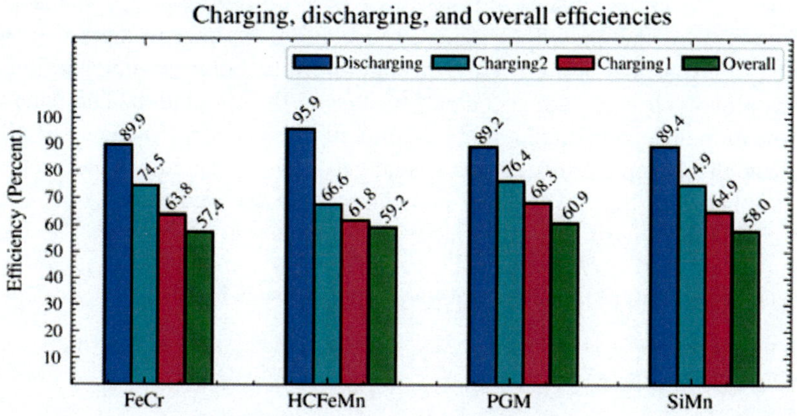


**Fig. 10** System charging and discharging efficiencies

observed. Figure 10 presents the first charging cycle efficiency, the second charging cycle efficiency, the discharging cycle efficiency, and the process overall efficiency. The process overall efficiency is the product of the charging and discharging cycle efficiencies.

The recovery of energy from the wall contributes to the high recovery efficiency. This is further enhanced by the stratification method. The energy required to pump the fluid through the packed bed was not accounted for in the evaluation of the charging and discharging cycle's efficiencies. The pumping energy requirement is dependent on the column height and the particle size distribution of the packing materials. Further demand result from the high superficial flow velocity through the packing due to fluid expansion at high temperatures. Based on the work of Hanchen et al. [20], the simulated thermal storage of this work would result in a fluid pumping energy requirement of no more than 2% of the overall energy input. This means no significant changes can result from including pumping energy demand on the overall system performance, however, experimentally studying the performance of the slag will confirm the accuracy of these assumptions.

## Conclusion

Four or more crystal structures were identified based on the present mineral phases from the FactSage simulation. This confirms the slag decrepitating potential which may compromise the long-term thermal stability of the slags. However, an experimental study of the slag thermal stability with respect to thermal disintegration must be carried out to quantify the possible number of thermal cycles each slag can be in use before replacement. The thermal capacity of all the slags is comparable to other filler materials reported in literature [9]. The advantage of pyrometallurgical slags is their availability and low cost.

The thermal storage capacity ratio of 45% is within an expected range given the performance of the encompassing wall insulation. Its improvement can move the capacity ratio to just above 50%. The stand-by time between the charging and the discharging cycles was not accounted for during the simulation. The benefit of the stratification arrangement can be seen through the recovery efficiency. Both the charging and discharging efficiencies are relatively low compared to values reported in literature [11, 18]. Significant improvement can be achieved by reducing energy lost to the wall and environment.

**Acknowledgements** This paper is published with the permission of MINTEK

## References

1. International Energy Agency (2014) Technology roadmap energy storage, IEA, France. Available: https://www.iea.org/reports/technology-roadmap-energy-storage
2. Gasia J, Miró L, Cabeza LF (2017) Review on system and materials requirements for high temperature thermal energy storage. Part 1: general requirements. Renew Sustain Energy Rev 75:1320–1338. https://doi.org/10.1016/j.rser.2016.11.119
3. Cabeza LF, Sole C, Castell A, Oro E, Gil A (2012) Review of solar thermal storage techniques and associated heat transfer technologies. Proc IEEE 100(2):525–538. https://doi.org/10.1109/JPROC.2011.2157883
4. Zanganeh G, Pedretti A, Zavattoni S, Barbato M, Steinfeld A (2012) Packed-bed thermal storage for concentrated solar power—pilot-scale demonstration and industrial-scale design. Sol Energy 86(10):3084–3098. https://doi.org/10.1016/j.solener.2012.07.019
5. Allen KG, Von Backström TW, Kröger DG (2014) Packed rock bed thermal storage in power plants: design considerations. Energy Procedia 49:666–675. https://doi.org/10.1016/j.egypro.2014.03.072
6. Khare S, Dell'Amico M, Knight C, McGarry S (1013) Selection of materials for high temperature sensible energy storage. Sol Energy Mater Sol Cells 115:114–122. https://doi.org/10.1016/j.solmat.2013.03.009
7. Agalit H, Zari N, Maaroufi M (2017) Thermophysical and chemical characterization of induction furnace slags for high temperature thermal energy storage in solar tower plants. Sol Energy Mater Sol Cells 172:168–176. https://doi.org/10.1016/j.solmat.2017.07.035
8. Tiskatine R et al (2016) Experimental evaluation of thermo-mechanical performances of candidate rocks for use in high temperature thermal storage. Appl Energy 171:243–255. https://doi.org/10.1016/j.apenergy.2016.03.061
9. Calderón-Vásquez I et al (2021) Review on modeling approaches for packed-bed thermal storage systems. Renew Sustain Energy Rev 143:110. https://doi.org/10.1016/j.rser.2021.110902
10. Ortega-Fernández I, Calvet N, Gil A, Rodríguez-Aseguinolaza J, Faik A, Aguanno BD (2015) Thermophysical characterization of a by-product from the steel industry to be used as a sustainable and low-cost thermal energy storage material. Energy 89:601–609. https://doi.org/10.1016/j.energy.2015.05.153
11. Grirate H, Agalit H, Zari N, Elmchaouri A, Molina S, Coutuorier R (2016) Experimental and numerical investigation of potential filler materials for thermal oil thermocline storage. Sol Energy 131:260–274
12. Esence T, Bruch A, Molina S, Stutz B, Fourmigué JF (2017) A review on experience feedback and numerical modeling of packed-bed thermal energy storage systems. Sol Energy 153:628–654. https://doi.org/10.1016/j.solener.2017.03.032

13. FactSage.com. https://www.factsage.com/. Accessed 20 Aug. 2022
14. Basson J, Curr TR, Gericke WA (2007) South Africa's ferro alloys industry- present status and future outlook, New Delhi, vo 11, p 6
15. Jones RT (2005) An overview of Southern African PGM smelting. In: 44th Annual conference of metallurgists, Calgary, Alberta, pp 147–178
16. Mills KC, Yuan L, Jones RT (2011) Estimating the physical properties of slags. J South Afr Inst Min Metall 111:649–658
17. Steenkamp JD, Tangstad M, Pistorius PC (2011) Thermal conductivity of solidified manganese-bearing slags—a preliminary investigation. J South Afr Inst Min Metall, pp 327–343
18. Sambo SN, Hockaday L, Seodigeng T (2021) Experimental study of packed bed heat transfer in a shaft Kiln to pre-heat manganese ore with hot air. SSRN Electron J. https://doi.org/10.2139/ssrn.3926617
19. Sambo SN (2022) The development of a heat transfer model for a Shaft Kiln to preheat manganese ore with hot air. Master's Thesis, Vaal University of Technology, Vanderbijl Park, South Africa
20. Hänchen M, Brückner S, Steinfeld A (2011) High-temperature thermal storage using a packed bed of rocks—heat transfer analysis and experimental validation. Appl Therm Eng 31(10):1798–1806. https://doi.org/10.1016/j.applthermaleng.2010.10.034

# $CO_2$ Free FeMn/Mn Production Through Molten Oxide Electrolysis

**Karen S. Osen, Halvor Dalaker, Ana Maria Martinez, Henrik Gudbrandsen, Zhaohui Wang, and Ida Kero**

**Abstract** Molten Oxide Electrolysis (MOE) is a new $CO_2$-free technology that could alleviate the environmental impact of ferroalloy production. In this work the objectives were to perform (i) experimental lab scale studies of MOE for Mn from MnO using synthetic raw materials and a commercial Ni–Cr–Fe alloy as oxygen-evolving anode, and (ii) theoretical mass and energy balances for different cases of Mn ore purification for a hypothetical industrial MOE FeMn process. Low carbon Mn metal was recovered at lab scale, and the feasibility of MOE manganese production proven. While the Ni–Fe–Cr anode performed satisfactory during the electrolysis, some dissolution of Ni, Fe and Cr was detected. Theoretical energy and mass balance calculations gave a total energy consumption of 6.5–7.6 MWh/tonne Mn depending on the purification scenario. The results showed that the purified ore greatly improves the process. Non-purified ore requires large amounts of flux and more hydrogen in the pre-reduction step, both aspects increasing the need for energy. Non-purified ore also generates large amounts of spent electrolyte. Despite the above, it is an open question whether the benefits of purifications would outweigh the costs.

**Keywords** Manganese · Molten oxide electrolysis (MOE) · Sustainability

## Introduction

Ferromanganese is an important alloying compound for the steel industry and is today produced by carbothermic reduction in Submerged Arc Furnaces (SAF) [1]. This is a productive and efficient process, but it emits carbon dioxide (1–1.6 kg $CO_2$/kg Mn) and thus contributes to global warming [2]. Methods for substituting fossil carbon as reductant is needed to reach the goal of the Paris agreement of keeping a global temperature rise this century below 2 °C [3]. Molten iron production by direct electrolysis of iron ore through Molten Oxide Electrolysis (MOE), has been identified by the American Iron and Steel Institute (AISI) as one of four possible breakthrough

K. S. Osen · H. Dalaker (✉) · A. M. Martinez · H. Gudbrandsen · Z. Wang · I. Kero
SINTEF Industry, 7465 Trondheim, Norway
e-mail: halvor.dalaker@sintef.no



C. Fleuriault et al. (eds.), *Advances in Pyrometallurgy*, The Minerals, Metals & Materials Series, https://doi.org/10.1007/978-3-031-22634-2_24

technologies able to alleviate the environmental impact of iron and steel production [4]. Electrons are then the reducing agent, and with an oxygen-evolving anode, the chemical energy in the carbon will be completely replaced with electrical energy. A molten oxide electrolysis process for ferromanganese from silica-based electrolytes, using a $CO_2$ evolving carbon anode has previously been realized at pilot scale (50 kA) at Sadacem in Gent, Belgium, proving the feasibility of Mn electrolysis from oxide melts [5, 6]. An important motivation for the mentioned work was to reduce the carbon content in the produced metal. A similar process, run with oxygen evolving anodes, could represent a new technology for low carbon manganese and ferromanganese production with potential to reach the commonly cited goal for zero $CO_2$ emissions in 2050 [7, 8].

This work investigates $CO_2$ free Mn/FeMn electrolysis in oxide melts, and the objective has been twofold: (i) Experimental lab scale studies to establish a proof of concept of Mn electrolysis in a molten oxide electrolyte using a commercial Ni–Cr–Fe metal oxygen-evolving anode and (ii) Theoretical mass and energy balances for comparisons between different cases of a hypothetical industrial process for FeMn production.

## Lab Scale MOE Electrolysis of Mn

The goal for these trials was to prove the concept of producing pure Mn metal, using a synthetic feed material for the sake of simplicity. In a real process, one must consider the available raw materials' composition and their influence on the operation. These aspects were omitted from the experiments but are discussed theoretically in the mass and energy balance section. Oxide melts possess lower conductivity and are therefore poorer electrolytes than the commonly used molten salts. In general, a high basicity is beneficial for an electrolytic process. The basicity is the ratio of $CaO/SiO_2 + Al_2O_3$ in the oxide melt and the higher it is, the higher the ionic conductivity of the melt. At the same time, as the CaO content increases, the electrolyte mix will also become more corrosive which is challenging with respect to materials choices. To avoid shuttle redox reactions and loss in current efficiency, it is beneficial to use MnO, rather than $MnO_2$ as feed to the electrolytic process. A hypothetical industrial process must therefore include a step for pre-reduction of Mn ore, which is explored theoretically in Section "Mass and Energy Balance". In the experimental work, the electrolyte composition was made from synthetic materials, mixing 40 wt% MnO, 9.9 wt% $Al_2O_3$, 23.4 wt% $SiO_2$ and 26.7 wt% CaO. The composition was selected from the following requirements: basicity 0.8, melting point below 1350 °C (to have a "not too high" operating temperature since in general the higher the temperature, the more demanding material requirements), and high initial content of MnO so that the experiments could be run continuously for a certain amount of time without feeding MnO. Other important parameters to control are the cathodic and anodic current density (cd). The higher the current density, the higher the production rate. At the same time, too high cd can give co-reduction of electrolyte constituents (Si, Al) at the cathode, and higher anode dissolution rate. At high temperatures the electrode kinetics are usually fast, so the limiting current density is set by the mass transport rate of the reactant towards the electrode surface, increasing with increasing temperature and concentration. Molten aluminium is processed at 0.8–1 A/cm$^2$ (both anode and cathode) at 960 °C. In this work, the current density was chosen to be 1–2 A/cm$^2$ and 0.3–0.6 A/cm$^2$ at the cathode and anode, respectively.

Cathode materials, cell components, and particularly inert oxygen-evolving anodes pose a demanding material challenge. For the Al electrolysis process taking place in cryolite melts, searches for inert anodes have been conducted over many years and different materials have been investigated including various cermets, ceramics and metals [9]. This research is still ongoing and has received renewed interest along with the general increase in awareness on global warming. Elysis™, a joint venture between Alcoa and Rio Tinto, has been able to produce carbon free aluminium in the industrial environment by using a cermet- type inert anode [9, 10]. Iridium has proven to work well at lab-scale as inert anodes for MOE iron production, but Ir would be expensive for a commercial process [6]. Boston Metal, a spin-off company from MIT, is currently developing a steel making MOE process, working with 2.5 kA semi-industrial cells using a chromium rich FeCr inert anode [11, 12].

The advantages of metals as inert anodes are their mechanical strength, electronic conductivity, resistance to thermal shock, and that they can easily be machined. They generally perform as oxygen-evolving anodes due to a thin protective oxide layer on the surface which remains stable during the operation despite the demanding conditions. Molten oxide melts might represent a "milder" corrosion threat than molten cryolite. Oxides also have another advantage over fluorides since there will always be a surplus of oxide ions to be discharged at the anode as opposed to in a fluoride melt where depletion of oxide at the anode surface can result in break-down of the passive oxide layer that protects the anode. But the higher operating temperature for MOE (above the melting point of the oxide mixture and the metal) rules out some of the Cu- and Ni bearing alloys suggested for Al electrolysis. In this work the anode of choice was a commercial Ni–Cr-Fe alloy with melting point around 1400 °C.

## *Experimental*

230 g of the selected electrolyte oxide mixture (40 wt% MnO, 9.9 wt% $Al_2O_3$, 23.4 wt% $SiO_2$ and 26.7 wt% CaO) was weighed from pure commercially available oxides and placed into a graphite crucible (200 mm high, 50 mm inner diameter). The cathode material used were 3 mm diameter rods of Mo. The oxygen evolving anode used was a 10 mm diameter commercial Ni–Cr–Fe Incoloy® alloy (Inconel 825) rod.

A principal sketch of the set-up is shown in Fig. 1.In most of the experiments, the crucible was lined with molybdenum foil in the bottom and along the walls. This was both to avoid direct contact between any produced metal and graphite as

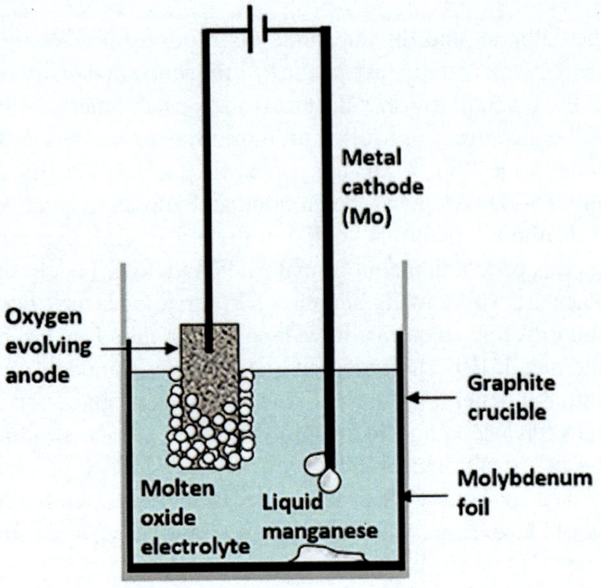


**Fig. 1** Principal sketch of the experimental set up

well as minimise the risk of spontaneous carbothermic reduction of Mn. To check the latter mechanism, reference experiments with direct contact between graphite and molten oxide were also carried out. The crucible containing the electrolyte mix as well as the electrode set up were inserted into a sealed vertical closed furnace, kept at 200 °C overnight to remove moisture, and heated to the working temperature ~1350 °C the next day, all during purging with argon gas (purity 99.999%). Electrolysis experiments were run by imposing a constant current of 3–4 A (giving the above-mentioned cd) for typically 100–300 min, using an HP 6031A system power supply. The working temperature was measured with a thermocouple type B (Pt30% Rh-Pt6% Rh) situated above the melt, and the cell voltage and temperature were recorded by means of a multichannel Keithley 2000 Multimeter. After shut-down, the bath was cooled, and the metal recovered was analysed by scanning electron microscope (SEM) with an energy dispersive spectroscopy (EDS) detector. The carbon content in the metal was analysed by combustion IR spectroscopy, using a CS844 combustion analyzer from LECO Corp.

## *Experimental Results and Discussion*

The results from the trials where graphite came into direct contact with the molten bath proved the necessity of a metal liner in the graphite crucible. Unwanted spontaneous carbothermic formation of Mn saturated with C was observed despite $\Delta G°$ being slightly positive for this reaction at the operating temperature, so the aim of a complete $CO_2$ free process must rule out carbon and graphite as materials in contact with the molten oxide.

Figure 2a shows a SEM picture and composition determined by EDS of the surface of metallic Mn as recovered from the bottom of the electrolytic cell. The carbon content was found to be 0.11 wt%. As seen in the figure, the metal contained Mo and traces of Al and Si. Figure 2b shows a Mo cathode after electrolysis, clearly consumed, which suggests that a liquid Mn-Mo alloy is formed when Mn is reduced. There are limited options for cathode materials in this process, and Mo was selected for its high thermal stability and inertness towards the oxide melt. The results show that it was not quite inert towards manganese.

The Faradaic current efficiency (i.e. the ratio between metal produced and the theoretical amount corresponding to the total electrical charge passed) was not so easily determined since some of the Mn produced could not be recovered. This is probably caused by Mn evaporation, a hypothesis supported by the Mn condensate found on the Mo cathode above the bath surface after several experiments, (see SEM/EDS of condensate shown in Fig. 2c). Although Mn boils at 2060 °C, pure Mn has a considerable vapor pressure of 0.008 atm at 1400 °C [13], whereas at 1400 °C it is reduced to ~0.002 atm above a molten Mn7Fe–$C_{sat}$ alloy. This is due to the lower Mn activity in this mixture (0.35) compared to pure Mn [1]. These observations imply that it could perhaps be more feasible to produce a Mn-Fe alloy directly in an MOE process rather that aiming for pure Mn.

Another aspect that should be kept in mind when developing a molten oxide process for Mn electrolysis is that due to the oxygen gas evolved at the anode and the lack of carbon in the system, the atmosphere will be much more oxidizing in such a reactor than in the SAF process. This can contribute to re-oxidation of Mn to MnO and/or MnO to $MnO_2$, mechanisms which would decrease the current efficiency and metal yield of the process.

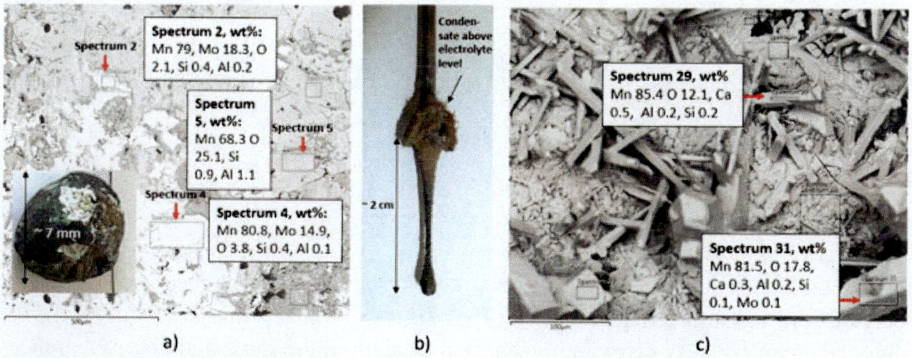


**Fig. 2** **a** Mn metal produced and corresponding SEM micrograph (back-scattered) with composition obtained from EDS point analysis, the lightest areas representing the metal, the darker areas remains of the oxide electrolyte, **b** consumed Mo cathode with condensate after electrolysis, **c** SEM picture with composition obtained from EDS analysis of the condensate recovered from the Mo cathode shown in **b**

**Fig. 3** **a** Ni–Cr–Fe anode after molten oxide electrolysis experiment (duration 360 min), **b** SEM/EDS analysis of the Mo foil in the cell bottom after electrolysis

The Ni–Cr–Fe-anode performed satisfactory during the trials in the sense that the cell voltage was relatively stable around 3–3.3 V. As shown in Fig. 3a, the anode had a certain wear, especially in the interface between the bath and the oxygen rich atmosphere above. This wear rate was calculated to be 0.13 g/h corresponding to 3.6 cm/year, which, if scalable, from an operational point of view probably would be satisfactory. However, as shown in Fig. 3b, up to 1.5% of Fe and traces of Cr and Ni were detected by EDS analysis of some of the metal gathered on the Mo foil in the crucible bottom, indicating that elements released from the anode may enter and pollute the cathode product over time. Iron dissolution from the anode and subsequent deposition into the metal is not a problem if FeMn is the targeted product. Typical impurity contents analyzed are 220–300 ppm Ni and 200–1200 ppm Cr and if these values become too high one might not obtain a marketable product. To verify the metal purity properly, other analysis methods must be applied, especially for metal obtained in long-term trials.

## Mass and Energy Balance

### *Approach and Methodology*

The experimental work discussed above was done with synthetic pure components. In an industrial process, the available raw materials' composition must be considered. The kind of raw materials used, and the type of pre-treatment processes employed will influence the mass and energy balances. In this section, the mass and energy balance are presented for two scenarios in which MOE of manganese could be realized:

1. Manganese ore with no purification
2. Manganese ore where 99% or 99.9% of non-Mn, non-Fe oxides have been removed.

**Table 1** Average composition of the Mn ore feedstock used for the calculations [1]

| $H_2O$ | $XH_2O$ | MnO | $MnO_2$ | $Fe_2O_3$ | $SiO_2$ | $Al_2O_3$ | MgO | CaO | BaO | $K_2O$ | P | $CO_2$ | Sum |
|---|---|---|---|---|---|---|---|---|---|---|---|---|---|
| 6.50 | 4.08 | 3.27 | 67.69 | 5.07 | 5.83 | 6.28 | 0.10 | 0.10 | 0.19 | 0.70 | 0.10 | 0.10 | 100 |

It should be noted that by nature of when considering immature technologies, the discussion will be based on several assumptions made with varying levels of confidence. Nonetheless, an analysis and discussion of the mass and energy balance can still be useful and illustrative. In this particular case, the comparison of the two scenarios will give insight into how much energy can be expended in an ore purification step before more gains are achieved by simply running the process with the non-purified ore as-is.

In high temperature electrochemical processes, it is beneficial to work with a baseline electrolyte with the required properties (appropriate melting point and thermal stability, sufficient ionic conductivity, chemical stability, etc.), and feed the raw material continuously with a speed matching the metal production rate set by the current applied. If the raw material contains substantial amounts of other compounds, this will alter the electrolyte properties with time, eventually causing operational problems and need for adjustment of electrolyte amounts and composition; tapping, additions etc. In the case of non-purified manganese ore, we see from the manganese ore composition in Table 1 that the electrolyte would accumulate $SiO_2$ and $Al_2O_3$, thus quickly lowering the basicity ($CaO/Al_2O_3 + SiO_2$) and changing other electrolyte properties as well. This problem would be diminished with a purified ore, though probably not eliminated entirely, since the purification step would not be 100% efficient.

An $MnO_2$-rich ore was selected as the feedstock for the calculations. Tangstad [1] gives the compositions of four different COMILOG (Compagnie Minière de l'Ogooué in Gabon) ores, the normalized average of which is presented in Table 1. In Scenario 1, this was the ore composition used as feed. In Scenarios 2a and b, an unspecified purification process was assumed to remove 99% and 99.9% of all non-Mn, non-Fe oxides, respectively. The details of the purification step are not discussed here but could be a combination of ore dressing and/or hydrometallurgical processes. The other process steps are outlined in Table 2 and described in more detail below. All thermochemical data is taken from the FactOxide and FSstel databases, and thermodynamic calculations are performed using Factsage [14].

## Pre-reduction Step

As mentioned in the experimental section, it is preferable to work with MnO rather than $MnO_2$. Therefore, in all scenarios it is assumed that a pre-reduction step using hydrogen converts all manganese oxides into MnO. It was also mentioned that the presence of iron in the metal product helps stabilize the Mn. In this theoretical work we have assumed that the iron oxide in the ore can be this source of iron for the final product. Thus, the iron oxides are assumed to be completely reduced in the

**Table 2** Scenarios and process steps

| Process step | Scenario | |
|---|---|---|
| | 1 | 2a, 2b |
| **Purification** | None | Removal of 99% (a) or 99.9% (b) of non-Mn/Fe-oxides |
| Pre-reduction | All Mn-oxides reduced to MnO, all iron oxides to Fe, by $H_2$ | All Mn-oxides reduced to MnO, all iron oxides to Fe, by $H_2$ |
| Electrolysis | Batch wise feeding and tapping CaO and $SiO_2$ electrolyte components added with ore | Continuous (≥1 year). Initial CaO and $SiO_2$ electrolyte, continuous feeding of purified ore |

pre-reduction step and carried forward to the electrolysis step as metallic iron. This may not be the most efficient option, but it has been chosen here for simplicity as it precludes the need for additional ferrous raw materials.

The pre-reduction is assumed to take place at 850 °C, with the material charged hot into the electrolysis step. Some energy loss is expected during transfer, so the pre-reduced material is assumed to be at 750 °C as it enters the electrolysis cell.

## Process-start Up and Run

For all scenarios, a large scale 300 kA electrolysis plant ("Hall Heroult like", with horizontal electrode configuration), is assumed, with an annual capacity of 400,000 tonnes of Mn and cathodic current density of 1 A/cm$^2$. The cell voltage and the current efficiency were assumed to be 3.3 V (ref experimental work in Section "Lab Scale MOE Electrolysis of Mn") and 80%, respectively. In all scenarios the starting electrolyte is built from (raw or purified) ore, CaO and $SiO_2$ to obtain the highest possible MnO-content (~39%) and a basicity of 0.8 while remaining fully liquid at 1350 °C. After start-up, the operation differs between the scenarios:

Scenario 1: As MnO converts into Mn, the properties of the electrolyte change due to depletion of MnO and no fresh feed is added. Here the basicity will not change. In this scenario, the metal and spent electrolyte must be tapped and re-charged twice a day. Longer operation would lead to formation of solids as the MnO-content drops and phases like $CaSiO_3$ are no longer soluble in the oxide electrolyte system.

Scenario 2a–b: The scenarios with purified ore envision a continuous process with a more stable electrolyte composition over time, where the purified ore is fed continuously as the Mn is reduced. Non-reduced oxides accumulate in the electrolyte, but the chemistry and volume of the electrolyte will change very slowly: For a 99.9% purification efficiency, solids only start to form after 18.6 years. For 99.0%, this happens after 1.9 years. It is likely that the process must be stopped for other purposes more often, so it is assumed that the cell is emptied and re-charged once per year for both purification efficiencies.

In addition to the formation of solids, the electrolyte volume and viscosity (calculated using the viscosity module of FactSage [14] were also considered as possible parameters that could reduce the lifetime of the electrolyte, but these were found to give less stringent limits than solid formation.

### Energy and Material Requirements

The mass balances presented here consider the ore required for one year of operation, the additional CaO and $SiO_2$ fluxes, as well as the hydrogen required for pre-reduction. The material outputs, in addition to metal product, are spent electrolyte and tailings from the purification process. No acids or other raw materials for, nor any wastes/by-products (beyond the oxides removed from the ore) from, the purification step are considered.

In addition to the electrical energy required to convert MnO to Mn, the process requires energy in the form of heat. In the pre-reduction step, the following contributions are considered:

- heating and evaporation (assumed to take place at 200 °C) of moisture and crystal water in the ore from 25 to 200 °C,
- enthalpy of reactions at 500 °C ($2MnO_2 + H_2 = Mn_2O_3 + H_2O$; $Fe_2O_3 + H_2 = 2FeO + H_2O$),
- enthalpy of reactions at 850 °C ($Mn_2O_3 + H_2 = 2MnO + H_2O$; $FeO + H_2 = Fe$), and
- heating of dry ore from 25 to 850 °C (with composition and heat capacity changing at 500 °C).

During electrolysis, the heat requirements are:

- heating of additional $SiO_2$ and CaO from 25 to 1350 °C,
- heating of pre-reduced ore and iron from 750 to 1350 °C,
- the enthalpies of melting and mixing of the oxides, and
- the enthalpy of dissolution of manganese in iron.

## *Discussion on Results from Mass and Energy Balance*

The overview of the results of the mass and energy balance is presented in Table 3: Generally, having a purified ore greatly improves the process. Without any purification step, large amounts of flux are required to keep the electrolysis going, which also means that more energy is required to heat the additional material. The figures obtained point to an energy gain of ≈1MWh per tonne of manganese in hydrogen production and heat input to the electrolysis process. For comparison, the Bayer

**Table 3** Overview of output data from the mass and energy balance for both scenarios

| | Raw ore, scenario 1 | Purified ore, scenario 2, purification efficiency | | | Raw ore, scenario 1 | Purified ore, scenario 2, purification efficiency | |
|---|---|---|---|---|---|---|---|
| | | 99.00% | 99.90% | | | 99.00% | 99.90% |
| **Material inputs [kg/tonne Mn]** | | | | **Energy inputs [MWh/tonne Mn]** | | | |
| Ore | 2563 | 1977 | 1977 | Electrolysis | 4.02 | 4.02 | 4.02 |
| $SiO_2$-flux | 1096 | 2.05 | 2.06 | Heating/melting ore and flux | 1.01 | 0.59 | 0.48 |
| CaO-flux | 1152 | 1.64 | 1.65 | Pre-reduction | 0.12 | 0.11 | 0.11 |
| $H_2$ | 50 | 39 | 39 | ($H_2$ production) | 2.53 | 1.95 | 1.95 |
| | | | | Total | 7.67 | 6.66 | 6.56 |
| **Waste outputs [kg/tonne Mn]** | | | | Total exlc. Hydrogen | 5.15 | 4.61 | 4.71 |
| Spent electrolyte | 3012 | 9.0 | 6.4 | $H_2$ share of energy (%) | 33 | 29 | 30 |
| Tailings | 0 | 294 | 291 | Mn yield (%) | 77 | 100 | 100 |

process consumes 1.9–5.8 MWh per tonne of refined alumina [15], which indicates that from energy considerations alone, it seems doubtful that the benefits of purifications would outweigh the costs.

Table 4 presents the electrolyte compositions as well as basicity and viscosity at the start and end for the three scenarios, and the numbers show that the batch-wise operation also has some other unfortunate consequences. While the fraction of remaining MnO in the spent electrolyte is higher in Scenario 2a–b, more ore is required per ton of product in Scenario 1 because so much more spent electrolyte is discarded. Since most of the discarded MnO originated as $MnO_2$, more hydrogen is also used in the pre-reduction step. Another aspect is that it is likely that the increase in viscosity and change in basicity seen in Table 4 for the non-purified (Scenario 1) and the ore purified to "only" 99% (Scenario 2a) will lead to lower ionic conductivity. It cannot be ruled out that the limits for how low this can be and still have a suitable electrolyte are already crossed. In that case, using Scenario 2b will be a better option.

The added flux in Scenario 1 ending up as spent electrolyte also gives massive volumes of by-product. The purification step in Scenario 2a & b also produces a by-product, in the form of tailings, but the total waste volumes are less than 10% of the amount produced in Scenario 1.

With significant amounts of both MnO and $SiO_2$, it is possible that the spent electrolyte could be used for silicomanganese production [1], and as it is tapped at 1350 °C it should also be possible to extract some of the high-temperature heat in useful form. Recovering some of this material and/or energy from the spent electrolyte would improve Scenario 1. In general, the proposed processes introduce several interesting

**Table 4** Electrolyte compositions [in wt%] at the start and end of electrolysis for the various scenarios

| | Scenario | MnO | $SiO_2$ | $Al_2O_3$ | MgO | CaO | BaO | $K_2O$ | Viscosity [poise] | Basicity, $CaO/(SiO_2 + Al_2O_3)$ |
|---|---|---|---|---|---|---|---|---|---|---|
| Start | All | 38.9 | 29.4–33.9 | 0.04–4.18 | 0–0.06 | 26.83–27.1 | 0–0.13 | 0–0.47 | 1.44–1.49 | 0.8 |
| End | 1 | 12.81 | 41.94 | 5.98 | 0.09 | 38.33 | 0.18 | 0.67 | 5.87 | 0.8 |
| | 2a | 26.35 | 37.20 | 15.46 | 0.23 | 18.55 | 0.47 | 1.73 | 8.74 | 0.35 |
| | 2b | 37.21 | 34.35 | 2.18 | 0.03 | 25.91 | 0.066 | 0.25 | 1.76 | 0.71 |

questions regarding circularity, but a detailed discussion of these are beyond the scope of this article.

It is also possible to use non-purified ore in a process more similar to Scenario 2, that is with continuous feeding. This could give a more beneficial mass balance but will introduce other operational challenges like fast electrolyte volume increase. Finally, it cannot be ruled out that Si would co reduce with Mn as the $SiO_2$/MnO ratio in the electrolyte increases. None of these options have been considered here.

Assuming that all Fe and P entering the electrolysis process end up in the final product, the metal composition will be 90.56% Mn, 9.19% Fe and 0.25% P for Scenario 1, when using non-purified ore. Operating the process according to Scenarios 2a–b, where the ore is purified, gives a metal composition of ~92.7% Mn and ~7.3% Fe, with a P level of 2–19 ppm depending on the degree of purification (99.9% contra 99%).

## Discussion and Conclusions

Small lab-scale experiments have shown the feasibility of manganese production using molten oxide electrolysis. There were some Mn losses due to evaporation and possibly re-oxidation, but these situations could likely be improved by producing an iron-containing alloy and improving the reactor design. The Mo cathode alloyed with the Mn produced in the reduction step. While the metallic Ni–Fe–Cr anode performed satisfactory from an operational point of view, some wear and "leakage" of Ni, Fe and Cr was detected. Its long-term behaviour and potential for metal product contamination has not been established. Giving the low carbon content obtained in the produced Mn metal, this might offer a direct route to a low carbon product. Manganese ore fines could be a suitable feed material for an electrolytic process, implying that raw material costs could be lower than for the SAF process, at least for the early movers. Advances in electrode materials development will be key to the further progress of molten oxide Mn electrolysis.

Mass and energy calculations puts the energy consumption at around 4 MWh/tonne of Mn for the electrolysis process (assuming a self-heating cell), and at 6.5–7.6 MWh/tonne when heating requirements and the electricity to produce hydrogen for pre-reduction are also taken into account. This can be compared to numbers for current carbothermic reduction, which can be extracted from the sustainability report of Eramet [16]. In 2018, they spent an average of 3.8 MWh of electrical energy per tonne of Mn alloy. Besides, they used energy-containing carbon materials corresponding to an additional 3.6 MWh per tonne.

Another factor to consider is the area efficiency. It is difficult to predict the exact area footprint of a future MOE-based Mn-process, but in aluminium electrolysis one standard cell of 48 $m^2$ will produce appr 2.2 tonne Al/day corresponding to 0.045 tonne/$m^2$/day. Correcting for the specifics of Mn (molar weight and number of electrons per mol metal), this corresponds to 0.135 tonne Mn/$m^2$/day. Thus, the area required will be much greater than for the current SAF which produces 2–300 tons/day on an area of 113 $m^2$. (1.8–2.7 tonne/$m^2$/day) [1]. This simple calculation involves the reduction unit only (electrolysis cell versus SAF furnace) and excludes surrounding infrastructure, but it still gives an impression of the likely order of magnitude difference.

Above all, the most important factor to consider is the $CO_2$ emissions. The traditional SAF method emits $CO_2$, while the proposed MOE method does not. If this fact does not have a high premium, it is unlikely that a consideration of any other numbers will lead to a favourable view of the electrolysis process. However, if the world is serious about reaching its climate targets, molten oxide electrolysis of manganese appears to be a promising technology.

**Acknowledgements** The mass balance calculations in this work have been performed as part of the project "Reduced $CO_2$ emissions in metal production", which is financed by the Norwegian Research Council (project number 280968), and the industry partners Elkem, Eramet, Wacker, Finnfjord, and TiZir. The experimental work was funded by SINTEF AS. Senior Research Scientist Samuel Senanu at SINTEF AS is kindly acknowledged for carrying out the SEM and EDS analysis.

## References

1. Olsen S, Tangstad M, Lindstad T (2007) Production of manganese ferroalloys. Tapir Academic press, Trondheim
2. Kero IT, Eidem PA, Ma Y, Indresand H, Aarhaug TA, Grådahl S (2019) Airborne emissions from Mn ferroalloy production. JOM 71(1):349–365
3. Kero IT, Dalaker H, Sende Osen K, Ringdalen E (2021) Some carbon-free technologies for manganese ferroalloy production. Available at SSRN 3926069
4. Cavaliere P (2019) Clean ironmaking and steelmaking processes; electrolysis of iron ores: most efficient technologies for greenhouse emissions abatement. Springer
5. Winand RFP (1970) Process and cell for the production of manganese of low carbon content by the means of a fused electrolytic bath. US patent 3,535,214, 20 October 1970
6. Allanore A (2015) Features and challenges of molten oxide electrolytes for metal extraction. J Electrochem Soc 162(1):E13–E22
7. European Commission (2021) A European Green Deal. https://ec.europa.eu/info/strategy/priorities-2019-2024/european-green-deal_en. Accessed 9 Sept 2022
8. United Nations (2020) Carbon neutrality by 2050: the world's most urgent mission. United Nations https://www.un.org/sg/en/content/sg/articles/2020-12-11/carbon-neutrality-2050-the-world%E2%80%99s-most-urgent-mission. Accessed 9 Sept 2022
9. He Y, Zhou K, Zhang Y, Xiong H, Zhang L (2021) Recent progress of inert anodes for carbon-free aluminium electrolysis: a review and outlook. J Mater Chem A 9:25272–25285
10. Elysis (2021) Recent progress of inert anodes for carbon-free aluminium electrolysis: a review and outlook https://www.elysis.com/en/carbon-free-aluminium-smelting-a-step-closer-elysis-advances-commercial-demonstration-and-operates. Accessed 2 Sept 2022
11. Boston Metal. Transforming metal production. https://www.bostonmetal.com/transforming-metal-production/. Accessed 2 Sept 2022
12. Allanore A, Yin L, Sadoway DR (2013) A new anode material for oxygen evolution in molten oxide electrolysis. Nature 497:353–356
13. Jones RT (1997) Thermodynamics and its applications—an overview 1997. Presented at SAIMM Pyrometallurgy School, Mintek, Randburg, 20–21 May 1997

14. Bale CW, Bélisle E, Chartrand P, Decterov SA, Eriksson G, Gheribi AE, Hack K, Jung IH, Kang YB, Melançon J, Pelton AD, Petersen S, Robelin C, Sangster J, Van Ende MA (2016) FactSage thermochemical software and databases, 2010–2016. Calphad 54:35–53. www.factsage.com
15. Scarsella AA, Noack S, Gasafi E, Klett C, Koschnick A (2015) Energy in alumina refining: setting new limits. In: Hyland M (ed) Light metals 2015. Springer, Cham. https://doi.org/10.1007/978-3-319-48248-4_24
16. Eramet Norway (2018). Eramet Norway sustainability report 2018. https://eramet.no/wp-content/uploads/Eramet-Sustainability-Report-2018-ENG-low.pdf. Accessed 5 Sept 2021

# Ferronickel Production from Nickel Laterite via Sulfide Chemistry

**Caspar Stinn and Antoine Allanore**

**Abstract** Nickel and ferronickel are critical components for production of products ranging from commodity metals to lithium-ion batteries. Current and future production rely on processing lateritic sources of nickel, which exhibit a range of technical and sustainability challenges. New sulfide-based process chemistries have been developed as platform technologies to decarbonize mining, materials separations, impurity management, and metals production. Herein, we utilize sulfide chemistry to produce carbon-free ferronickel. We first demonstrate selective sulfidation of mixed laterite feedstocks to form an iron-nickel sulfide matte. We then explore the thermodynamics of vacuum thermal treatment processes to enrich the matte in nickel via selective oxidation of iron. Finally, we employ aluminothermic reduction via reactive vacuum distillation to produce ferronickel from iron-nickel sulfide. These results lay the groundwork for an autothermal pathway to manufacture ferronickel from lateritic ore without direct greenhouse gas emissions.

**Keywords** Ferronickel · Nickel · Laterite · Sulfidation · Decarbonization · Vacuum

## Introduction

Nickel is crucial for a wide range of applications spanning from super alloys to lithium-ion battery cathodes. However, the largest consumer of nickel remains the stainless steel industry [1], where ferronickel and ferrochromium are blended with other alloying agents to reach a target product composition [2]. Ferronickel grades of 5–25% nickel are commonly employed [3]. Nickel is predominantly mined from sulfide and laterite deposits, with the latter accounting for approximately 72% of global production [4]. While secondary sources of nickel such as lithium ion battery cathode recycling remain important [5], the majority of demand growth is expected to be met by further development of laterite resources [4].

C. Stinn · A. Allanore (✉)
Department of Materials Science and Engineering, Massachusetts Institute of Technology, Cambridge, MA, USA
e-mail: allanore@mit.edu



C. Fleuriault et al. (eds.), *Advances in Pyrometallurgy*, The Minerals, Metals & Materials Series, https://doi.org/10.1007/978-3-031-22634-2_25

Nickel laterite may be divided into three main fractions within a given deposit [6]. Each deposit and fraction exhibit unique chemistry and processing challenges [7]. Heavily weathered iron-rich limonite generally constitutes the uppermost layer, where nickel and iron are usually present at grades of 0.8 wt% to 1.5 wt% and 40 wt% to 50 wt% respectively [1]. Deeper in the deposit where less weathering has occurred, nickel and iron are present in hydrated and magnesium silicate phases at grades of 1.5–4 wt% and 10–25 wt% respectively [1]. These regions are termed saprolite. Limonite and saprolite layers are often separated by a layer of smectite clay which exhibits intermediate iron and nickel levels [6, 7].

Due to differences in iron and gangue contents, nickel is conventionally extracted from limonite and saprolite via separate pathways. The iron to nickel ratio in saprolite is similar to that employed in ferronickel, allowing it to undergo direct pyrometallurgical co-reduction of iron and nickel species [3]. The nickel content may be further upgraded by sulfidation of the ferronickel to matte for subsequent iron/nickel separation via matte smelting. Sulfidation may be conducted during reduction roasting (PT Inco) or after ferronickel refining (Le Nickel) [1].

The iron to nickel ratio of limonite is higher than in saprolite, complicating its conversion to ferronickel. Direct reduction of low grade limonite and chromite to stainless steel has been conducted, however this approach is often challenged by low yields, mass transport limitations, high energy usage, and excessive slag generation [8]. Most limonite is processed hydrometallurgically via leaching with sulfuric acid at temperatures of 250 °C and pressures of 40 atm [1]. Nickel from limonite is dissolved while iron is stabilized as insoluble hematite or jarosite, enabling iron/nickel separation and ultimately production of high purity nickel/cobalt metal or compounds. The presence of magnesium oxide mineral phases greatly increases acid consumption however, causing the hydrometallurgical extraction of nickel from saprolite to be uneconomical [4]. Alternatively, ammonia leaching via the Caron process may be employed, albeit with higher energy usage and lower nickel yields [1].

In practice limonite, smectite, and saprolite layers are often blended within a laterite deposit [6], complicating nickel extraction. When conventional processes are used to handle mixed laterite feedstocks, higher energy and reagent usage are needed. Hydrometallurgical and pyrometallurgical processing of nickel both remain energy intensive at 52 kWh and 43 kWh respectively per kg of nickel from laterite at a nickel grade of 1 wt% [9]. Prior to smelting, classification by size may be used to separate finer limonite particles from coarser saprolite grains [10]. However, each fraction would then still require separate pathways for subsequent smelting. These factors motivate the search for a low energy process capable to handing mixed laterite feeds. Atmospheric heap leaching processes employing nitric acid and hydrochloric acid in place of sulfuric acid for treatment mixed laterites have been piloted, yet remain undeployed at an industrial scale [4].

Sulfidation chemistry is a promising avenue for sustainable and economical extraction of nickel from complex laterite feedstocks. Previously, Harris et al. demonstrated the selective sulfidation of limonite to produce a nickel-rich sulfide concentrate [11, 12]. Meanwhile, sulfidation of reduction roasted saprolite has been practiced industrially at the PT Inco facility since the late 1970s [13]. More recently, a combined thermodynamic/kinetic framework has been established to better control sulfidation selectivity [14]. Technoeconomic and life cycle assessment have likewise shown that pyrometallurgical selective sulfidation can reduce environmental impacts and capital costs by 60–90% versus legacy hydrometallurgy [14]. Low cost, carbon free electrolytic and vacuum thermal methods for reduction of sulfides are concurrently emerging [15–17]. Together, these innovations lay the groundwork for expanded utilization of sulfidation chemistry in nickel and ferronickel production from laterite. Herein, we explore the applicability of sulfidation for nickel extraction from complex nickeliferous laterite feedstocks. We then model the thermodynamics of vacuum thermal treatments for subsequent nickel upgrading, iron removal via oxidation, and sulfur recovery. Finally, we demonstrate aluminothermic reduction of iron-nickel sulfide for the synthesis of ferronickel alloy. With these processing operations, we establish the foundation for a new, minerals agnostic system chemistry for carbon-free nickel and ferronickel production.

## Sulfidation of Nickel Laterite

Sulfidation of limonite has been extensively explored by Harris et al. with promising results [11, 12], yet processing challenges remained. Previous work was directed toward producing a sulfide concentrate for flotation, where nickel and iron species in the limonite were sulfidized at conversions up to 80–90% and 40–50% respectively. A tradeoff was observed between the conversion of nickel oxide to sulfide and the sulfidation selectivity of nickel versus iron oxide. Sulfidation was conducted by heating a mixed briquette of sulfur and laterite, potentially hindering control of sulfidation kinetics and mass transport [14]. Operating temperatures below 950 °C produced sub-micron sulfide phases that were not amenable to liberation and separation. At temperatures above 950 °C sulfide particles on the order of 1–10 microns were able to be liberated, but proved challenging to float. Ultimately, nickel recoveries of 35% to 45% were achieved.

Recent advances may by leveraged to improve sulfidation performance. The use of gaseous elemental sulfur enables better control of sulfidation kinetics, mass transport, and nucleation/growth versus condensed or briquetted sources [14]. Meanwhile, conducting sulfidation at higher temperatures may enable the formation of immiscible liquid matte and slag products, improving separation and recovery and eliminating the burden of subsequent fine grinding [14]. Subsequent matte smelting or vacuum thermal treatment processes may be employed for further iron-nickel separation as needed. In this section, we explore the selective sulfidation of nickeliferous laterite to form immiscible liquid matte and slag phases amenable to subsequent processing.

## *Experimental Methods*

A nickeliferous lateritic ore from Indonesia was employed as a precursor for sulfidation. Sampling was conducted using the cone and quarter method. Prior to sulfidation, the ore was ground using a mortar and pestle to pass through a 212 micron sieve. A sample was then mounted in epoxy and polished for microscopy. Composition and morphology were analyzed via energy dispersive X-ray spectroscopy (SEM/EDS, JEOL JSM-6610LV JEOL Ltd., Sirius SD detector, SGX Sensortech Ltd.). The average composition of the nickeliferous laterite is reported in Table 1, with spatial distributions of some metallic elements illustrated in Fig. 1.

For a given element mapped in Fig. 1, cooler colors denote lower concentrations whereas warmer colors denote higher concentrations. Iron oxides and hydroxides were observed to be the most prevalent species, corresponding to a significant limonite fraction in the laterite. Mixed oxides of silicon, aluminum, and magnesium were also present, signifying the presence of smectite and saprolite fractions. In iron-rich phases, nickel is contained at a grade of about 1 wt%. In silica-rich, magnesia-poor phases, nickel content was observed to be lower at approximately 0.3 wt%. In mixed silica-magnesia phases, nickel content was higher at about 3 wt%. Occasional manganese-rich phases were found, which exhibited elevated levels of nickel of around 10 wt%.

The nickeliferous laterite was sulfidized in an alumina packed bed reactor at a temperature of 1300 °C and pressure of 1 atm for 1 h using procedures and equipment reported previously [14]. Elemental sulfur was employed as the sulfidizing agent at a partial pressure of 0.5–0.7 atm. An argon carrier gas was employed at a flowrate of 500 sccm for a system pressure of 1 atm. Following sulfidation, a portion of the product was mounted in epoxy for subsequent polishing and microscopy.

## *Results and Discussion*

During grinding and sieving of the nickeliferous laterite, limonite and saprolite phases were found to exhibit differences in particle size. Iron-rich oxide/hydroxide particles were generally observed between 10 and 50 microns in size. Silica, alumina, and magnesia-rich particles were larger at 100 to 200 microns in size. This trend is consistent with observations by others, suggesting that limonite and saprolite may in principle be separated via classification [10]. Following separation, conventional processing pathways differ greatly for limonite and saprolite. They must generally be treated at separate facilities for smelting to be economical [1, 4]. This motivates the search for a nickel extraction technology better suited to treat mixed limonite/saprolite feeds. Sulfidation is a promising candidate for a minerals agnostic process.

Sulfidation of the nickeliferous laterite at 1300 °C resulted in the formation of immiscible liquid sulfide matte and oxide slag phases. Sulfur/oxygen ratios and

**Table 1** Average bulk composition of the laterite determined via SEM/EDS

| Ni | Fe | Si | Al | Mg | Cr | Mn | O | S | Other |
|---|---|---|---|---|---|---|---|---|---|
| 1.1 wt% (±<0.1) | 37.0 wt% (±2.0) | 12.4 wt% (±1.7) | 6.5 wt% (±0.9) | 2.5 wt% (±0.4) | 1.6 wt% (±0.4) | 0.9 wt% (±0.4) | 35.0 wt% (±1.7) | 2.5 wt% (±0.2) | 0.4 wt% (±0.1) |

Standard deviations are reported in parenthesis

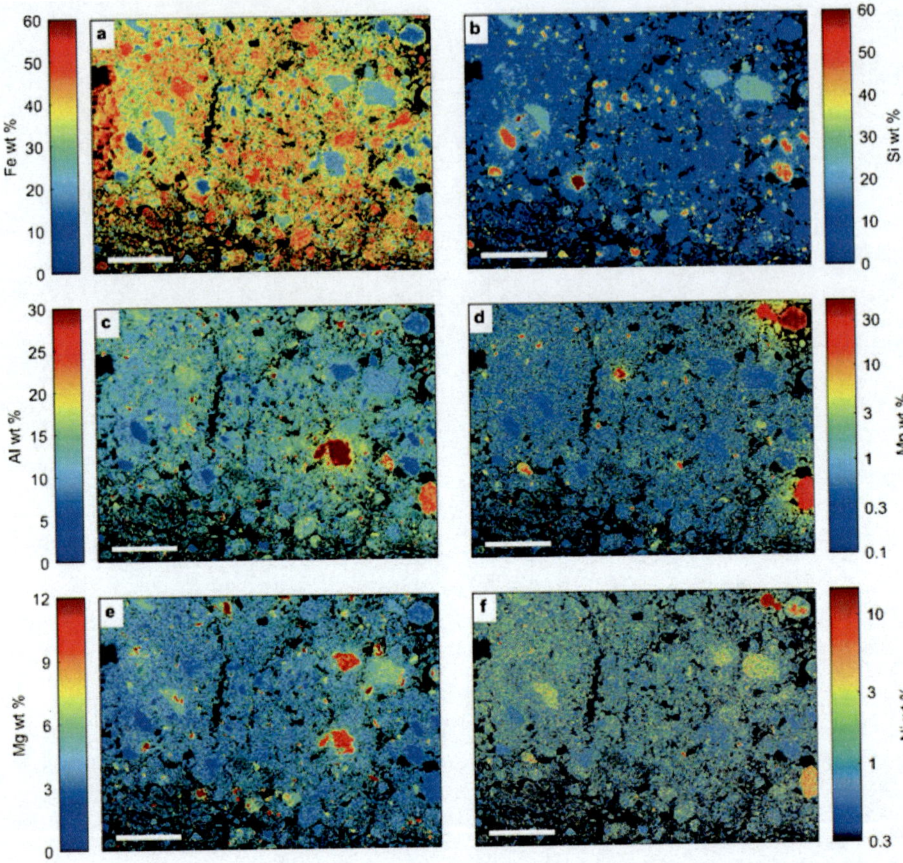


**Fig. 1** Distribution of iron (**a**), silicon (**b**), aluminum (**c**), manganese (**d**), magnesium (**e**), and nickel (**f**) in the laterite feedstock. Nickel is observed to be primarily distributed between iron-rich phases at a grade of around 1 wt% and magnesium-silicon-rich phases at a grade of around 3 wt%. Some manganese-rich phases are observed with elevated nickel contents on the order of 10 wt%. Silicon-rich magnesium-poor phases exhibit the lowest nickel trades at around 0.3 wt% or less (SEM/EDS map, scale bars: 500 μm)

metallic element weight fractions are mapped in Figs. 2 and 3 via SEM/EDS. For the sulfur/oxygen ratio, warmer colors denote higher sulfur contents. For metallic element maps, warmer colors denote higher relative weight fractions of a given element. Four major phases were observed: an iron-rich sulfide matte phase, and silica-rich, alumina-rich phase, and iron-rich oxide phases in the slag. Some matte droplets were entrained in the slag phase and vice versa, as shown in Fig. 3. This indicates that the time required for coalescence of the liquid phases exceeded that of sulfidation. Operating parameters such as temperature may be optimized to favorably modify viscosity for matte and slag coalescence.

Compositions and phase fractions following sulfidation are reported in Table 2. Limonite phases fully reacted, with nickel and some iron entering the sulfide matte

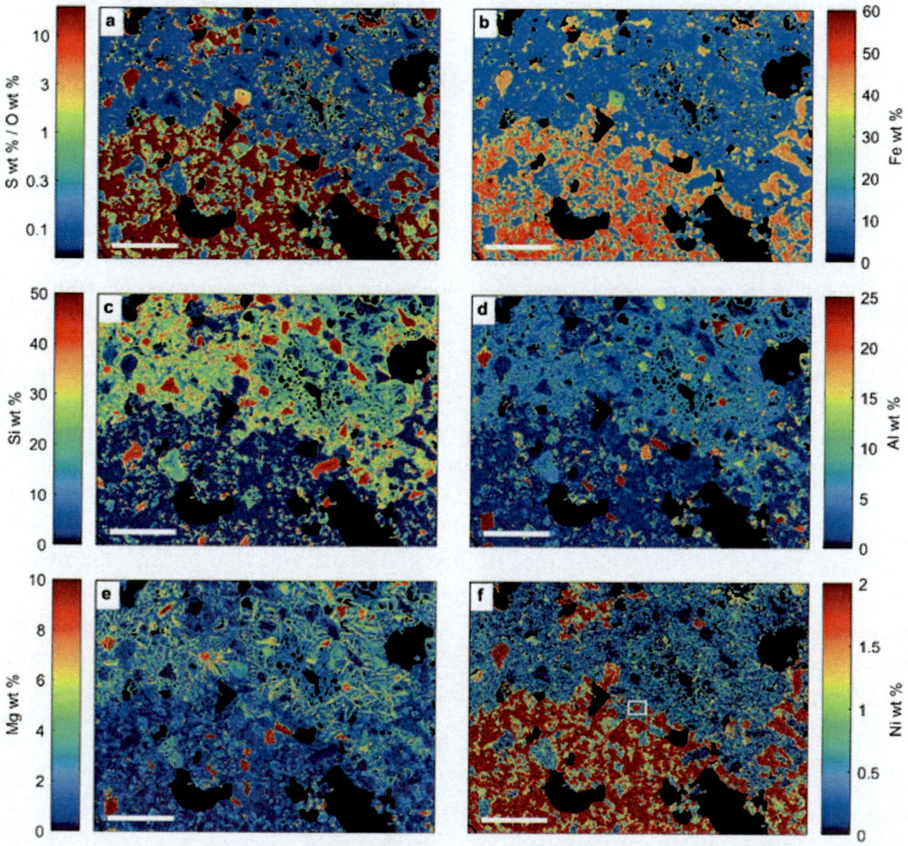


**Fig. 2** Ratio of sulfur to oxygen (**a**) and distribution of iron (**b**), silicon (**c**), aluminum (**d**), magnesium (**e**), and nickel (**f**) in the sulfidized laterite. Sulfidation at 1300 °C resulted in the formation of an iron-rich matte and silicon–aluminum–magnesium–rich slag phase. A nickel content in the matte of approximately 1.8 wt% was obtained, corresponding to a nickel sulfidation conversion of 89%. Within the slag, nickel contents remained highest in magnesium-rich phases and lowest in silicon-rich phases. Some entrainment of matte droplets in the slag and vice versa was observed. The boxed region shown in panel **f** corresponds to the area depicted in Fig. 3 (SEM/EDS map, scale bars: 500 μm)

phase. The remaining iron from limonite entered the slag oxide phases. The nickel contents of all slag phases were lower following sulfidation. Within the slag oxide, silica-rich acidic phases retained lower concentrations of nickel than magnesia-rich basic phases. Overall, nickel partitioned to the sulfide matte phase at a conversion of 89% for a grade of 1.8 wt%. Iron partitioned to the sulfide phase at a conversion of 83% for a grade of 54 wt%. These results indicate that sulfidation is capable of recovering nickel in an iron sulfide matte from both limonite and saprolite laterite phases.

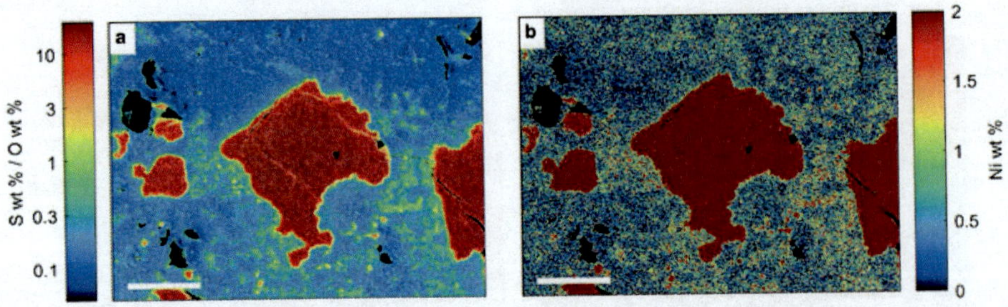


**Fig. 3** Ratio of sulfur to oxygen (**a**) and distribution of nickel (**b**) around an entrained matte particle within the slag phase. The boxed region shown in Fig. 2f corresponds to the area depicted here (SEM/EDS map, scale bars: 30 μm)

A tradeoff exists between nickel grade and recovery. Nevertheless, co-sulfidation of iron oxide may be beneficial. The sulfidation of iron oxide is exothermic [14], indicating that higher rates of iron oxide sulfidation lead to lower energy consumption at a given process temperature, ultimately enabling autothermal reactor operation. When sulfidation results in the formation of immiscible matte and slag phases, the resultant slag chemistry may be modified to enable subsequent smelting operations for nickel upgrading. Vacuum thermal treatments are particularly promising for production of nickel or ferronickel alloys from iron-nickel sulfides, which are explored thermodynamically in the following section.

## Nickel Sulfide Matte Upgrading and Sulfur Recovery

Depending on the iron to nickel ratio in the sulfidized product, several pathways are available for the production of low carbon nickel or ferronickel products from sulfidized laterite. Since sulfidation may be used to generate a liquid iron-nickel sulfide matte, melt based methods are the most attractive. One option is matte smelting, which employs oxygen addition to selectively convert iron sulfide to oxide while enriching the matte in nickel [1]. Product sulfur dioxide may be then used to produce sulfuric acid. While technically feasible, the attractiveness of this approach for iron removal from sulfidized laterite will be dependent on the economic landscapes of sulfuric acid production and elemental sulfur consumption. Alternatively, vacuum thermal treatments may be employed during matte smelting to recover a significant fraction of the sulfur in elemental form for reuse during sulfidation.

**Table 2** Average SEM/EDS composition of phases following sulfidation

| Phase | Ni | Fe | Si | Al | Mg | Cr | O | S | Other |
|---|---|---|---|---|---|---|---|---|---|
| Fe-rich sulfide | 1.8 wt% (±0.2) | 54.3 wt% (±1.4) | 1.9 wt% (±0.3) | 1.0 wt% (± 0.1) | 0.5 wt% (±<0.1) | 1.2 wt% (±0.1) | 2.4 wt% (±0.7) | 36.8 wt% (±0.2) | 0.2 wt% (±0.1) |
| Al-rich oxide | 0.4 wt% (±0.1) | 15.1 wt% (±1.5) | 4.2 wt% (±1.2) | 23.9 wt% (±2.4) | 9.5 wt% (±0.8) | 10.0 wt% (±2.9) | 33.6 wt% (±1.3) | 3.2 wt% (±0.6) | 0.2 wt% (±0.1) |
| Fe-rich oxide | 0.2 wt% (±0.1) | 21.7wt% (±0.7) | 4.4 wt% (±0.2) | 7.3 wt% (±0.5) | 4.1 wt% (±0.6) | 29.3 wt% (±0.9) | 29.9 wt% (±0.9) | 2.8 wt% (±0.2) | 0.3 wt% (±0.2) |
| Si-rich oxide | 0.1 wt% (±<0.1) | 8.0 wt% (±0.7) | 44.9 wt% (±1.2) | 1.9 wt% (±0.3) | 0.6 wt% (±0.1) | 0.9 wt% (±0.3) | 40.7 wt% (±0.9) | 2.7 wt% (±0.3) | 0.1 wt% (±<0.1) |

Standard deviations are reported in parenthesis

## *Thermodynamic Framework*

In conventional matte smelting, iron sulfide from the matte is selectively reacted with oxygen to form iron oxide, producing sulfur dioxide as described in the following reaction. Silica is added to produce fayalite, fluxing iron oxide while increasing selectivity and conversion of iron sulfide to oxide.

$$\frac{2}{3}\mathrm{FeS}\ (\mathrm{l}) + \mathrm{O_2}\ (\mathrm{g}) + \frac{1}{3}\mathrm{SiO_2}(\mathrm{s,l}) = \frac{1}{3}\mathrm{Fe_2SiO_4}\ (\mathrm{l}) + \frac{2}{3}\mathrm{SO_2}\ (\mathrm{g}) \quad (1)$$

At elevated temperatures and lower pressures, iron sulfide thermally decomposes to iron metal and sulfur gas. The following reactions may also occur to generate elemental sulfur in place of sulfur dioxide.

$$2\mathrm{FeS} = 2\mathrm{Fe}\ (\mathrm{l\ in\ matte}) + \mathrm{S_2}\ (\mathrm{g}) \quad (2)$$

$$2\ \mathrm{Fe}(\mathrm{l\ in\ matte}) + \mathrm{O_2}\ (\mathrm{g}) + \mathrm{SiO_2}\ (\mathrm{s,l}) = \mathrm{Fe_2SiO_4}\ (\mathrm{l}) \quad (3)$$

Sulfide matte fluxes iron [18], allowing a sulfur deficient matte phase to form (Reaction 2). Iron may then be removed from the matte via selective oxidation (Reaction 3).

To compare the relative favorability of Reactions 1 versus 2 and 3 under ambient pressure (1 atm) and industrial vacuum (0.001 atm), Gibbs energy minimization was conducted with the FactSage 8.0 Equilib module using the FactPS, FToxid, and FSstel databases. A feedstock matte phase consisting of 98 wt% iron(II) sulfide and 2 wt% nickel (II) sulfide at 1300 °C was modelled as a single phase using the FSstel-Liqu solution phase. The silica feedstock was modelled as quartz at 25 °C. The matte was modelled to react with silica in a 5:3 mass ratio at 1500 °C with varying levels of oxygen addition under either ambient pressure or industrial vacuum. The possible formation of liquid matte (FSstel-Liqu, single phase), pyrrhotite solid solution (FSstel-PYRR), metal disulfide solid solution (FSstel-MeS2), pentlandite solid solution (FSstel-Pent), liquid slag (FToxid-SLAGA, single phase), spinel solid solution (FToxid-SPINA), metal monoxide solid solution (FToxid-MeO_A), clinopyroxene solid solution (FToxid-cPyrA), and olivine solid solution (FToxid-OlivA) were considered, along with pure solid compounds and gaseous compounds present in the FactPS, FToxid, and FSstel databases.

Gibbs energy minimization results for the reaction of the matte with silica and oxygen under ambient and industrial vacuum conditions are presented in Fig. 4. Matte, slag, and unreacted silica in the form of cristobalite were calculated to be the thermodynamically stable condensed product phases at 1500 °C. The product gas phase was found to be dominated by elemental sulfur and sulfur dioxide. Figure 4a and b illustrate the predicted yields and grades of iron and nickel in the matte phase at ambient pressure (1 atm) and industrial vacuum (0.001 atm) respectively. The discontinuity under ambient conditions at an oxygen feed of 0.2 kg/kg of sulfide corresponds to phase separation of the liquid into matte and slag phases. This separation occurs at lower oxygen feeds under vacuum due to greater depletion of sulfur from the matte from vacuum thermal decomposition of iron and nickel sulfides.

Figure 4c and d illustrate the predicted yields of gaseous elemental sulfur and the ratios of sulfur to sulfur dioxide in the gas phase as a function of oxygen addition at ambient pressure and industrial vacuum respectively. Under ambient conditions, sulfur is evolved through exchange reactions upon oxygen addition. Under vacuum, additional sulfur is also evolved due to vacuum thermal decomposition of the matte, increasing the yield of sulfur and the ratio of sulfur to sulfur dioxide in the gas phase. Under the conditions considered herein, up to 80% of sulfur in the matte may be recovered via vacuum thermal treatment and selective oxidation.

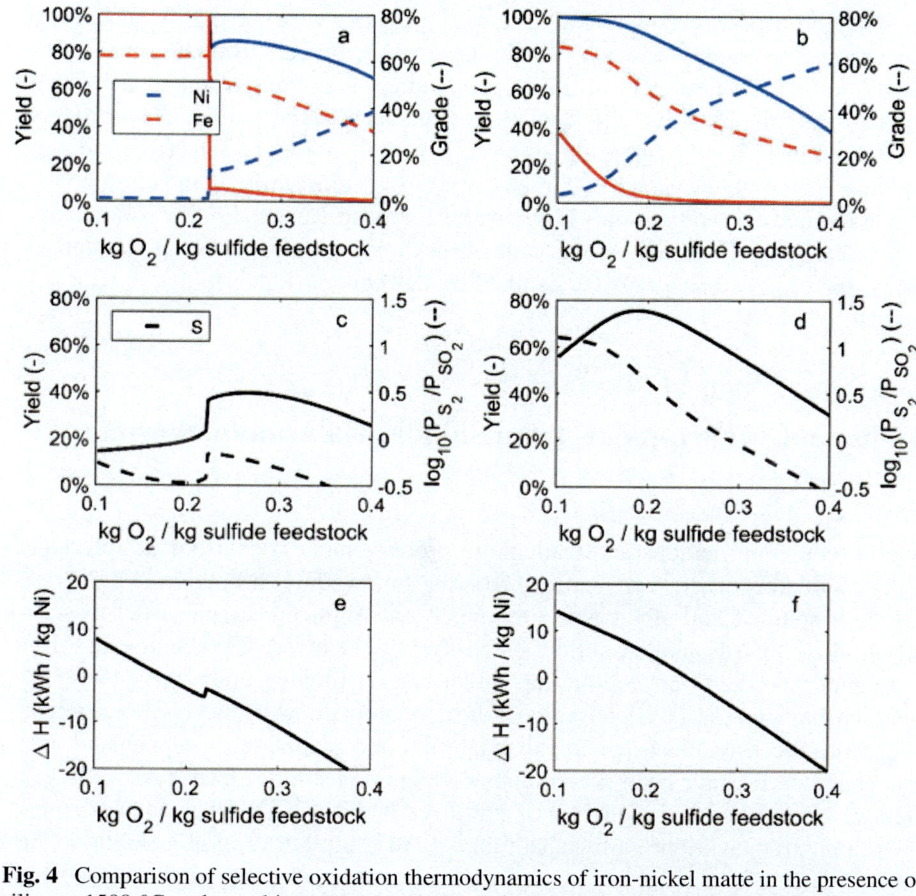


**Fig. 4** Comparison of selective oxidation thermodynamics of iron-nickel matte in the presence of silica at 1500 °C under ambient pressure (1 atm, left panels) and industrial vacuum (0.001 atm, right panels). Iron and nickel yields and grades in the matte sulfide as a function of oxygen feed are shown in panels **a** and **b**. Panels **c** and **d** illustrate the yields of sulfur from the matte to the gas phase and the ratios of sulfur to sulfur dioxide generated during selective oxidation. The enthalpies of the processes are depicted in panels **e** and **f**

Figures 4e and f illustrate the predicted enthalpy changes from initial conditions to final conditions at ambient pressure and industrial vacuum respectively. Under ambient conditions, the reaction becomes autothermal at a lower oxygen input (>0.17 kg/kg of sulfide) than under vacuum conditions (>0.24 kg/kg of sulfide). Heating is largely supplied by the exothermic mixing of iron oxide with silica and the combustion of sulfur. The extents of these reactions are both related to the rate of oxygen inlet. Since additional sulfur is vaporized under vacuum, additional oxygen is needed to achieve autothermal operation. Sequential vacuum and ambient pressure treatments could be used to optimize sulfur recovery and energy usage.

Nickel upgrading and sulfur recovery via selective oxidation under vacuum would likely use reactor geometries and process controls that are more similar to a basic oxygen furnace (BOF) or Ruhrstahl-Heraeus (RH) degasser than a Pierce-Smith converter. Slag chemistries will need to be optimized to maximize selectivity in the oxidation of iron versus nickel. The quality of this thermodynamic analysis is subject to the accuracy of available solution models. Process evaluation via Gibbs energy minimization assumes equilibrium operation; non-equilibrium behavior may provide favorable or unfavorable deviations from calculated behavior. Nevertheless, these preliminary results suggest that through the utilization of a selective oxidation stage conducted under vacuum, the majority of the sulfur utilized in sulfidation to form nickel and iron sulfides may be recovered for reuse. Residual sulfur dioxide may be converted to sulfuric acid along with sulfur dioxide captured during sulfidation. Once a matte has been upgraded to a suitable nickel content, it may undergo reduction to metal.

## Production of Ferronickel from Sulfides via Aluminothermic Reduction

Several reduction methods are available to produce nickel or ferronickel products from iron-nickel sulfide. Conventional roasting followed by carbothermic reduction may be employed [1], albeit with significant environmental impacts [9]. A low-carbon alternative is molten sulfide electrolysis, which recovers elemental sulfur as an anode product and exhibits no direct carbon dioxide emissions [19]. This approach has previously been explored for the production of nickel-cobalt alloys [17]. More recently, electrochemical separation and reduction of copper and iron from chalcopyrite have been accomplished via molten sulfide electrolysis [15]. For selective electrochemical reduction of sulfidized laterite, the Wagner-Allanore thermodynamic framework provides a path forward to control the iron/nickel ratio in the metal product [20]. The lack of phase stability, solution, viscosity, and conductivity models for mixed sulfides however hinders optimization of supporting electrolytes [19, 21].

Another promising low-carbon approach is aluminothermic reduction via reactive vacuum distillation, which has recently been explored for production of aluminum master alloys [16]. Aluminothermic reduction of molten iron-nickel sulfide matte to liquid ferronickel via reactive vacuum distillation occurs via the following reactions:

$$\text{FeS (l, matte)} + \frac{2}{3}\text{Al (l)} = \text{Fe(l, alloy)} + \frac{1}{3}\text{Al}_2\text{S}_3\text{ (g)} \tag{4}$$

$$\text{NiS (l, matte)} + \frac{2}{3}\text{Al (l)} = \text{Ni(l, alloy)} + \frac{1}{3}\text{Al}_2\text{S}_3\text{ (g)} \tag{5}$$

However, previous analysis [16] suggests that under an industrial vacuum at 0.001 atm and an aluminum sulfide partial pressure of 0.001 atm, incomplete conversion is possible. Likewise, liquid ferronickel is fluxed by both liquid iron-nickel sulfide matte and aluminium [18, 22], leading to challenges with residual aluminum and sulfur removal. To achieve higher purity ferronickel, the aluminum sulfide partial pressure must be minimized. One avenue to lower the aluminum sulfide partial pressure is to react the gaseous aluminum sulfide product with a slag to form a matte that is immiscible with ferronickel. From the sulfidation series of oxides [14], aluminum sulfide is expected to sulfidize calcia. When aluminothermic reduction is conducted in the presence of a calcia-containing slag, the following reaction is proposed:

$$\frac{1}{3}\text{Al}_2\text{S}_3\text{ (g)} + \text{CaO (l, slag)} = \text{CaS(l, matte)} + \frac{1}{3}\text{Al}_2\text{O}_3\text{ (l, slag)} \tag{6}$$

Depending on the matte and slag chemistry, solution effects may shift the equilibrium of Reaction 6. Calcium sulfide is a conventional product from ferronickel desulfurization processes and is immiscible with ferronickel [1]. Therefore, the use of a calcia containing slag during aluminothermic reduction of iron-nickel matte is a promising pathway to minimize aluminum and sulfur impurities in the ferronickel alloy product.

## *Experimental Methods*

For aluminothermic reduction of iron-nickel sulfide, 3 g of iron(II) sulfide (FeS, 99.9%, Lot # 19872600, Strem Chemicals) and nickel subsulfide ($Ni_3S_2$, 99.7%, Lot # MKCF2714, Aldrich) mixed in a 2:1 mass ratio served as the feedstock. The ratio of iron to nickel ratio in the feed is expected to be obtainable via selective oxidation or matte smelting (see above). For the alumina-calcia-silica slag, 6 g aluminum oxide ($Al_2O_3$, 99.9%, Lot # Z30F064, Alfa Aesar), silicon oxide ($SiO_2$, 99.5%, Lot # Z17A018, Alfa Aesar), and calcium oxide (CaO, Certified Powder, Lot # 197017, Fischer Chemical) were mixed in a 3:2:1 mass ratio. For the reductant, 1 g of aluminum (Al, 99.9%, Lot # MKCH1235, Sigma-Aldrich) was utilized. The aluminum, mixed sulfide, and mixed oxide were then sequentially added to an alumina crucible. Aluminothermic reduction via reactive vacuum distillation at a temperature of 1500 °C and pressure of 0.001 atm was conducted using procedures and equipment reported previously [16]. Following vacuum thermal treatment, volatized products were collected from the top of the crucible and furnace tube. The crucible was broken for recovery of metal, matte, and slag phases. Products were then mounted in epoxy and polished for subsequent SEM/EDS analysis.

## *Results and Discussion*

Aluminothermic reduction of iron-nickel sulfide via reactive vacuum distillation resulted in the formation of a matte, a slag, a metallic pool, and a volatile distillate which collected in the top of the crucible and furnace tube. These products are shown in Fig. 5. SEM/EDS compositions of metal, matte, and slag phases are included in Table 3. The bulk ferronickel product contained and iron to nickel ratio of 1.8, with a sulfur content of approximately 1 wt%, an aluminum content of approximately 0.3 wt%, and a calcium content of approximately less than <0.1 wt%. Analytical methods that are more sensitive for light and dilute elements, such as combustion analysis or spectroscopy, are needed to refine these quantifications. Within the ferronickel, nickel rich (light) and iron rich (dark) metallic phases were detected. The silicon content of the ferronickel was 14 wt%. However, excess silicon is usually present in crude ferroalloys and is removed industrially via oxygen purge [3]. The slag was found to be rich in aluminum and silicon, whereas the matte was found to rich in aluminum and depleted in silicon. The matte showed unexpectedly high levels of oxygen (Table 3, *); it remains unclear whether the matte was an oxysulfide liquid or oxidized during polishing and analysis. In both the slag and matte phases, only trace levels of nickel were detected. Iron levels were higher in the matte at 4 wt%, indicating that reduction showed some selectivity for nickel. The condensed distillate product was shown to be dominated by aluminum–silicon–calcium–oxygen–sulfur species. Iron and nickel contents of the distillate were below the detectable limit at <0.1 wt%. A mass balance reveals >95% conversion of nickel sulfide to ferronickel.

For the calcia-alumina-silica slag, matte, and aluminum feedstock ratios employed herein, at 1500 °C the aluminothermic reduction (Reactions 4–6) to produce ferronickel is calculated to be exothermic at −600 kWh/tonne of nickel. This excess heat may be used to melt aluminum and slag feedstocks. Further work is necessary to determine the optimal amount and chemistry of the slag to achieve sufficient ferronickel purity while maintaining autothermal reactor operation.

While analytical methods more suitable for light elements, such as combustion analysis, are needed to better quantify sulfur content, preliminary analysis suggests that little to no desulfurization may be necessary. Industrially, FeNi20 alloy is marketable at nickel contents of 15–25% and sulfur contents of 0.4 wt%[3]. Removal of silicon via oxidation and dilution of our ferronickel alloy with high purity iron to 15 wt% nickel would lower our sulfur content to <0.5 wt%, near the vicinity of typical ferronickel FeNi20 products [3]. Further optimization of slag chemistry and operating conditions may lead to lower levels of sulfur. Our results suggest that aluminothermic reduction via reactive vacuum distillation is a promising pathway

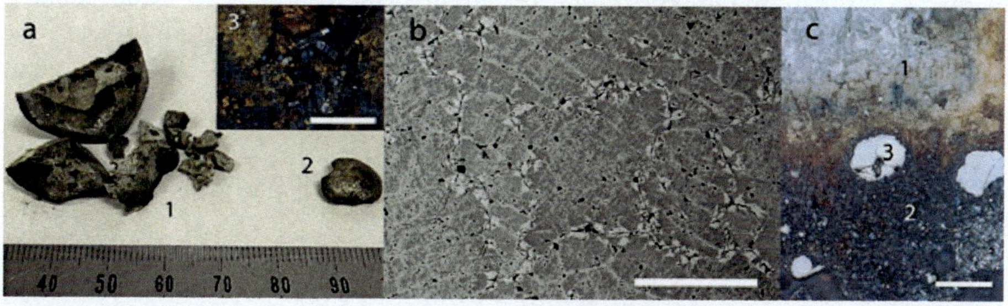

**Fig. 5** Products of aluminothermic reduction of iron-nickel sulfide via reactive vacuum distillation. Panel **a** (optical, inset scale bar: 1 mm) corresponds to slag/matte (1), ferronickel (2), and condensed distillate (3) products. Panel **b** (SEM backscatter electron composite, scale bar: 200 μm) shows the microstructure of the solidified ferronickel. Lighter phases were found to be richer in nickel. Panel **c** (optical, scale bar: 500 μm) shows a cross section of the slag (1)/ matte (2) interface, with dispersed droplets of ferronickel (3)

**Table 3** Average SEM/EDS composition of metal, matte, and slag following aluminothermic reduction via reactive vacuum distillation

| Phase | O | Al | Si | S | Ca | Fe | Ni |
|---|---|---|---|---|---|---|---|
| Bulk metal | 1.5 wt% (±0.1) | 0.3 wt% (±<0.1) | 14.0 wt% (±<0.1) | 1.0 wt% (±<0.1) | <0.1 wt% (±<0.1) | 53.7 wt% (±0.2) | 29.5 wt% (±0.2) |
| Bulk matte | 37.9 wt%* (±0.4) | 36.5 wt% (±0.8) | 1.8 wt% (±0.9) | 13.0 wt%* (±0.9) | 6.6 wt% (±0.7) | 4.0 wt% (±1.7) | 0.2 wt% (±0.1) |
| Bulk slag | 39.0 wt% (±0.3) | 31.1 wt% (±0.5) | 14.2 wt% (±0.2) | 6.5 wt% (±0.5) | 8.6 wt% (±0.3) | 0.4 wt% (±0.9) | 0.1 wt% (±<0.1) |

Standard deviations are reported in parenthesis

to produce carbon-free ferronickel from iron-nickel matte, with a potentially lower refining burden than conventional methods. In practice, scrap aluminum could be utilized for reduction to lower costs, yet the behavior of scrap impurities would also require analysis.

## Conclusion

A sulfide-based processing pathway is presented for ferronickel production from laterite. Analyses of laterite feedstock and sulfidation products suggest that sulfidation is an omnivorous technology capable of recovering nickel from mixed limonite and saprolite phases into an iron sulfide matte. Compared to conventional hydrometallurgical routes that are net consumers of sulfuric acid, sulfidation can enable the production of sulfuric acid as a byproduct via capture and abatement of sulfur dioxide. Vacuum thermal treatments for selective oxidation are thermodynamically promising to upgrade the nickel content of the matte while recovering elemental sulfur for recycling to sulfidation. Finally, iron-nickel sulfide matte was reduced to ferronickel using aluminothermic reduction via reactive vacuum distillation. With optimization, ferronickel produced via this pathway may require less refining than conventional reduction routes. Laterite sulfidation, matte upgrading, and aluminothermic reduction to ferronickel are all exothermic processes, and may enable ferronickel production via a fully autothermal process that is free of direct greenhouse gas emissions.

**Acknowledgements** The authors wish to thank Prof. Zaki Mubarok and Dr. Taufiq Hidayat for their insight and for the laterite ore sample.

## References

1. Crundwell FK, Moats MS, Ramachandran V, Robinson TG, Davenport WG (2011) Extractive metallurgy of nickel, cobalt, and platinum-group metals. Elsevier, Oxford, UK
2. Norgate TE, Jahanshahi S, Rankin WJ (2004) Alternative routes to stainless steel a life cycle approach. In: Tenth international ferroalloys congress, pp 693–704
3. Polyakov O (2013) Technology of ferronickel. In: Gaslikeditor M (ed) Handbook of ferroalloys. Elsevier, Oxford, UK
4. Stanković S, Kamberović Ž, Friedrich B, Stopić S R, Sokić M, Marković B, Schippers A (2022) Options for hydrometallurgical treatment of Ni-Co lateritic ores for sustainable supply of nickel and cobalt for European battery Industry from South-Eastern Europe and Turkey. Metals (Basel) 12(5). https://doi.org/10.3390/met12050807
5. Harper G, Sommerville R, Kendrick E, Driscoll L, Slater P, Stolkin R, Walton A, Christensen P, Heidrich O, Lambert S, Abbott A, Ryder K, Gaines L, Anderson P (2019) Recycling lithium-ion batteries from electric vehicles. Nature 575(7781):75–86. https://doi.org/10.1038/s41586-019-1682-5
6. König U (2021) Nickel laterites—mineralogical monitoring for grade definition and process optimization. Minerals 11(11). https://doi.org/10.3390/min11111178
7. Gleeson SA, Butt CRM, Elias M (2003) Nickel laterites: a review. Soc Econ Geol Newsl 45:11–18. https://doi.org/10.5382/SEGnews.2003-54.fea
8. Keskinkilic E (2019) Nickel laterite smelting processes and some examples of recent possible modifications to the conventional route. Metals (Basel) 9(974):1–16
9. Norgate T, Jahanshahi S (2010) Low grade ores—smelt, leach or concentrate? Miner Eng 23(2):65–73. https://doi.org/10.1016/j.mineng.2009.10.002
10. Seo J, Kim K, Bae I, Lee J, Kim H (2016) A study on classification of limonite and saprolite from nickel laterite ores. J Korean Inst Resour Recycl 25(1):40–47. https://doi.org/10.7844/kirr.2016.25.1.40
11. Harris CT, Peacey JG, Pickles CA (2011) Selective sulphidation of a nickeliferous lateritic ore. Miner Eng 24:651–660. https://doi.org/10.1016/j.mineng.2010.10.008
12. Harris CT, Peacey JG, Pickles CA (2013) Selective sulphidation and flotation of nickel from a nickeliferous laterite ore. Miner Eng 54:21–31
13. Diaz CM, Landolt CA, Vahed A, Warner AEM, Taylor JC (1988) A Review of nickel pyrometallurgical operations. JOM 40(9):28–33. https://doi.org/10.1007/BF03258548
14. Stinn C, Allanore A (2022) Selective sulfidation of metal compounds. Nature 602:78–83. https://doi.org/10.1038/s41586-021-04321-5
15. Daehn KE, Stinn C, Rush L, Benderly-Kremen E, Wagner ME, Boury C, Chmielowiec B, Gutierrez C, Allanore A (2022) Liquid copper and iron production from chalcopyrite, in the absence of oxygen. Metals (Basel) 12(9):1440. https://doi.org/10.3390/met12091440

16. Stinn C, Allanore A (2022) Aluminothermic reduction of sulfides via reactive vacuum distillation. Light Metals 2022:681–688
17. Stinn C, Allanore A (2021) Selective sulfidation and electrowinning of nickel and cobalt for lithium ion battery recycling. In: Andersoneditor C (ed) Ni–Co 2021: The 5th international symposium on nickel and cobalt. Springer Nature, Cham, pp 99–110
18. Waldner P, Sitte W (2008) Thermodynamic modeling of Fe–Ni pentlandite. J Phys Chem Solids 69(4):923–927. https://doi.org/10.1016/j.jpcs.2007.10.011
19. Daehn K, Allanore A (2020) Electrolytic production of copper from chalcopyrite. Curr Opin Electrochem 22:110–119. https://doi.org/10.1016/j.coelec.2020.04.011
20. Wagner M-E, Allanore A (2022) Electrochemical separation of $Ag_2S$ and $Cu_2S$ from molten sulfide electrolyte. J Electrochem Soc 169(6):063511. https://doi.org/10.1149/1945-7111/ac7101
21. Stinn C, Nose K, Okabe T, Allanore A (2017) Experimentally determined phase diagram for the barium sulfide-copper(I) sulfide system above 873 K (600 °C). Metall Mater Trans B. https://doi.org/10.1007/s11663-017-1107-5
22. Bradley AJ (1949) Microscopical studies on the iron-nickel-aluminum system. Part 1 – $\alpha + \beta$ alloys and isothermal sections of the phase equilibrium diagram. J Iron Steel Inst 163:19–30

# Author Index





C. Fleuriault et al. (eds.), *Advances in Pyrometallurgy*, The Minerals, Metals & Materials Series, https://doi.org/10.1007/978-3-031-22634-2

# Subject Index





C. Fleuriault et al. (eds.), *Advances in Pyrometallurgy*, The Minerals, Metals & Materials Series, https://doi.org/10.1007/978-3-031-22634-2